Topics in
Modern Analysis

PRENTICE-HALL INTERNATIONAL, INC., *London*
PRENTICE-HALL OF AUSTRALIA, PTY, LTD., *Sydney*
PRENTICE-HALL OF CANADA, LTD., *Toronto*
PRENTICE-HALL OF INDIA (PRIVATE) LTD., *New Delhi*
PRENTICE-HALL OF JAPAN, INC., *Tokyo*

Topics in
Modern Analysis

Robert G. Kuller

Department of Mathematics
Northern Illinois University

Prentice-Hall, Inc., Englewood Cliffs, N.J.

To Janet

Preface

This book is intended as a text for a one-year course in analysis. It can follow an advanced calculus course, but it should be suitable for any student who has had some introduction to rigorous analysis. One of the principal aims of the book is to bring the reader within reach of a serious study of functional analysis as quickly as possible; and with this goal in view various basic types of functions are treated in the context of normed vector spaces. The text includes a thorough treatment of Lebesgue theory and, in the last chapter, an introduction to dual spaces.

Some effort has been made to keep the book self-contained. Thus, discussions of basic set theory, the real-number system, and elementary vector space theory are included. The student will find many exercises, ranging from insultingly easy to difficult.

Thanks are due to Jack R. Britton, who first encouraged me to under-take the preparation of this material; to the several groups of students who were subjected to preliminary versions; and to William Vekovius, who read and commented on the final draft. It is my hope that some of the enthusiasm of my teacher, T. H. Hildebrandt, who first introduced me to the subject, comes through to the reader. I would also like to acknowledge the fine work of Betty Surzyn in typing the manuscript, and the excellent cooperation of the editors of Prentice-Hall in producing the finished textbook.

R. G. KULLER

Contents

1 Preliminaries

Section 1 LOGICAL NOTATION

We assume that the reader has had experience with formal proofs in mathematics; so the purpose of this section is largely to set down some notations we shall use.

Let P and Q be statements. By $P \Rightarrow Q$ we mean "*P implies Q*" or "*if P then Q*"; by $P \Leftrightarrow Q$ we mean "*P is equivalent to Q*" or "*P if and only if Q.*" Other expressions synonymous with $P \Rightarrow Q$ are "*P is sufficient for Q*" and "*Q is necessary for P,*" while $P \Leftrightarrow Q$ may be read "*Q is necessary and sufficient for P.*" By $\sim P$ we mean the *negation* of P.

Many mathematical theorems are assertions that statements of the type $P \Rightarrow Q$ are true. The statement $\sim Q \Rightarrow \sim P$ is logically equivalent to $P \Rightarrow Q$ and is known as its *contrapositive*. Also, for any R, the statement

$$(P \text{ and } \sim Q) \Rightarrow (R \text{ and } \sim R)$$

is logically equivalent to $P \Rightarrow Q$. Hence, to prove the truth of $P \Rightarrow Q$, we can proceed directly, contrapositively, or by the third route, that of deducing a *contradiction*. Unfortunately, contradictions can be produced by erroneous arguments more easily than by any other way, so that proofs by this method should be avoided whenever possible.

It is sometimes considerably simpler to prove $\sim Q \Rightarrow \sim P$ than to show $P \Rightarrow Q$. For this reason the reader should get in the habit of automatically formulating the contrapositives of propositions he wishes to prove. Even if the direct approach turns out to be better, insight is frequently gained by

1

this procedure. It is not always obvious how to write down $\sim P$ from P, especially if P is complicated. But in our work, only relatively few types of statements will occur, and the reader will soon become proficient in forming $\sim P$ from P.

EXAMPLE. Let P be the statement, "For every x there exists a y such that Q is true." In this case the statement $\sim P$ reads, "there exists an x such that, for all y, Q is false." (Q is a statement involving x and y.)

The statement $Q \Rightarrow P$ is known as the *converse* of the statement $P \Rightarrow Q$.

Section 2 SETS

A set is a collection of objects or elements. Any collection of sets we consider will be a collection of subsets of some mathematically well-defined set. With this precaution, there is no danger of meeting any of the paradoxes which can arise by consideration of collections of sets which are "too large" as, for example, the collection of *all* sets. The symbol $\varnothing$ will be used to denote the empty set, that is, the set containing no members.

Let S be any set, and let $A, B, C, \ldots$ be subsets of S. We adopt the following definitions.

$A \cup B$, the *union* of A and B, is the subset of S consisting of all members of S which are in either A or B (or both).

$A \cap B$, the *intersection* of A and B, is the subset of S consisting of all members of S which are in both A and B.

$A - B$ is the set consisting of those members of A which are not in B.

$A \triangle B$, the *symmetric difference* of A and B, is defined as $(A - B) \cup (B - A)$.

$\tilde{A}$, the *complement* of A, is $S - A$. This will also be written $\sim A$.

$A \subset B$ means that A is contained in B, but is not equal to B.

$A \subseteq B$ means that A is a subset of B, and may indeed be the same as B.

The operations $\cup$, $\cap$, and $\sim$ satisfy the identities

$$A \cup A = A \qquad\qquad A \cap A = A$$
$$A \cup B = B \cup A \qquad\qquad A \cap B = B \cap A$$
$$A \cup (B \cup C) = (A \cup B) \cup C \qquad A \cap (B \cap C) = (A \cap B) \cap C$$

together with the distributive laws

$$A \cup (B \cap C) = (A \cup B) \cap (A \cup C)$$
$$A \cap (B \cup C) = (A \cap B) \cup (A \cap C).$$

Further, if $\{A_\alpha\}$ is any family of subsets of S, then the following complementation laws hold.

$$\sim\left(\bigcap_\alpha A_\alpha\right) = \bigcup_\alpha \tilde{A}_\alpha$$

$$\sim\left(\bigcup_\alpha A_\alpha\right) = \bigcap_\alpha \tilde{A}_\alpha.$$

Although all of these identities require formal proof, only the proof of the first complementation law will be given here. We will need the following additional notation:

$a \in A$ means that a is a member of the set A,   while

$a \notin A$ means that $a \in \tilde{A}$.

To prove $\sim\left(\bigcap_\alpha A_\alpha\right) = \bigcup_\alpha \tilde{A}_\alpha$, we show (i) $\sim\left(\bigcap_\alpha A_\alpha\right) \subseteq \bigcup_\alpha \tilde{A}_\alpha$, and (ii) $\sim\left(\bigcap_\alpha A_\alpha\right) \supseteq \bigcup_\alpha \tilde{A}_\alpha$. Now, if $a \in \sim\left(\bigcap_\alpha A_\alpha\right)$, then $a \notin \bigcap_\alpha A_\alpha$, and there exists α_0 such that $a \in \tilde{A}_{\alpha_0}$. Since $\tilde{A}_{\alpha_0} \subseteq \bigcup_\alpha \tilde{A}_\alpha$, we see that $a \in \bigcup_\alpha \tilde{A}_\alpha$. Hence, every member of $\sim\left(\bigcap_\alpha A_\alpha\right)$ is also a member of $\bigcup_\alpha \tilde{A}_\alpha$, and therefore $\sim\left(\bigcap_\alpha A_\alpha\right) \subseteq \bigcup_\alpha \tilde{A}_\alpha$. Conversely, if $a \in \bigcup_\alpha \tilde{A}_\alpha$, then there exists α_0 such that $a \notin A_{\alpha_0}$. Since $A_{\alpha_0} \supseteq \bigcap_\alpha A_\alpha$, we see that $a \in \sim\left(\bigcap_\alpha A_\alpha\right)$. Thus every member of $\bigcup_\alpha \tilde{A}_\alpha$ is also a member of $\sim\left(\bigcap_\alpha A_\alpha\right)$, and hence $\bigcup_\alpha \tilde{A}_\alpha \subseteq \sim\left(\bigcap_\alpha A_\alpha\right)$. This finishes the proof.

Definition 1.2.1. A collection $\mathscr{B}$ of subsets of a set S is called an *algebra* of subsets of S if

(a) $A \cup B \in \mathscr{B}$ whenever $A \in \mathscr{B}$ and $B \in \mathscr{B}$,
(b) $A \cap B \in \mathscr{B}$ whenever $A \in \mathscr{B}$ and $B \in \mathscr{B}$,
(c) $\tilde{A} \in \mathscr{B}$ whenever $A \in \mathscr{B}$.

Conditions (a), (b), and (c) are called *conditions of algebraic closure*, and by using this terminology, we can rephrase the definition to read, "A collection $\mathscr{B}$ of subsets of a set S is called an algebra of subsets of S if $\mathscr{B}$ is closed under the operations $\cup$, $\cap$, and $\sim$." We make two comments on this definition.

1.   If a collection $\mathscr{B}$ of subsets of S is closed under $\cup$ and $\sim$, or closed under $\cap$ and $\sim$, then $\mathscr{B}$ is an algebra of sets. (This follows from the identities $A \cap B = \sim[\tilde{A} \cup \tilde{B}]$ and $A \cup B = \sim[\tilde{A} \cap \tilde{B}]$.)

2.   If $\mathscr{B}$ is a nonempty algebra of subsets of S, then $\varnothing \in \mathscr{B}$ and $S \in \mathscr{B}$. (For, if $A \in \mathscr{B}$, then $\tilde{A} \in \mathscr{B}$ and $S = A \cup \tilde{A} \in \mathscr{B}$. Hence, also, $\varnothing = \tilde{S} \in \mathscr{B}$.)

EXAMPLE 1. The collection of all subsets of a given set S is an algebra of subsets of S.

EXAMPLE 2. Let I^+ be the set of all positive integers, and let $\mathscr{B}$ contain $\varnothing$, I^+, all finite subsets of I^+, and all complements of finite subsets of I^+. Then $\mathscr{B}$ is an algebra of subsets of I^+. (Note that $\mathscr{B}$ is not the collection of all subsets of I^+.)

Section 3 RELATIONS

Definition 1.3.1. The *Cartesian product* of two sets $A = \{a_\alpha\}$ and $B = \{b_\beta\}$ is the set of all ordered pairs (a_α, b_β), where $a_\alpha \in A$, $b_\beta \in B$; it is denoted by $A \times B$.

The Cartesian product of n sets is defined similarly.

Definition 1.3.2. A *binary relation* $*$ on a set S is defined by a subset G of $S \times S$. The element a is said to stand in the relation $*$ to the element b, that is, $a * b$, if and only if the ordered pair (a, b) is in G.

Without some further restrictions on the nature of G, the notion of binary relation is too general to be of much use. We now study certain special kinds of binary relations.

Definition 1.3.3. An equivalence relation on a set S is a binary relation $*$ on S which satisfies the following three conditions:

(a) $a * a$ for all $a \in S$ (reflexivity)
(b) $a * b$ implies $b * a$ (symmetry)
(c) $a * b$ and $b * c$ imply $a * c$ (transitivity).

In terms of G, we are requiring that $(a, a) \in G$ for all $a \in S$, $(a, b) \in G$ if and only if $(b, a) \in G$, and $(a, b) \in G$, $(b, c) \in G$ imply $(a, c) \in G$.

EXAMPLE 1. Consider the set I of all integers, and define $a * b$ to mean that $a - b$ is even, i.e., divisible by 2. The $*$ relation is reflexive, symmetric, and transitive. The subset G of $I \times I$ associated with $*$ is $G = \{(a, b) \in I \times I : a - b \text{ is even}\}$.†

In applications we often identify a and b if $a * b$. To make this identification process clearer, we define

†This is read "the set of all (a, b) in $I \times I$ such that $a - b$ is even."

$$\lfloor a \rfloor = \{x \in S; \; x * a\}$$

and call it the *equivalence class* of *a*.

Lemma 1.3.4. If $*$ is an equivalence relation on S, then

(a) $a \in \lfloor a \rfloor$ for all a,
(b) $\lfloor a \rfloor = \lfloor b \rfloor \Leftrightarrow a * b$,
(c) for any $\lfloor a \rfloor$ and $\lfloor b \rfloor$, either $\lfloor a \rfloor = \lfloor b \rfloor$ or $\lfloor a \rfloor \cap \lfloor b \rfloor = \varnothing$.

Proof. (a) follows from $a * a$. To prove (b) we note that $\lfloor a \rfloor = \lfloor b \rfloor$ implies $a \in \lfloor b \rfloor$, and hence $a * b$. Conversely, if $x \in \lfloor a \rfloor$, then $x * a$ which, combined with $a * b$, gives $x * b$, or $x \in \lfloor b \rfloor$. Hence $\lfloor a \rfloor \subseteq \lfloor b \rfloor$ and, in like manner, we can show that $\lfloor b \rfloor \subseteq \lfloor a \rfloor$. It follows that $\lfloor a \rfloor = \lfloor b \rfloor$ as desired. To prove (c) we show that if there is an x in $\lfloor a \rfloor \cap \lfloor b \rfloor$, then $\lfloor a \rfloor = \lfloor b \rfloor$. Now, $x \in \lfloor a \rfloor \cap \lfloor b \rfloor$ implies $x * a$ and $x * b$. Applying symmetry and transitivity, we obtain $a * b$ which, by (b), yields $\lfloor a \rfloor = \lfloor b \rfloor$.

The lemma shows that an equivalence relation on a set S serves to partition the set into disjoint equivalence classes. In the preceding example, the set I of all integers is partitioned by the given equivalence relation into exactly two equivalence classes, $\lfloor 0 \rfloor = \{$all even integers$\}$ and $\lfloor 1 \rfloor = \{$all odd integers$\}$.

Another important instance of the general notion of a binary relation is that of an *order* relation.

Definition 1.3.5. A relation $\le$ on a set S is called a *partial* order relation if

(a) $a \le a$ for all a (reflexivity)
(b) $a \le b$ and $b \le a$ imply $a = b$ (antisymmetry)
(c) $a \le b$ and $b \le c$ imply $a \le c$ (transitivity).

A partial order relation $\le$ on a set S is called a *simple* ordering if,

(d) for any a and b, either $a \le b$ or $b \le a$.

An example of a simply ordered set is the set I of all integers under its natural ordering.

EXAMPLE 2. Let S be the set of all real numbers, and consider the Cartesian plane $S \times S$. For (a, b) and (c, d) in $S \times S$ define $(a, b) \le (c, d) \Leftrightarrow [a \le c$ and $b \le d]$. The relation $\le$ partially orders $S \times S$, but it is not a simple ordering of $S \times S$. The reader should find a geometrical interpretation for this relation.

In exercises 8 and 9 the notion of a partial ordering is treated further.

Section 4 FUNCTIONS

A binary relation $*$ on a set S may have the property

$$a * b, \ a * c \Rightarrow b = c.$$

If so, then this relation defines a (single-valued) *function, f,* by the rule

$$f(a) = b \Leftrightarrow a * b,$$

and the subset G of $S \times S$ defining $*$ is called the *graph* of f.

More generally, a subset, G, of a Cartesian product, $S_1 \times S_2$, such that

$$[(a, b) \in G, (a, c) \in G] \Rightarrow [b = c]$$

defines a single-valued function g by the rule

$$g(a) = b \Leftrightarrow (a, b) \in G.$$

Again, G is called the graph of g. The *domain* $\mathscr{D}(g)$ of g and the *range* $\mathscr{R}(g)$ of f are defined by

$$\mathscr{D}(g) = \{a \in S_1 : (a, b) \in G \quad \text{for some } b \in S_2\}$$
$$\mathscr{R}(g) = \{b \in S_2 : (a, b) \in G \quad \text{for some } a \in S_1\}.$$

In other words, $\mathscr{D}(g)$ is the projection of G onto S_1, and $\mathscr{R}(g)$ is the projection of G onto S_2.

The function g is an *onto* function if $\mathscr{R}(g) = S_2$, and it is *one-to-one* if $[g(a_1) = g(a_2)] \Rightarrow [a_1 = a_2]$.

For $A \subseteq S_1$ and $B \subseteq S_2$ we define

$$g(A) = \{g(a): a \in A\}.$$

the *image* of A under g, and

$$g^{-1}(B) = \{a \in S_1 : g(a) \in B\},$$

the *inverse image* of B under g. Note that $g(A) \subseteq S_2$ and $g^{-1}(B) \subseteq S_1$. Unless g is one-to-one, g^{-1} is not a function.

The following formulas will be left to the reader to prove (see exercise 11).

$$g(\bigcup_\alpha A_\alpha) = \bigcup_\alpha g(A_\alpha) \qquad g[g^{-1}(B)] \subseteq B$$

$$g(\bigcap_\alpha A_\alpha) \subseteq \bigcap_\alpha g(A_\alpha) \qquad g^{-1}[g(A)] \supseteq A$$

(I)

$$g^{-1}(\bigcup_\alpha B_\alpha) = \bigcup_\alpha g^{-1}(B_\alpha)$$

$$g^{-1}(\bigcap_\alpha B_\alpha) = \bigcap_\alpha g^{-1}(B_\alpha)$$

The operations of addition and multiplication of real numbers are special types of functions known as binary operations. Specifically, a *binary operation* on a set S is a function with domain $S \times S$ and range in S.

A situation frequently met in mathematics is as follows. We are given a binary operation $a \circ b$ on S and an equivalence relation $*$ on S. It is natural to try to define a binary operation $\cdot$ on the set of equivalence classes by

$$\lfloor a \rfloor \cdot \lfloor b \rfloor = \lfloor a \circ b \rfloor.$$

But this is not always possible, since the equivalence class $\lfloor a \circ b \rfloor$ may depend on the representatives a and b rather than on the classes $\lfloor a \rfloor$ and $\lfloor b \rfloor$. We need to know that, if a $a' \in \lfloor a \rfloor$ and $b' \in \lfloor b \rfloor$, then $\lfloor a' \circ b' \rfloor$ is the same as $\lfloor a \circ b \rfloor$.

Definition 1.4.2. An equivalence relation $*$ on a set S is said to have the *substitution property* with respect to the binary operation $\circ$ on S if $[a' * a, b' * b]$ implies $[(a' \circ b') * (a \circ b)]$.

The reader can verify that, if $*$ does have the substitution property over $\circ$, then the definition $\lfloor a \rfloor \cdot \lfloor b \rfloor = \lfloor a \circ b \rfloor$ is proper.

EXAMPLE. Let I be the set of all integers, and let $*$ be defined by $a * b$ if and only if $b - a$ is divisible by p (p is a fixed nonzero integer). It is easy to verify that $*$ has the substitution property over addition, and hence, that we can define addition of equivalence classes. In this example there are exactly p equivalence classes, $\lfloor k \rfloor = \{k \pm np: n = 0, 1, 2, \ldots\}$, $k = 0, 1, 2, \ldots, p - 1$.

EXERCISES

1. Prove that, for any family $\{A_\alpha\}$ of subsets of a set S, $\sim[\bigcup_\alpha A_\alpha] = \bigcap_\alpha \tilde{A}_\alpha$.

2. Let S be the set of all quadratic polynomials with real coefficients. Define

$$(ax^2 + bx + c) * (Ax^2 + Bx + C)$$

if the first polynomial can be made identical with the second by replacing x by $x + d$ for some d. Show that $*$ is an equivalence relation, and give a geometrical interpretation.

3. Give examples of relations which are

    a. reflexive and symmetric, but not transitive,
    b. reflexive and transitive, but not symmetric,
    c. symmetric and transitive, but not reflexive.

4. On a set S, let $*$ be a relation which is reflexive and transitive, but not symmetric. Define

$$x \circ y \Leftrightarrow [x * y \text{ and } y * x]$$

    a. Show that "$\circ$" is an equivalence relation.
    b. Show how $*$ induces a partial ordering of the equivalence classes in a natural way.

5. Let I^+ be the positive integers, and define $m * n \Leftrightarrow [m$ is a factor of $n]$. What can you say about the relation "$*$"?

6. Let $F[a, b]$ be the set of all real-valued functions on an interval $[a, b]$. Define $f \leq g \Leftrightarrow \{f(t) \leq g(t)$ for all $t \in [a, b]\}$. Show that $F[a, b]$ is a partially ordered set under this relation, and interpret the ordering geometrically.

7. Let $\mathscr{B}_n$ be the algebra of all subsets of a set of n elements. How many sets does $\mathscr{B}_n$ contain?

8. Let P be a partially ordered set. An element z in P is called the *least upper bound* of elements x and y in P if and only if

$$z \geq x \text{ and } z \geq y, \quad \text{and} \quad w \geq x, w \geq y \Rightarrow w \geq z.$$

    a. Show that, if x and y have a least upper bound z, then it is unique. [HINT: Let z' be another least upper bound for x and y, and show $z = z'$.]
    b. Define the greatest lower bound of two elements x and y in P. Show that if it exists, then it is unique.

9. A *lattice* is a partially ordered set in which every two elements have a least upper bound and a greatest lower bound. Show that $F[a, b]$ is a lattice. (See exercises 6 and 8.)

10. Prove that the $\mathscr{B}$ of example 2 of section 2 is an algebra of sets.

11. a. Prove the general formulas (I) of section 4.
b. Give examples to show that equality may not hold in the three formulas in (I) where it is not specified.
c. Show that, in general, no connection exists between $g(\tilde{A})$ and $\sim[g(A)]$.

Section 5 ELEMENTARY RESULTS ON CARDINALITY

It is easy to compare the sizes of two finite sets, but more care is required in the case of infinite sets. The notion of a one-to-one correspondence is fundamental.

Definition 1.5.1

1. Two sets A and B are said to have the same *cardinality* if there exists a one-to-one function which maps A onto B. In this case we write $A \sim B$.
2. The cardinality, or cardinal number, $c(A)$ of a set A is the set of all sets which have the same cardinality as A.

Clearly, the relation $\sim$ is reflexive, symmetric, and transitive, and it is therefore an equivalence relation on any well-defined class of sets S. Unfortunately, we cannot take S as the set of all sets, for there are contradictions inherent in this notion. We now define an ordering on the cardinal numbers, noting first that $c(A) = c(B) \Leftrightarrow A \sim B$.

Definition 1.5.2. $c(A) \le c(B)$ if and only if A can be put into a one-to-one correspondence with a subset of B.

The relation $\le$ is reflexive and transitive. The fact that it is also antisymmetric is somewhat deeper, and is equivalent to the following theorem.

Theorem 1.5.3. If A can be put into one-to-one correspondence with a subset B' of B, and if B can be put into one-to-one correspondence with a subset A' of A, then there is a one-to-one correspondence between A and B.

Proof. Let the given correspondences be denoted by f and g, that is:

$$f: \quad A \longleftrightarrow B' \subseteq B$$
$$g: \quad B \longleftrightarrow A' \subseteq A.$$

The restriction g' of g to B' can be composed with f to yield a one-to-one mapping

$$h: \quad A \overset{f}{\longleftrightarrow} B' \overset{g'}{\longleftrightarrow} A_1 \subseteq A'$$

between A and a subset A_1 of A'. Now, if we could find a one-to-one corres-pondence between A and A', then the scheme

$$A \longleftrightarrow A' \longleftrightarrow B$$

would yield the final desired result.

Theorem 1.5.3 is then a consequence of the following assertion:

Let $M \supseteq M_1' \supseteq M_1$ and let p be a one-to-one correspondence between M and M_1. Then there exists a one-to-one correspondence between M and M_1'.

To prove this assertion, we let $M_0 = M$ and define inductively $M_{k+1} = p(M_k)$ for $k \geq 0$; similarly, we define $M_{k+1}' = p(M_k')$ for $k \geq 1$. The simple fact that $A \supseteq B \Rightarrow p(A) \supseteq p(B)$ then gives $M_0 \supseteq M_1' \supseteq M_1 \supseteq M_2' \supseteq M_2 \supseteq \cdots \supseteq M_k' \supseteq M_k \supseteq \cdots$, with $M_k' \longleftrightarrow M_{k+1}'$, and $M_k \longleftrightarrow M_{k+1}$ for all k. These correspondences, in turn give $(M_k - M_{k+1}') \longleftrightarrow (M_{k+1} - M_{k+2}')$ for all k. Letting $N = \bigcap_{k=0}^{\infty} M_k$, we have

$$M = (M_0 - M_1') \cup (M_1' - M_1) \cup (M_1 - M_2') \cup (M_2' - M_2) \cup \cdots \cup N$$

and

$$M_1' = (M_1' - M_1) \cup (M_1 - M_2') \cup (M_2' - M_2)$$
$$\cup (M_2 - M_3') \cup \cdots \cup N,$$

where the sets in each union are disjoint.

The desired one-to-one correspondence between M and M_1' is then obtained by using the identity mapping on N and on the sets of type $(M_k' - M_k)$, and the mapping p between the sets $(M_k - M_{k+1}')$ and $(M_{k+1} - M_{k+2}')$. This completes the proof.

Corollary. If $L \supseteq M \supseteq N$ and $c(L) = c(N)$, then $c(L) = c(M) = c(N)$.

EXAMPLE 1. If $a \leq x \leq b$ and $c \leq y \leq d$ are any two closed intervals in R, the real numbers system, then the function $y = c + \dfrac{d-c}{b-a}(x-a)$ gives a one-to-one correspondence between them.

EXAMPLE 2. We introduce the notations

$$[a, b] = \{x: \quad a \leq x \leq b\}$$
$$[a, b) = \{x: \quad a \leq x < b\}$$

$$(a, b] = \{x: \quad a < x \le b\}$$
$$(a, b) = \{x: \quad a < x < b\}$$

Clearly $[0, 1] \supset [0, 1) \supset [0, \frac{1}{2}]$, and by example 1, $[0, 1] \sim [0, \frac{1}{2}]$. The corollary then yields $[0, 1] \sim [0, 1)$.

Our work so far shows that the $\le$ relation between cardinal numbers is a partial ordering. *It is actually a simple ordering*, since it can be proved that, for any two sets A and B, either $A \sim B_1 \subseteq B$ or $B \sim A_1 \subseteq A$. We shall use this fact, but relegate its demonstration to the appendix, where an additional set theoretic principle will be introduced; that principle is essential for the proof.

Definition 1.5.4. For a set A we say that

(a) A is *finite* if, for some positive integer n, $A \longleftrightarrow \{1, 2, \ldots, n\}$.
(b) A is *infinite* if it is not finite.
(c) A is *denumerable* if it can be put in a one-to-one correspondence with I^+, the set of all positive integers.
(d) A is *countable* if it is either finite or denumerable.
(e) A is *uncountable* if it is not countable.

EXAMPLE 3. The set I of all integers is denumerable. A one-to-one correspondence between I^+ and I is given by

$$
\begin{array}{ccccccc}
1 & 2 & 3 & 4 & 5 & 6 & 7 \quad \cdots \\
\updownarrow & \updownarrow & \updownarrow & \updownarrow & \updownarrow & \updownarrow & \updownarrow \\
0 & 1 & -1 & 2 & -2 & 3 & -3 \quad \cdots
\end{array}
$$

EXAMPLE 4. The rational numbers Q are denumerable. To see this, note that for any n there is only a finite number $N(n)$ of rationals p/q such that $0 \le p/q < 1$ and $0 < q \le n$. The set of rationals p/q, $0 \le p/q < 1$, is then the union of a denumerable number of finite sets and is therefore denumerable. The set of all rationals, as a denumerable union of denumerable sets, is also denumerable. (We are using theorem 1.5.5, proven below.)

EXAMPLE 5. Let A be the set of all real numbers which are roots of polynomials with rational coefficients. By multiplying such a polynomial by the least common multiple of the denominators of its coefficients, we obtain a polynomial with integer coefficients; hence, it is equivalent to consider polynomials of this type. For such a polynomial, $p(t) = a_n t^n + a_{n-1} t^{n-1} + \cdots + a_1 t + a_0$, we define $N(p) = |a_n| + \cdots + |a_0| + n$, the *index* of $p(t)$. Now, there are only a finite number of polynomials (with integer coefficients) of index m, and hence, only a finite number $M(m)$ of algebraic numbers (roots of polynomials with integer coefficients) arise from this set of polynomials. Thus the set A of all algebraic numbers is a denumerable union of finite sets, and it is therefore denumerable.

Theorem 1.5.5. A countable union of countable sets is countable.

Proof. Let the sets $\{A_n\}$ be countable. Then we have

$$A_1 = \{a_{11}, a_{12}, a_{13}, \ldots, a_{1n}, \ldots\},$$
$$A_2 = \{a_{21}, a_{22}, a_{23}, \ldots, a_{2n}, \ldots\},$$
$$A_3 = \{a_{31}, a_{32}, a_{33}, \ldots, a_{3n}, \ldots\},$$

$$\cdot \qquad \cdot \qquad \cdot \qquad \qquad \vdots$$
$$\cdot \qquad \cdot \qquad \cdot \qquad \qquad \cdot$$

$$A_n = \{a_{n1}, a_{n2}, a_{n3}, \ldots, a_{nn}, \ldots\}, \ldots .$$

and the set $\bigcup_n A_n$ can be enumerated as follows:

$$a_{11}; \quad a_{21}, a_{12}; \quad a_{31}, a_{22}, a_{13}; \quad a_{41}, a_{32}, a_{23}, a_{14}; \quad \ldots$$

(If some of the A_n's are finite, then certain terms will be omitted.) This completes the proof.

The reader may have noticed that an infinite set may be in one-to-one correspondence with a proper subset of itself. For example, the positive integers are in one-to-one correspondence with the even positive integers. This peculiarity is characteristic of infinite sets, for it can be proved that a set is infinite if and only if it can be put into a one-to-one correspondence with a proper subset of itself. (See exercise 8.)

We come now to the problem of exhibiting an uncountable set.

EXAMPLE 6. The set R of all real numbers is uncountable. It suffices to show that $[0, 1)$ is uncountable. Now, a number in $[0, 1)$ is an infinite decimal (we will discuss this in the next chapter), and if we agree not to permit decimals to end in an infinite sequence of 9's, then there is a one-to-one correspondence between $[0, 1)$ and the set of decimals $.b_1 b_2 \cdots b_n \cdots .$

Assume that a one-to-one function from I^+ into $[0, 1)$ is given as follows:

$$1 \longrightarrow a_1 = .a_{11} a_{12} \cdots a_{1n} \cdots$$
$$2 \longrightarrow a_2 = .a_{21} a_{22} \cdots a_{2n} \cdots$$

$$\cdot \qquad \cdot \qquad \cdot$$
$$\cdot \qquad \cdot \qquad \cdot$$
$$\cdot \qquad \cdot \qquad \cdot$$

$$n \longrightarrow a_n = .a_{n1} a_{n2} \cdots a_{nn} \cdots$$

$$\cdot \qquad \cdot \qquad \cdot$$
$$\cdot \qquad \cdot \qquad \cdot$$
$$\cdot \qquad \cdot \qquad \cdot$$

We define

$$\left\{\begin{matrix} b_n = 1 & \text{if} & a_{nn} \neq 1 \\ b_n = 2 & \text{if} & a_{nn} = 1 \end{matrix}\right\} \quad \text{for } n = 1, 2, \ldots.$$

Then the number $b = .b_1 b_2 b_3 \cdots b_n \cdots$ is different from all the a_n's (its decimal disagrees with theirs). It follows that f is not an *onto* function, and that there is no one-to-one mapping of I^+ onto $[0, 1)$. Hence $[0, 1)$ is uncountable.

Theorem 1.5.6. The cardinal number of the set of all subsets of I^+ is $c(R)$.

Proof. Let S be the set of all subsets of I^+. We determine a member of S by going down the list of positive integers and checking off those to be included. Alternatively, we associate a 1 with those integers to be kept and a 0 with those to be discarded. Hence S has the same cardinality as the set of all infinite sequences of 0's and 1's.

Now, R is in one-to-one correspondence with $[0, 1)$ (see the exercises), and if the numbers in $[0, 1)$ are written as decimals in the binary system, then $[0, 1)$ is equivalent to a subset of S. (To avoid ambiguities, we do not permit binary decimals to end in an infinite sequence of 1's.) Hence R is equivalent to a subset of S.

Finally, S is equivalent to the set of all numbers in $[0, 1)$ whose base-ten decimals contain only zeros and ones. Hence S is equivalent to a subset of R, and an application of theorem 1.5.3 completes the proof.

EXERCISES

1.   Prove that the set of irrational numbers is uncountable.

2.   Give a direct proof that $[0, 1)$ and $[0, 1]$ are equivalent (i.e., do not use theorem 1.5.3).

3.   Show that $R \sim (0, 1)$.

4.   Show that $[0, 1] \times [0, 1]$ is equivalent to $[0, 1]$. [Use the mapping $(.a_1 a_2 \cdots a_n \cdots, .b_1 b_2 \cdots b_n \cdots) \rightarrow (.a_1 b_1 a_2 b_2 \cdots a_n b_n \cdots)$.]

5.   Show that $R \sim R \times R$.

6.   Let S be the set of all sequences of real numbers. Show that S has cardinality $c(R)$.

7. Let S be the set of all sequences of elements from a countable set A. Show that $c(S) = c(R)$.

8. a. Show that a denumerable set can be put into a one-to-one correspondence with a proper subset of itself.
    b. Show that an uncountable set has a proper denumerable subset.
    c. Show that an uncountable set can be put into a one-to-one correspondence with a proper subset of itself.

9. Let T be the set of all sequences of nonnegative integers having only finitely many nonzero entries. Show that $c(T) = c(I^+)$.

10. Let B be the set of all finite subsets of a countable set A. Show that $c(B) = c(A)$.

11. Let A be any infinite set and I a denumerable set disjoint from A. Show that $C(A) = C(I \cup A)$.

2 Real Numbers

In chapter 1 we tacitly assumed a knowledge of the real-number system, R. In this chapter we explore the nature of R much more deeply.

Section 1 FIELDS

The real-number system has aspects which are algebraic and aspects which deal with the notion of convergence. In this section and the next, we treat mostly the former, and we begin by defining an algebraic structure known as a field.

Definition 2.1.1. A field F is a set containing at least two members and having two binary operations called addition and multiplication. These two operations

$$(a, b) \longrightarrow a + b$$
$$(a, b) \longrightarrow ab$$

are assumed to satisfy the following axioms.

1. $a + b = b + a$ $ab = ba$ (commutative laws)
2. $a + (b + c) = (a + b) + c$ $(ab)c = a(bc)$ (associative laws)
3. $a(b + c) = ab + ac$ (distributive law)
4. There exists an element 0 such that

15

$$a + 0 = a \qquad \text{for all } a \text{ in } F.$$

5. There exists an element 1 such that

$$a1 = a \qquad \text{for all } a \text{ in } F.$$

6. For every element a in F there exists an element $(-a)$ such that

$$a + (-a) = 0.$$

7. For every element a different from 0 in F there exists an element

$$a^{-1} \quad \text{such that} \quad aa^{-1} = 1.$$

Additional relations valid in any field are given in the following lemma.

Lemma 2.1.2

(a) (Uniqueness of 0) If $a + 0' = a$ for all $a \in F$, then $0' = 0$.
(b) (Uniqueness of 1) If $a1' = a$ for all $a \in F$, then $1' = 1$.
(c) (Uniqueness of additive inverses) If $a + b = 0$, then $b = -a$.
(d) (Uniqueness of multiplicative inverses) If $ab = 1$, $a \neq 0$, then $b = a^{-1}$.
(e) $0a = 0$ for all $a \in F$.
(f) $-a = (-1)a$ for all $a \in F$.
(g) $(-1)(-1) = 1$.
(h) $ab = 0 \Rightarrow (a = 0 \quad \text{or} \quad b = 0)$.

Proof

(a) For $a = 0$ we have $0 + 0' = 0$, but also $0 + 0' = 0' + 0 = 0'$; hence $0 = 0'$.
(b) For $a = 1$, we have $11' = 1$, but also $11' = 1'1 = 1'$; hence $1 = 1'$.
(c) From $a + b = 0$, we obtain $b = (-a) + a + b = -a$, and $b = -a$.
(d) From $ab = 1$, $a \neq 0$, we obtain $b = a^{-1}ab = a^{-1}$, or $b = a^{-1}$.
(e) $0a = (0 + 0)a = 0a + 0a$; addition of $-(0a)$ to both sides yields $0 = 0a$.
(f) $a + (-1)a = 1a + (-1)a = [1 + (-1)]a = 0a = 0$, and the uniqueness of $-a$ yields $(-1)a = -a$.
(g) $0 = 0(-1) = [1 + (-1)](-1) = 1(-1) + (-1)(-1)$
$= (-1) + (-1)(-1)$.
Addition of 1 to both sides yields $1 = (-1)(-1)$.
(h) If $ab = 0$ and $a \neq 0$, then there exists a^{-1}, and $b = a^{-1}ab = a^{-1}0 = 0$.

This completes the proof. Henceforth we will write $a - b$ for $a + (-b)$.

EXAMPLE. Let I be the set of all integers. Define on I the relation

$$a * b \Leftrightarrow (p \text{ divides } a - b),$$

where p is a fixed prime number. As indicated in the example in section 4 of chapter 1, $*$ is an equivalence relation on I, and it has the substitution property with respect to addition. It also has the substitution property with respect to multiplication. Thus the rules

$$\lfloor a \rfloor \lfloor b \rfloor = \lfloor ab \rfloor$$
$$\lfloor a \rfloor + \lfloor b \rfloor = \lfloor a + b \rfloor$$

define two binary operations on the finite set of equivalence classes

$$\lfloor 0 \rfloor, \lfloor 1 \rfloor, \lfloor 2 \rfloor, \ldots, \lfloor p - 1 \rfloor.$$

This set, with the two operations, is called J_p; proof that J_p is a field is left for the exercises. Although J_m is well-defined for any positive integer m, it is not a field unless m is prime, for if $m = cd$, then $\lfloor 0 \rfloor = \lfloor c \rfloor \lfloor d \rfloor$, in contradiction to part (h) of lemma 2.1.2.

The preceding example shows that the field axioms are quite general. The property of *order* has not yet been incorporated into our discussion. We now do this.

Definition 2.1.3. An ordered field F is a field in which there is a subset P (called the positive elements of F) such that

(a) F is *partitioned* by the sets $\{0\}$, P, and $-P$. ($-P$ is the set of all additive inverses of the elements of P.)

(b) $a \in P, b \in P \Rightarrow a + b \in P$.

(c) $a \in P, b \in P \Rightarrow ab \in P$.

If we now define $y \le x \Leftrightarrow (x - y) \in P \cup \{0\}$, we immediately see that we have a *simple order* relation defined on F. The proof is trivial and is left to the reader, as is the verification of the implications

$$a \le b \Rightarrow a + c \le b + c$$
$$a \le b, c \ge 0 \Rightarrow ac \le bc$$
$$a \le b, c \le 0 \Rightarrow ac \ge bc.$$

We also have

$$1 \in P$$

$$a \in P \Rightarrow a^{-1} \in P.$$

To prove the first, we need only show that 1 is not in $-P$, since we know $1 \neq 0$. Now, if $1 \in -P$, then $-1 \in P$ and $(-1)(-1) = 1 \in P$; but we cannot have both $1 \in P$ and $1 \in -P$. Hence $1 \notin -P$ and $1 \neq 0$; so we conclude $1 \in P$. Similarly, if $a \in P$ and $a^{-1} \in -P$, then $a(-a^{-1}) \in P$ or $-1 \in P$. Since this is false, we must have $a^{-1} \in P$.

We now make the assumption that at least one ordered field exists. This assumption could be avoided by giving an explicit construction of the rational number system Q from more primitive hypotheses, but this is a direction in which we do not wish to go. (See, however, exercises 11 and 12.)

Theorem 2.1.4. Every ordered field F contains a smallest subset F' such that F' is itself an ordered field under the operations of F.

Proof. Any subset G of F which is itself an ordered field must contain 0 and 1, and hence, all sums of 1's, i.e., all elements of the form $1, 1 + 1$, $1 + 1 + 1, \ldots$. (See exercise 14.) These sums must be distinct, for otherwise, some sum of 1's would be 0; but no sum of members of P can lie outside P. Using the usual notation for sums of 1's, we see that G must contain

$$I = \{\ldots, -n, \ldots, -2, -1, 0, 1, 2, \ldots, n, \ldots\},$$

the multiplicative inverses of all nonzero members of I, and all products mn^{-1}, where $m, n \in I$. Denoting this set by F', we have $G \supseteq F'$. It is now a routine matter to verify that F' is an ordered field. (We refer the reader to the exercises for details.) The proof is now complete, for F' is necessarily the smallest subfield of F.

Now, if F' is the minimal subfield of the ordered field F and if G' is the minimal subfield of the ordered field G, then there is a unique one-to-one correspondence h between F' and G' such that

$$h(a + b) = h(a) + h(b)$$
$$h(ab) = h(a)h(b)$$
$$a \leq b \Rightarrow h(a) \leq h(b).$$

The detailed proof is left for the exercises. Because of this correspondence, we think of Q, the rational numbers, as the smallest ordered field. The mapping h is called an *order isomorphism* between F' and G', and the latter are said to be *order isomorphic*.

We now state the *density* property of ordered fields.

(D) If $a < b$ then there is an element c such that $a < c < b$. (By the strict inequality $a < b$ we mean $a \leq b$ but $a \neq b$.)

A corollary to (D) is the property that, between any two members of F, there are denumerably many others. For proof of (D) and this corollary, we refer the reader to the exercises.

The simple order in Q has another important property which we now define.

Definition 2.1.5. The order in an ordered field is said to be *Archimedian* if, for any two elements $a > 0$, $b > 0$, there is a positive integer n such that $na > b$.

The proof that the order in Q is Archimedian is left to the exercises. There the reader will also meet an ordered field which is not Archimedian. In the next section we introduce the key property of the real-number field which characterizes this field in the class of all ordered fields.

EXERCISES

1. a. Let a and b be two fixed integers and let $S = \{xa + yb : x, y \in I\}$, where I is the set of all integers. Let d be the smallest positive integer in S. Show that S is the set of all integral multiples of d.
    b. Show that d is the greatest positive common divisor of a and b.
    c. Show that if 1 is the greatest common divisor of a and b, then there exist integers p and q such that $pa + qb = 1$.
    d. Show that J_p is a field whenever p is prime.

2. Show that, in any field,
    a. $(a^{-1})^{-1} = a$;
    b. $-(-a) = a$;
    c. $-(a + b) = (-a) + (-b)$;
    d. $(ab)^{-1} = a^{-1} b^{-1}$ $(a, b \neq 0)$.

3. Complete the proof of theorem 2.1.4 by showing that F' is an ordered field.

4. Verify that an ordered field F is simply ordered by the relation

    $$y \leq x \Leftrightarrow [x - y \in P \cup \{0\}].$$

5. Prove the implications $a \leq b \Rightarrow a + c \leq b + c$, $a \leq b$, $c \geq 0 \Rightarrow ac \leq bc$, and $a \leq b$, $c \leq 0 \Rightarrow ac \geq bc$.

6. Show that $a < b \Rightarrow [a < 2^{-1}(a + b) < b]$, and hence, that the density property (D) and its corollary are valid in any ordered field.

7. Prove that the order in Q is Archimedian.

8. Prove that no choice of P could make J_p an ordered field.

9. Let G be the set of all real numbers of the form $a + b\sqrt{3}$, where a and b are rational. Show that G is an Archimedian ordered field under the usual definitions of addition, multiplication, and order. (Assume you know what $\sqrt{3}$ is.)

10. Let F be the set of all quotients of polynomials in x which have rational coefficients. Show that F is a field under the usual definitions of addition and multiplication. Let P be the subset of F consisting of all functions in F which become and remain positive as $x \longrightarrow +\infty$. Show that F is an ordered field under the ordering induced by P. Show that this order is not Archimedian.

11. In this problem we assume I^+, the set of positive integers, is completely known, together with the operations of addition and multiplication on I^+, and their properties, including the cancellation laws:

$$\text{CA}: \quad a + b = a + c \Rightarrow b = c$$
$$\text{CM}: \quad ab = ac \Rightarrow b = c.$$

(Subtraction and division would lead outside of I^+ and cannot be assumed.)

a. Let $\sim$ be defined on $I^+ \times I^+$ by $(a, b) \sim (c, d) \Leftrightarrow a + d = b + c$. Show that $\sim$ is an equivalence relation on $I^+ \times I^+$. Define

$$(a, b) + (c, d) = (a + c, b + d), \quad \text{and}$$
$$(a, b) \cdot (c, d) = (ac + bd, bc + ad)$$

and show that $\sim$ has the substitution property with respect to both of these operations.

b. Let I be the set of equivalence classes in $I^+ \times I^+$. Show that the addition and multiplication in I [induced by the operations defined in (a)] are both commutative and associative.

c. Show that the cancellation law of addition is valid in I.

d. Show that the general equation $\lfloor (a, b) \rfloor + \lfloor (x, y) \rfloor = \lfloor (c, d) \rfloor$ has a unique solution.

e. Show that I has a unique additive zero element, $\lfloor (a, a) \rfloor$.

f. Prove the cancellation law of multiplication in I.

g. Show that I has a unique multiplicative identity element.

h. Show that $\lfloor (a, b) \rfloor \lfloor (c, d) \rfloor = \lfloor (e, e) \rfloor$ implies that either $\lfloor (a, b) \rfloor = \lfloor (e, e) \rfloor$ or $\lfloor (c, d) \rfloor = \lfloor (e, e) \rfloor$.

    i. Prove the distributive law of multiplication over addition.

    j. Show that $\lfloor (a, b) \rfloor \lfloor (c, c) \rfloor = \lfloor (c, c) \rfloor$.

    k. Show that there are equations of the form $\lfloor (a, b) \rfloor \lfloor (x, y) \rfloor = \lfloor (c, d) \rfloor$ which have no solutions.

12. Use the familiar notation $\{\ldots, -n, \ldots, -2, -1, 0, 1, 2, \ldots, n, \ldots\}$ for the set I of exercise 11.

    a. In $I \times (I - \{0\})$, define $(p, q) * (r, s) \Leftrightarrow ps = qr$, and show that $*$ is an equivalence relation having the substitution property with respect to the operations

$$(p, q) + (r, s) = (ps + rq, qs)$$
$$(p, q)(r, s) = (pr, qs).$$

    b. Let Q be the set of equivalence classes with the operations of addition and multiplication induced by the definitions in (a). Show that Q is a field.

13. Complete the discussion of the correspondence h at the end of the section. (Prove first that $h(0) = 0$ and $h(1) = 1$.)

14. (Refer to the proof of theorem 2.1.4.) It is conceivable that elements $\hat{0}$ and $\hat{1}$, distinct from 0 and 1, serve as identity elements in a proper subfield G of F, and that 0 and 1 are not in G. Prove that this cannot happen.

Section 2 THE REAL NUMBERS

In the previous section, the rational-number system Q was characterized as the smallest ordered field. We now introduce a notion which will single out the real-number system in the class of all ordered fields.

Definition 2.2.1. An ordered field F is *complete* if every subset of F which has an upper bound has a least upper bound.

The terms upper bound and least upper bound are defined as follows. An element a is an *upper bound* for a set S if $a \geq s$ for all $s \in S$. An upper bound b is the *least upper bound* for S if $a \geq b$ for every upper bound a of S. If a set S has a least upper bound a, then a is unique, for if a' were a second least upper bound for S, then we would have $a \leq a'$ and $a' \leq a$, or $a = a'$. The terms lower bound and greatest lower bound are similarly defined and we again have uniqueness for the latter.

In a complete ordered field, any subset S which has a lower bound must have a greatest lower bound, for the set $-S = \{-s: s \in S\}$ is bounded above and, hence, has a least upper bound a; the element $-a$ is then the greatest lower bound of S.

We will use the notation

$$b = \sup A$$
$$a = \inf A$$

to designate that b is the least upper bound of A and a is the greatest lower bound of A. The abbreviations *sup* and *inf* stand for *supremum* and *infimum*, respectively. It will be convenient to write

$$\sup A = \infty$$

when A is not bounded above and

$$\inf A = -\infty$$

when A is not bounded below. The symbols ∞ and $-\infty$ are *not* to be interpreted as members of F.

Lemma 2.2.2. A complete ordered field F is Archimedian.

Proof. Let a and b be two members of P, the set of positive elements of F, and suppose $na \leq b$ for all positive integers n. Then the set $\{na\}$ has a least upper bound c. Further, since $a > 0$, we have $c - a < c$; thus, if $na \leq c - a$ for all n, the element c would not be the *least* upper bound of $\{na\}$. Hence, for some integer m, we have $ma > c - a$, and $(m + 1)a > c$, which is also contradictory to the definition of c. Therefore, our supposition $na \leq b$ for all n must be false, and we must have $Na > b$ for some N. The lemma is now proved.

To see that Q is not complete, consider the set $S = \{p/q: p^2/q^2 < 2\}$. It is bounded above (by $\frac{3}{2}$ for example); we now show that it has no least upper bound.

Suppose the rational number x is the least upper bound of S. Then exactly one of the following must hold:

(a) $x^2 = 2$
(b) $x^2 < 2$
(c) $x^2 > 2$.

We first show that $x^2 = 2$ is impossible. Suppose $x^2 = 2$ and $x = r/s$, where r and s have no common factor. Then $r^2/s^2 = 2$ and $r^2 = 2s^2$, which implies that 2 is a factor of r^2 and thus also of r. Hence 2^2 is a factor of r^2, and 2 must therefore be a factor of s, contrary to the assumption that r and s have no common factor.

Now suppose $x^2 < 2$. We will find a rational number y such that $x^2 < y^2 < 2$, in contradiction to the definition of x. It is natural to try to find a small rational number r such that $y = x + r$ satisfies $(x + r)^2 < 2$, or $x^2 + 2rx + r^2 = x^2 + r(2x + r) < 2$. If we require $r < 1$, then $x^2 + r(2x + r) < x^2 + r(2x + 1)$, and we see that the choice $r < \min\left(1, \dfrac{2 - x^2}{2x + 1}\right)$ will suffice. In like manner, it can be shown that $x^2 > 2$ is impossible.

It follows that the set S has no least upper bound, and hence, that the Archimedian ordered field Q is not complete.

At this point we have no guarantee that any complete ordered field exists. We will not give here the classical construction due to Dedekind; instead, we will simply *assume* that there exists as least one complete ordered field R.

Lemma 2.2.3

(a) If $a < b$ in R, then there exists an element $r \in Q$ such that $a < r < b$.

(b) For every $a \in R$, $a = \sup\{r \in Q : r < a\}$.

Proof

(a) Without loss we can assume $a > 0$. By the Archimedian property there is an integer N such that $1/N < (b - a)$. Now let m be an integer such that $(m - 1)/N \leq a < (m/N)$. The rational $r = m/N$ satisfies $a < r < b$.

(b) The set $S = \{r \in Q : r < a\}$ is bounded above by a; hence, it has a least upper bound $a' \leq a$. If $a' < a$, then a rational r can be found such that $a' < r < a$. This is impossible, since $r \in S$ and $a' = \sup S$. It follows that $a' = a$, and the proof is finished.

We are now ready to demonstrate that there is essentially only one complete ordered field.

Theorem 2.2.4. If R and R' are any two complete ordered fields, then there exists a one-to-one mapping f of R onto R' such that

$$f(a + b) = f(a) + f(b)$$
$$f(ab) = f(a)f(b)$$
$$a \leq b \Rightarrow f(a) \leq f(b).$$

In other words, R and R' are order isomorphic.

Sketch of the Proof. If Q and Q' denote the rational subfields of R and R', respectively, then by the results of the preceding section, there is exactly one order isomorphism f between Q and Q'. Let S' be the image under f of $\{r \in Q : r < a\}$ where a is in R but not in Q. It is easy to see that S' is bounded above. Thus we can define $f(a) = \sup S'$ and, in this way, extend f to a mapping from R into R'. The proof that f, thus extended, is an order isomorphism between R and R' is somewhat lengthy but not difficult. We omit the demonstration. (See, however, the exercises.)

The (unique) complete ordered field R will be henceforth referred to as the *real-number system*.

In the preceding chapter, we identified real numbers with infinite decimals. We now elaborate on this. By the Archimedian property of the order, any real number a can be written $n + b$, where n is an integer and $0 \leq b < 1$; hence it suffices to obtain a decimal representation for numbers b satisfying $0 < b < 1$. Now, $0 < 10b < 10$, and we let b_1 be the largest integer less than or equal to $10b$. The existence of such an integer is again guaranteed by the Archimedian property of R. Then $.b_1 < b < 1$ and $0 \leq b - .b_1 < .1$. If $b = .b_1$, we are finished; otherwise $0 < b - .b_1 < .1$ and $0 < 10^2(b - .b_1) < 10$. Let b_2 be the largest integer less than or equal to $10^2(b - .b_1)$. Continuing in this way, we obtain an infinite decimal $.b_1b_2b_3 \cdots b_n \cdots$ such that $0 \leq b - .b_1 \cdots b_n < 10^{-n}$ for all n. By $b = .b_1b_2 \cdots b_n \cdots$, we mean simply that b is the least upper bound of the numbers $.b_1, .b_1b_2, \ldots, .b_1b_2 \cdots b_n, \ldots$.

The notion of *absolute value* is essential in our subsequent work.

Definition 2.2.5. The absolute value of a real number a is defined by

$$|a| = \begin{cases} a, & \text{if } a \geq 0 \\ -a, & \text{if } a < 0. \end{cases}$$

The absolute-value function has the following properties.

$$|x| \geq 0; \qquad |x| = 0 \Leftrightarrow x = 0,$$
$$|xy| = |x| \cdot |y|, \quad \text{and}$$
$$|x + y| \leq |x| + |y|.$$

The first two follow at once from our definitions, while the third is obvious if $xy \geq 0$, in which case it becomes an equality. If $xy < 0$, we can assume $x > 0$, $y < 0$, in which case $x + y < x < x + (-y) = |x| + |y|$. Also,

$$-(x + y) = -x - y < -y < -y + x = |x| + |y|.$$

Combining these inequalities, we obtain $|x + y| < |x| + |y|$. We also have

$$|x - y| = |x + (-y)| \leq |x| + |-y| = |x| + |y|.$$

The complex-number system is now easily treated.

Definition 2.2.6. The complex-number system C is the set $R \times R$ with the operations of addition and multiplication defined by

$$(x_1, y_1) + (x_2, y_2) = (x_1 + x_2, y_1 + y_2)$$
$$(x_1, y_1)(x_2, y_2) = (x_1 x_2 - y_1 y_2, x_1 y_2 + x_2 y_1)$$

The proof that C is a field is straightforward. The additive identity element in C is $(0, 0)$, while the multiplicative identity is $(1, 0)$. The subset of C consisting of all elements of the form $(a, 0)$ is itself a field and is, of course, isomorphic to R. The element $(0, 1)$ corresponds to the more familiar $i = \sqrt{-1}$. In future work with C, we will use the usual notation $a + ib$, where $a, b \in R$, instead of (a, b).

It is interesting that under no choice of P, the set of positive elements, does C become an ordered field. To prove this we suppose that C has been ordered, in which case we know that $(1, 0) > (0, 0)$. A routine calculation yields

$$(0, 1)^2 = -(1, 0) \quad \text{and} \quad [-(0, 1)]^2 = -(1, 0).$$

If $(0, 1) > (0, 0)$, we see that $-(1, 0) > (0, 0)$ or $(1, 0) < (0, 0)$, a contradiction. Similarly, $(0, 1) < (0, 0)$ leads to the same contradiction. Hence the element $(0, 1)$ is not in $P \cup \{0\} \cup (-P)$, and the assumed ordering is illegitimate.

Definition 2.2.7. The absolute value of a complex number $z = a + bi$ is defined as $|z| = \sqrt{a^2 + b^2}$.

The absolute-value function has the same properties in C as in R, namely,

$$|z| \geq 0; \quad |z| = 0 \Leftrightarrow z = 0,$$
$$|wz| = |w| \cdot |z|, \quad \text{and}$$
$$|w \pm z| \leq |w| + |z|.$$

The proof is found in the exercises, as is the proof of the following inequality, which is valid in both R and C.

$$||x| - |y|| \leq |x \pm y|$$

EXERCISES

1. Complete the proof of theorem 2.2.4.

2. Discuss the problem of representing real numbers by infinite decimals if the base 2 is used instead of the base 10.

3. Prove that the complex-number system C is a field under the addition and multiplication given in definition 2.2.6.

4. For a complex number (a, b), define $\overline{(a, b)} = (a, -b)$, the complex conjugate of (a, b). Prove
a. $\overline{w + z} = \bar{w} + \bar{z}$,
b. $\overline{wz} = \bar{w}\bar{z}$,
c. $|z|^2 = z\bar{z}$,
d. $z = \bar{z} \Leftrightarrow z$ is real.

5. Prove that the relations
$$|wz| = |w||z|$$
$$|w + z| \le |w| + |z|$$
hold for any two complex numbers w and z. Exactly when does the second relation become an equality?

6. Let a be a positive, real number, let n be a fixed, positive integer, and define $c = \sup \{b: b^n < a\}$. Show that $c^n = a$. (HINT: Show that $c^n < a$ and $c^n > a$ are both impossible.) Now show that $d^n = a, d > 0$, imply $d = c$. [This proves that any positive, real number a has a unique positive nth root for every positive integer n.]

7. Prove that between any real numbers a and b there is an irrational number c.

8. Prove that the numbers represented by terminating decimals are exactly those rational numbers p/q, where q is an integer of the form $2^m \cdot 5^n$ (p and q are assumed to have no common factor). (A terminating decimal is one which ends in an infinite sequence of zeros.)

9. Prove that every rational number in $(0, 1)$ is represented by a repeating decimal and conversely. (A repeating decimal is one of type $.a_1 \cdots a_k b_1 \cdots b_n b_1 \cdots b_n b_1 \cdots b_n \cdots$, where the block of digits $b_1 \cdots b_n$ repeats indefinitely.)

10. There are two types of nonterminating repeating decimals, those which repeat from the start, e.g., $.b_1 \cdots b_n b_1 \cdots b_n \cdots$, and those which repeat "after a while," e.g., $.a_1 \cdots a_k b_1 \cdots b_n b_1 \cdots b_n \cdots$, $k > 0$. Show that the first type represents rationals of the form p/q, where the greatest common divisor of 10 and q is 1. (Assume p and q have no common factor.)

11. a. Let A be a set of real numbers with sup $A = a$. Show that for any real number $\epsilon > 0$ there is a number a_1 in A such that $a - \epsilon \leq a_1 < a$. (If a happens to be in A, then we can take $a_1 = a$.)
    b. Formulate and prove an analogous statement about greatest lower bounds.

12. Prove that $||x| - |y|| \leq |x \pm y|$ holds in both R and C.

Section 3   PROPERTIES OF R

In the preceding section we characterized the real-number system R as a complete ordered field. We now introduce a number of concepts which will be useful in studying R.

Definition 2.3.1. Let A be a nonempty subset of R.

(a) A point x_0 is a *cluster point* of A if, for every $h > 0$, the interval $(x_0 - h, x_0 + h)$ contains at least two members of A.
(b) A', the *derived* set of A, is the set of all cluster points of A. Note that $x_0 \in A'$ if and only if, for every $h > 0$, the interval $(x_0 - h, x_0 + h)$ contains infinitely many members of A. (Why?)

Definition 2.3.2. Let A be a nonempty subset of R.

(a) The *closure* $\bar{A}$ of A is $A \cup A'$.
(b) A is *closed* if $A = \bar{A}$, i.e., $A' \subseteq A$.
(c) The interior $A°$ of A is the set of all points x for which there exists an $h > 0$ such that $(x - h, x + h) \subseteq A$; thus $A \supseteq A°$.
(d) A is *open* if $A = A°$.

We define the empty set $\varnothing$ to be both open and closed.

The reader should establish that $\bar{A} \supseteq A \supseteq A°$, and that the following remarks hold.

31734

1. R is both open and closed.
2. (a, b) is open.
3. $[a, b]$ is closed.
4. $[a, b)$ is neither open nor closed.
5. Any finite set is closed.
6. A' is closed.
7. $\bar{A}$ is closed.
8. $A°$ is open.

The relation between open and closed sets is brought out by the following theorem.

Theorem 2.3.3. A set A is closed (open) if and only if $\tilde{A}$ is open (closed).

Proof. If A is closed, then $A \supseteq A'$ and every $x \in \tilde{A}$ is also in $\sim(A')$; thus, for every such x there is an h_x such that $(x - h_x, x + h_x) \subseteq \tilde{A}$, which implies that $\tilde{A}$ is open. Conversely, if $\tilde{A}$ is open, then no point of $\tilde{A}$ can be in A', so that $A \supseteq A'$, i.e., A is closed.

We now list some additional elementary properties of open and closed sets, leaving all proofs as exercises.

1. For any A, $A°$ is the union of all open subsets of A, i.e., $A°$ is the largest open subset of A.
2. For any A, $\bar{A}$ is the intersection of all closed sets containing A, i.e., $\bar{A}$ is the smallest closed set containing A.
3. If $\{O_\alpha\}$ is any collection of open sets, then $\bigcup_\alpha O_\alpha$ is open. If $O_1, \ldots, O_n$ is a finite collection of open sets, then $\bigcap_{i=1}^{n} O_i$ is open.
4. If $\{K_\alpha\}$ is any collection of closed sets, then $\bigcap_\alpha K_\alpha$ is closed. If $K_1, \ldots, K_n$ is a finite collection of closed sets, then $\bigcup_{i=1}^{n} K_i$ is closed.

The preceding discussion centers on certain subsets of R and their special attributes. Vital to the continued study of R is the notion of a *sequence*.

Definition 2.3.4. A sequence in R is a function from I^+, the positive integers, into R.

Instead of using the functional notation $f(n)$, $n \in I^+$, for a sequence, we will write $\{x_n\}$, where $x_n = f(n)$. But in so doing, it is important not to think of a sequence simply as a subset of R. For example, we can speak of a one-to-one sequence $[n \neq m \Rightarrow f(n) \neq f(m)]$, whereas it makes no sense to talk of a one-to-one subset. Other useful special sequences are the monotone sequences, defined as follows.

nondecreasing:	$x_{n+1} \geq x_n$	for all n
nonincreasing:	$x_{n+1} \leq x_n$	for all n
decreasing:	$x_{n+1} < x_n$	for all n
increasing:	$x_{n+1} > x_n$	for all n.

Definition 2.3.5. A sequence $\{x_n\}$ in R is said to *converge* to a point x_0 if for any $\epsilon > 0$ there is an $N(\epsilon) \in I^+$ such that $|x_n - x_0| \leq \epsilon$ whenever $n \geq N(\epsilon)$. In this case we write limit $x_n = x_0$, or $x_n \longrightarrow x_0$.
$$n \to \infty$$

Lemma 2.3.6. Let A be a nonempty subset of R. Then $x_0 \in A'$ if and only if there is a one-to-one sequence $\{x_n\}$ of points of A such that $x_n \longrightarrow x_0$.

Proof. If $\{x_n\}$ is a one-to-one sequence of points in A converging to x_0, then for any $\epsilon > 0$ there is an $N(\epsilon)$ such that $|x_n - x_0| \leq \epsilon$ whenever $n \geq N(\epsilon)$. Thus the interval $(x_0 - \epsilon, x_0 + \epsilon)$ contains infinitely many distinct points of A, and $x_0 \in A'$. Conversely, if $x_0 \in A'$, then the interval $(x_0 - 1, x_0 + 1)$ contains a point $x_1 \in A$, $x_1 \neq x_0$. In like manner, the interval $(x_0 - |x_0 - x_1|, x_0 + |x_0 - x_1|)$ contains a point $x_2 \in A$, $x_2 \neq x_0$, x_1. Continuing in this way, we obtain a one-to-one sequence of points in A converging to x_0. This completes the proof.

Corollary. $x_0 \in \bar{A}$ if and only if there is a sequence of points in A converging to x_0.

A sequence $\{x_n\}$ is said to be *bounded* above (below) if $x_n \leq M$ ($x_n \geq m$) for some $M(m)$ and all n. A basic criterion for deciding when some sequences converge is as follows.

Theorem 2.3.7. Every bounded monotone sequence converges.

Proof. It suffices to consider the case where $\{x_n\}$ is nondecreasing and bounded. Let x_0 be the supremum of the range of the sequence. Then for any $\epsilon > 0$, there is an $N(\epsilon)$ such that $x_0 - \epsilon \leq x_{N(\epsilon)} \leq x_0$. (See exercise 11 of the preceding section.) The monotonicity of $\{x_n\}$ now yields $x_0 - \epsilon \leq x_{N(\epsilon)} \leq x_n \leq x_0$ for $n \geq N(\epsilon)$, and the convergence is proved.

We will also need the notion of a *subsequence* of a sequence.

Definition 2.3.8. Let $n(k)$ be an increasing function from I^+ into I^+ [i.e., $n(k + 1) > n(k)$ for all k], and let $f(n)$ define the sequence $\{x_n\}$. The function $f[n(k)]$ then defines a sequence which is a *subsequence* of $\{x_n\}$; it is designated $\{x_{n(k)}\}$.

For example consider the sequence $\{1, 1/2^2, 1/3^2, 1/4^2, \ldots, 1/n^2, \ldots\}$ defined by $f(n) = 1/n^2$. If $n(k) = 2k$, then $f[n(k)] = 1/4k^2$ and the corresponding subsequence is $\{\frac{1}{4}, \frac{1}{16}, \frac{1}{36}, \frac{1}{64}, \ldots\}$.

We come now to a second and very fundamental application of sequences to the study of the structure of R (the first was lemma 2.3.6).

Theorem 2.3.9. (Nested Intervals Theorem.) Let $\{I_n\} = \{[a_n, b_n]\}$ be a denumerable set of nonempty closed intervals such that $I_n \supseteq I_{n+1}$ for all n, and $\lim_{n \to \infty} |b_n - a_n| = 0$. Then the set $\bigcap_{n=1}^{\infty} I_n$ consists of a single point.

Proof. The inclusion $I_n \supseteq I_{n+1}$ yields $b_1 \geq b_2 \geq \cdots \geq b_n \geq \cdots \geq a_1$ and $a_1 \leq a_2 \leq \cdots \leq a_n \leq \cdots \leq b_1$ so that from theorem 2.3.7 we obtain numbers a_0 and b_0 with $a_0 = \lim_{n \to \infty} a_n$ and $b_0 = \lim_{n \to \infty} b_n$. Letting $n \longrightarrow \infty$ in the equality

$$|a_0 - b_0| \leq |a_0 - a_n| + |a_n - b_n| + |b_n - b_0|,$$

we see that $a_0 = b_0$. Thus

$$a_1 \leq a_2 \leq \cdots \leq a_n \leq \cdots \leq a_0 = b_0 \leq \cdots \leq b_n \leq \cdots \leq b_2 \leq b_1$$

and $a_0 = b_0 = \bigcap_{n=1}^{\infty} I_n$ as desired.

Continuing, we have the following equally important theorem.

Theorem 2.3.10. (Bolzano-Weierstrass.) A bounded infinite set of real numbers has at least one cluster point.

Proof. Suppose that A is infinite and that $m \leq a \leq M$ for all $a \in A$. At least one of the intervals $[m, \frac{1}{2}(m + M)]$, $[\frac{1}{2}(m + M), M]$ must contain infinitely many points of A. Let $[a_1, b_1]$ be such an interval. Similarly, one of the intervals $[a_1, \frac{1}{2}(a_1 + b_1)]$, $[\frac{1}{2}(a_1 + b_1), b_1]$ contains infinitely many points of A, and we designate it by $[a_2, b_2]$ again making a choice if necessary. Continuing in this way, we obtain a sequence $\{I_n\} = \{[a_n, b_n]\}$ where $I_n \supset I_{n+1}$ and $b_n - a_n = 2^{-n}(M - m)$. By the nested intervals theorem, $\bigcap_{n=1}^{\infty} I_n$ consists of a single point x_0. We show that x_0 is a cluster point of A. Given $\epsilon > 0$, choose n such that $b_n - a_n < \epsilon$; then $I_n \subset (x_0 - \epsilon, x_0 + \epsilon)$ and the interval $(x_0 - \epsilon, x_0 + \epsilon)$ contains infinitely many points of A. This completes the proof.

Having used sequences to prove the Bolzano-Weierstrass theorem, we reverse our tactics and use the latter to obtain a characterization of convergent sequences. The difficulty with the definition of convergence is that knowledge of the limit is involved. A criterion for convergence not requiring this knowledge will now be given.

Definition 2.3.11. A sequence $\{x_n\}$ is a *Cauchy sequence* if, for every $\epsilon > 0$, there exists an $N(\epsilon)$ such that $|x_m - x_n| \le \epsilon$ whenever $m, n \ge N(\epsilon)$.

Now if $x_n \longrightarrow x_0$, then $|x_n - x_0| \le \epsilon/2$ whenever $n \ge N(\epsilon/2)$. Thus $|x_m - x_n| \le |x_m - x_0| + |x_0 - x_n| \le \epsilon$ for $m, n \ge N(\epsilon/2)$, and convergent sequences are necessarily Cauchy. We have proved half of the following theorem.

Theorem 2.3.12. A sequence $\{x_n\}$ converges if and only if it is Cauchy.

Proof. Let $\{x_n\}$ be Cauchy; then $|x_m - x_n| \le \epsilon$ whenever $m, n \ge N(\epsilon)$ and, in particular, $|x_n - x_{N(\epsilon)}| \le \epsilon$ for $n \ge N(\epsilon)$. Hence only finitely many members of $\{x_n\}$ lie outside the interval $[x_{N(\epsilon)} - \epsilon, x_{N(\epsilon)} + \epsilon]$, and $\{x_n\}$ is bounded. If the sequence $\{x_n\}$ contains only finitely many distinct members, then it is ultimately constant, i.e., for some N it must happen that $x_n = x_m$ whenever $m, n \ge N$. (Otherwise, $\{x_n\}$ could not be a Cauchy sequence.) In this case it obviously converges. If $\{x_n\}$ contains infinitely many distinct members, then the Bolzano-Weierstrass theorem can be applied to yield a cluster point x_0. No other point x_0' can be a cluster point, for if we choose $\epsilon = \frac{1}{4}|x_0 - x_0'|$, and determine the corresponding $N(\epsilon)$, then $|x_m - x_n| \le \epsilon$ whenever $m, n \ge N(\epsilon)$. But some x_m must satisfy $|x_0 - x_m| \le \epsilon$ and some x_n must satisfy $|x_0' - x_n| \le \epsilon$ with $m, n \ge N(\epsilon)$. Thus

$$|x_0 - x_0'| \le |x_0 - x_m| + |x_m - x_n| + |x_n - x_0'| \le 3\epsilon = \tfrac{3}{4}|x_0 - x_0'|,$$

which is a contradiction. Hence, for any $\epsilon > 0$, only finitely many numbers of the sequence $\{x_n\}$ can lie outside the interval $(x_0 - \epsilon, x_0 + \epsilon)$, and the sequence converges. This completes the proof.

Our next objective is the Borel covering theorem, which will have an important generalization in the next chapter.

Theorem 2.3.13. Let K be any closed, bounded set in R and let $\{O_\alpha\}$ be a family of open sets such that $K \subset \bigcup_\alpha O_\alpha$. Then there is a finite subset of $\{O_\alpha\}$ whose union contains K.

Proof. Since K is bounded, there exist numbers m and M such that $m \le k \le M$ for all k in K. It now suffices to show that the closed interval $I_0 = [m, M]$ can be covered by $\tilde{K}$ together with a finite number of members of $\{O_\alpha\}$. Now, if I_0 cannot be covered by finitely many of these open sets, then one of the intervals $[m, \frac{1}{2}(m + M)]$, $[\frac{1}{2}(m + M), M]$ cannot be so covered; we denote it by I_1 and proceed (as in the proof of the Bolzano-Weierstrass theorem) to construct a set of closed intervals $\{I_n\}$ such that $I_n \supset I_{n+1}$, the length of I_n is $2^{-n}(M - m)$, and no I_n is covered by finitely many open sets from the prescribed collection. By the nested intervals theorem, there is a point x_0 in $\bigcap_{n=1}^{\infty} I_n$. This point is in one of the open sets, O, of the covering of I_0 and, hence, for some $\epsilon > 0$, the interval $(x_0 - \epsilon, x_0 + \epsilon)$ is in O. Also, for N large enough to guarantee $2^{-N}(M - m) < \epsilon$, the interval I_N is contained in $(x_0 - \epsilon, x_0 + \epsilon)$, and then $I_N \subset O$, in contradiction to the constructed property of the members of $\{I_N\}$. This completes the proof.

For our last result in this direction, we prove a theorem giving an explicit description of the nature of a general open set in R.

Theorem 2.3.14. Every open set in R can be uniquely written as a countable union of disjoint open intervals.

Proof. Let O be an open set. For any x in O there are rational numbers r and s such that $x \in (r, s) \subseteq O$. Letting $I_x = (r, s)$, we have $O = \bigcup_x I_x$, and since there are only countably many open intervals with rational endpoints, the collection $\{I_x\}$ contains only countably many distinct members. Changing our notation, we can write $O = \bigcup_{n=1}^{\infty} I_n$, and O is represented as a countable union of open intervals which are not, however, necessarily disjoint. We now define on $\{I_n\}$ a relation $*$ by

$$I_i * I_j \Leftrightarrow [\text{There exists a finite set of intervals } I_{k(1)}, \ldots, I_{k(n)} \text{ in } \{I_n\}$$
$$\text{such that the sets } I_i \cap I_{k(1)}, I_{k(1)} \cap I_{k(2)}, \ldots, I_{k(n)} \cap I_j$$
$$\text{are all nonempty}].$$

This is easily seen to be an equivalence relation. Let $\{J_k\}$ be the collection of sets each of which consists of the union of the members of an equivalence class. Clearly $O = \bigcup_k J_k$, and we now show

(i) $J_k \cap J_j = \varnothing$ for $k \neq j$;
(ii) J_k is an open interval for every k.

To see (i), note that $x_0 \in J_k \cap J_j$ implies $x_0 \in I_p \cap I_q$, where $I_p \subseteq J_k$, $I_q \subseteq J_j$; but in this case $I_p * I_q$, contrary to the construction. To prove (ii), let $b_k = \sup \{x : x \in J_k\}$ and $a_k = \inf \{x : x \in J_k\}$; it is clear that $J_k \subseteq (a_k, b_k)$. Conversely, if $x \in (a_k, b_k)$, then there exist numbers a_k' and b_k' in J_k such that $a_k < a_k' < x < b_k' < b_k$. Now $a_k' \in I_p$ and $b_k' \in I_q$ for some constituent intervals I_p and I_q of J_k, and $I_p * I_q$. Thus I_p and I_q are bridged by a finite chain of overlapping intervals, one of which must contain x. Hence $x \in J_k$, and the proof is complete, except for the uniqueness question, which is left for the exercises.

Corollary. Any closed set in R is the complement of the union of countably many disjoint open intervals.

We close this section with a few more remarks on sequences.

Definition 2.3.15

$$\overline{\lim_{n\to\infty}} \, x_n = \inf_n \, [\sup_m \, \{x_m : m \geq n\}]$$

$$\underline{\lim_{n\to\infty}} \, x_n = \sup_n \, [\inf_m \, \{x_m : m \geq n\}].$$

These expressions are read "the upper limit of $\{x_n\}$" and "the lower limit of $\{x_n\}$," respectively. Insight into exactly how $\overline{\lim_{n\to\infty}} \, x_n$ and $\underline{\lim_{n\to\infty}} \, x_n$ are determined by $\{x_n\}$ is furnished by the following lemma.

Lemma 2.3.16. The numbers $\bar{x} = \overline{\lim_{n\to\infty}} \, x_n$ and $\underline{x} = \underline{\lim_{n\to\infty}} \, x_n$ are characterized by (a) and (b):

(a) For any $\epsilon > 0$, there are only finitely many x_n's for which $x_n \geq \bar{x} + \epsilon$, but infinitely many x_n's for which $x_n \leq \bar{x} + \epsilon$.

(b) For any $\epsilon > 0$, there are only finitely many x_n's for which $x_n \leq \underline{x} - \epsilon$, but infinitely many x_n's for which $x_n \geq \underline{x} - \epsilon$.

The proof is left for the exercises. The lemma makes it clear that $\bar{x}$ and $\underline{x}$ are unaffected by the removal of finitely many terms from the sequence, just as the limit of a sequence (if it has one) is unaffected. In fact, we always have

$$\underline{\lim_{n\to\infty}} \, x_n \leq \overline{\lim_{n\to\infty}} \, x_n$$

and

$$\underline{\lim_{n\to\infty}} \, x_n = \overline{\lim_{n\to\infty}} \, x_n \Leftrightarrow \lim_{n\to\infty} x_n = x_0,$$

in which case $x_0 = \underline{x} = \bar{x}$.

Again, the details are relegated to the exercises.

EXERCISES

1. Prove the following:
    a. A' is closed for any $A \subseteq R$.

b. $\bar{A}$ is closed for any $A \subseteq R$.
c. A° is open for any $A \subseteq R$.

2.   Show that for any $A \subseteq R$
a. A° is the union of all open subsets of A.
b. $\bar{A}$ is the intersection of all closed sets containing A.

3.   a. Show that the union (intersection) of any number of open (closed) sets is open (closed).
b. Show that the union (intersection) of finitely many closed (open) sets is closed (open).

4.   We know that $(A^{\circ})^{\circ} = A^{\circ}$ and $(\bar{\bar{A}}) = \bar{A}$ for any $A \subseteq R$. (Why?) Is there any relation, in general, between A' and $(A')'$?

5.   Prove
a. $(A \cup B)' = A' \cup B'$.
b. $\overline{A \cup B} = \bar{A} \cup \bar{B}$.
c. These equalities can be extended to all finite unions, i.e.,

$$\bigcup_{i=1}^{n}(A_i') = (\bigcup_{i=1}^{n} A_i)' \quad \text{and} \quad \bigcup_{i=1}^{n} \bar{A}_i = \overline{\bigcup_{i=1}^{n} A_i}.$$

6.   Do the results of exercise 5 extend to infinite unions?

7.   Investigate exercises 5 and 6 if $\cup$ is replaced by $\cap$.

8.   a. Show that any open interval is the countable union of closed intervals.
b. Show that any open set is a countable union of closed sets.
c. Show that any closed set is a countable intersection of open sets.

9.   Prove the uniqueness part of theorem 2.3.14.

10.   a. Prove that no finite interval (a, b) can contain an uncountable family of nonempty disjoint open intervals. [HINT: If $\bigcup_{\alpha} I_\alpha \subseteq$ (a, b), then for any n, at most $n - 1$ I_α's can have length greater than $(1/n)(b - a)$.]
b. Prove that no uncountable family of nonempty disjoint open intervals exists in R.

11.   From the sequence $\{x_n\}$ define sequences $\{y_n\}$ and $\{z_n\}$ by

$$y_n = \sup_{m} \{x_m : m \geq n\}$$

$$z_n = \inf_{m} \{x_m : m \geq n\}$$

a. Show that $\{y_n\}$ and $\{z_n\}$ are monotone.
b. Show that $\underline{x} = \varliminf_{n \to \infty} x_n \leq \varlimsup_{n \to \infty} x_n = \bar{x}$.
c. Show that $\bar{x} = \underline{x}$ if and only if $\{x_n\}$ converges.
d. Prove lemma 2.3.16.
e. Show that subsequences of $\{x_n\}$ can be found which converge to $\bar{x}$ and $\underline{x}$.

12. Let $\{x_n\}$ and $\{y_n\}$ be two bounded sequences of nonnegative numbers. Show

$$\varlimsup_{n \to \infty} (x_n y_n) \leq (\varlimsup_{n \to \infty} x_n)(\varlimsup_{n \to \infty} y_n).$$

13. Show that

$$\varliminf_{n \to \infty} x_n + \varliminf_{n \to \infty} y_n \leq \varliminf_{n \to \infty} (x_n + y_n)$$

$$\leq \varliminf_{n \to \infty} x_n + \varlimsup_{n \to \infty} y_n \leq \varlimsup_{n \to \infty} (x_n + y_n)$$

$$\leq \varlimsup_{n \to \infty} x_n + \varlimsup_{n \to \infty} y_n.$$

14. By $\inf_{n} x_n$ ($\sup_{n} x_n$) we mean the inf (sup) of $\{x_n\}$ considered as a subset of R. Show that

$$\inf_{n} x_n + \inf_{n} y_n \leq \inf_{n} (x_n + y_n) \leq \inf_{n} x_n + \sup_{n} y_n$$

$$\leq \sup_{n} (x_n + y_n) \leq \sup_{n} x_n + \sup_{n} y_n.$$

3 Metric Spaces

The notions of convergence encountered in the real line, the complex plane, Euclidean 3-space, and various types of convergence of functions (e.g., uniform convergence and mean convergence) can be abstracted and studied in the general context of metric spaces. In this chapter we define the notion of metric space and develop some basic properties of these spaces.

Section 1 FUNDAMENTAL DEFINITIONS AND EXAMPLES

Definition 3.1.1. A *metric space* is a set M with a real-valued function $d(x, y)$ on $M \times M$ called a *distance function*. This function is required to have the properties:

(a) $d(x, y) \geq 0$; $d(x, y) = 0 \Leftrightarrow x = y$.
(b) $d(x, y) = d(y, x)$.
(c) $d(x, z) \leq d(x, y) + d(y, z)$ (the triangle inequality).

An immediate consequence of this definition is the property

(d) $|d(x, y) - d(x, z)| \leq d(y, z)$.

The proof is left as an exercise.

We omit the formal definition of sequences and subsequences in metric spaces. They are copies with obvious modifications of the definitions in R.

36

Definition 3.1.2. A sequence of points $\{x_n\}$ in M is said to *converge* to the point x_0 if $\lim\limits_{n\to\infty} d(x_n, x_0) = 0$. In this case we write $x_n \longrightarrow x_0$.

It is easy to show that if $x_n \longrightarrow x_0$, $x_n \longrightarrow y_0$, then $x_0 = y_0$; that is, limits are *unique*.

EXAMPLE 1. R, the set of all real numbers, is a metric space under the definition $d(x, y) = |x - y|$. [(a) and (b) are obvious, while (c) is proved as follows: $|x - z| = |(x - y) + (y - z)| \leq |x - y| + |y - z|$.]

EXAMPLE 2. $R \times R$ is a metric space under

$$d[(x_1, y_1), (x_2, y_2)] = \sqrt{(x_1 - x_2)^2 + (y_1 - y_2)^2}.$$

Relations (a) and (b) are obvious; (c) is a special case of a more general inequality proved in example 5.

EXAMPLE 3. The function

$$d_1[(x_1, y_1), (x_2, y_2)] = \max\left[|x_1 - x_2|, |y_1 - y_2|\right]$$

is also a metric on $R \times R$. Again (a) and (b) are clear, while (c) is proved as follows:

$$|x_1 - x_3| \leq |x_1 - x_2| + |x_2 - x_3| \leq \max\left[|x_1 - x_2|, |y_1 - y_2|\right]$$
$$+ \max\left[|x_2 - x_3|, |y_2 - y_3|\right]$$
$$|y_1 - y_3| \leq |y_1 - y_2| + |y_2 - y_3| \leq \max\left[|x_1 - x_2|, |y_1 - y_2|\right]$$
$$+ \max\left[|x_2 - x_3|, |y_2 - y_3|\right].$$

From these inequalities we see that

$$\max\left[|x_1 - x_3|, |y_1 - y_3|\right] \leq \max\left[|x_1 - x_2|, |y_1 - y_2|\right]$$
$$+ \max\left[|x_2 - x_3|, |y_2 - y_3|\right]$$

as desired.

EXAMPLE 4. A third metric on $R \times R$ is given by $d_2[(x_1, y_1), (x_2, y_2)] = |x_1 - x_2| + |y_1 - y_2|$. Again, (a) and (b) are obvious, while (c) is left as an exercise.

We now present an example which generalizes example 2 in several ways.

EXAMPLE 5. Let $C^{(n)} = C \times C \times \cdots \times C$ (n times) and define

$$d[(x_1, \ldots, x_n), (y_1, \ldots, y_n)] = \left\{ \sum_{k=1}^{n} |x_k - y_k|^2 \right\}^{1/2}.$$

As usual, (a) and (b) are clear, but the triangle inequality requires discussion. We begin by proving the Cauchy-Schwarz inequality:

$$\text{(C-S)} \quad \left| \sum_{k=1}^{n} x_k \bar{y}_k \right|^2 \le \sum_{k=1}^{n} |x_k|^2 \sum_{k=1}^{n} |y_k|^2$$

Clearly, for *any* complex λ, we have

$$\sum_{k=1}^{n} |x_k + \lambda y_k|^2 \ge 0.$$

This is equivalent to

$$\sum_{k=1}^{n} |x_k|^2 + \lambda \sum_{k=1}^{n} x_k \bar{y}_k + \bar{\lambda} \sum_{k=1}^{n} \overline{x_k \bar{y}_k} + |\lambda|^2 \sum_{k=1}^{n} |y_k|^2 \ge 0,$$

since

$$|x_k + \lambda y_k|^2 = (x_k + \lambda y_k)(\bar{x}_k + \bar{\lambda} \bar{y}_k) = |x_k|^2 + \lambda \bar{x}_k y_k + \bar{\lambda} x_k \bar{y}_k + |y_k|^2 |\lambda|^2.$$

If $\sum_{k=1}^{n} |y_k|^2 \ne 0$, we take

$$\lambda = -\left(\sum_{k=1}^{n} |y_k|^2 \right)^{-1} \left(\sum_{k=1}^{n} x_k \bar{y}_k \right),$$

and the C-S inequality follows. If $\sum_{k=1}^{n} |y_k|^2 = 0$ and $\sum_{k=1}^{n} |x_k|^2 \ne 0$, we can interchange the roles of $(x_1, \ldots, x_n)$ and $(y_1, \ldots, y_n)$. If both are zero, the inequality reduces to $0 = 0$.

We now employ the C-S inequality as follows:

$$\sum_{k=1}^{n} |x_k + y_k|^2 = \sum_{k=1}^{n} |x_k|^2 + \left[\sum_{k=1}^{n} x_k \bar{y}_k + \sum_{k=1}^{n} \bar{x}_k y_k \right] + \sum_{k=1}^{n} |y_k|^2$$

$$\le \sum_{k=1}^{n} |x_k|^2 + \left| \sum_{k=1}^{n} x_k \bar{y}_k \right| + \left| \sum_{k=1}^{n} \bar{x}_k y_k \right| + \sum_{k=1}^{n} |y_k|^2$$

$$\le \sum_{k=1}^{n} |x_k|^2 + 2\left\{ \sum_{k=1}^{n} |x_k|^2 \sum_{k=1}^{n} |y_k|^2 \right\}^{1/2} + \sum_{k=1}^{n} |y_k|^2.$$

Hence we have

$$\left\{ \sum_{k=1}^{n} |x_k + y_k|^2 \right\}^{1/2} \le \left\{ \sum_{k=1}^{n} |x_k|^2 \right\}^{1/2} + \left\{ \sum_{k=1}^{n} |y_k|^2 \right\}^{1/2},$$

which is called the Minkowski inequality.

The triangle inequality follows at once from the Minkowski inequality if we replace x_k by $x_k - y_k$ and y_k by $y_k - z_k$.

EXAMPLE 6. The complex number system C is a metric space under $d(z_1, z_2) = |z_1 - z_2|$. It is also a metric space under

$$d_1(z_1, z_2) = \frac{|z_1 - z_2|}{\{(1 + |z_1|^2)(1 + |z_2|^2)\}^{1/2}}.$$

Only the triangle inequality need be proved. From the identity

$$(z_1 - z_3)(1 + z_2\bar{z}_2) = (z_1 - z_2)(1 + \bar{z}_2 z_3) + (z_2 - z_3)(1 + z_1\bar{z}_2),$$

we obtain

$$|z_1 - z_3|(1 + |z_2|^2) \le |z_1 - z_2| \cdot |1 + \bar{z}_2 z_3| + |z_2 - z_3| \cdot |1 + z_1\bar{z}_2|.$$

Since, by C-S,

$$|1 + \bar{z}_2 z_3| \le (1 + |z_2|^2)^{1/2}(1 + |z_3|^2)^{1/2} \quad \text{and}$$

$$|1 + z_1\bar{z}_2| \le (1 + |z_1|^2)^{1/2}(1 + |z_2|^2)^{1/2}$$

we can extend this inequality to

$$|z_1 - z_3|(1 + |z_2|^2) \le |z_1 - z_2|(1 + |z_2|^2)^{1/2}(1 + |z_3|^2)^{1/2}$$
$$+ |z_2 - z_3|(1 + |z_1|^2)^{1/2}(1 + |z_2|^2)^{1/2}.$$

Division by $(1 + |z_2|^2)(1 + |z_1|^2)^{1/2}(1 + |z_3|^2)^{1/2}$ now yields the desired triangle inequality. This metric is useful in complex variable theory.

EXAMPLE 7. Let M be *any* set, and define

$$d(x, y) = \begin{cases} 1, & \text{if} \quad x \ne y \\ 0, & \text{if} \quad x = y. \end{cases}$$

We now define a number of terms which we will need in our discussions of metric spaces.

Definition 3.1.3. Let M be a metric space with metric $d(x, y)$.

1. An *open sphere* with center x_0 and radius $a > 0$ is the set $S(x_0; a) = \{x \in M : d(x_0, x) < a\}$. For brevity, we shall usually use "sphere" for "open sphere."
2. The point x_0 is a *cluster* point of the subset A of M if every open sphere $S(x_0; a)$ contains a point x of A with $x \ne x_0$.

3. The *derived set* A' of A is the set of all cluster points of A.
4. The *closure* $\bar{A}$ of A is $A \cup A'$.
5. A is *closed* if $A = \bar{A}$, or if $A = \varnothing$.
6. The point x_0 is an *interior point* of A if, for some $a > 0$, $S(x_0; a) \subseteq A$.
7. The set of all interior points of A is the *interior* of A and is written A°.
8. A set A is *open* if $A = A^\circ$, or if $A = \varnothing$.
9. A set B is *dense* in the set A if $B \subseteq A$, $\bar{B} = A$.
10. A set A is *bounded* if it is contained in some sphere.

We list some immediate consequences of the definition:

(i) M is both open and closed.
(ii) $x_0 \in A' \Leftrightarrow$ there is a one-to-one sequence $\{x_n\}$ of points in A such that $x_n \longrightarrow x_0$.
(iii) $x_0 \in \bar{A}$ if and only if there is a sequence $\{x_n\}$ of points in A such that $x_n \longrightarrow x_0$.
(iv) A is open if and only if it is a union of open spheres.
(v) $\bar{A} \supseteq A \supseteq A^\circ$.

The following four lemmas have proofs which are very similar to the proofs already given for corresponding results in R; hence, the proofs are omitted.

Lemma 3.1.4. A set A is open (closed) if and only if $\tilde{A}$ is closed (open).

Lemma 3.1.5. For any A, A° is open while A' and $\bar{A}$ are closed. A° is the union of all open sets contained in A, while $\bar{A}$ is the intersection of all closed sets containing A.

Lemma 3.1.6. The union of any family of open sets is an open set. The intersection of any family of closed sets is a closed set. Finite intersections of open sets are open, and finite unions of closed sets are closed.

Lemma 3.1.7. For any A and B, $(A \cup B)' = A' \cup B'$ and $\overline{A \cup B} = \bar{A} \cup \bar{B}$. Also, $A \subseteq B \Rightarrow \bar{A} \subseteq \bar{B}$.

We now define the *closed sphere* with center x_0 and radius a to be $\{x \in M : d(x_0, x) \leq a\}$. The terminology is justified by the fact that *closed spheres are closed sets* (see the exercises). It is then clear that $\overline{S(x_0, a)} \subseteq \{x : d(x_0, x) \leq a\}$, but the inclusion may be proper. For example, let M consist of the nonnegative real numbers and the number -1, with the same metric as in R. Then $\{x : d(0, x) \leq 1\} = \{-1\} \cup [0, 1]$, but $\overline{S(0; 1)} = [0, 1]$.

Definition 3.1.8. A metric space M is *separable* if it contains a countable dense subset.

For example, R is separable since the rationals are a countable dense subset. Also, $R \times R$ under the Euclidean metric is separable since the points having both coordinates rational constitute a countable dense subset.

Definition 3.1.9. A family of open sets $\{O_\alpha\}$ is called a *basis* for the open sets of a metric space M if every open set in M is a union of sets from this family.

An alternative formulation of this definition reads: $\{O_\alpha\}$ is a *basis* for the open sets of M if, for any point $x \in M$ and open set O containing x, there is an O_α such that $x \in O_\alpha \subseteq O$.

For most purposes it suffices to work with basic open sets rather than arbitrary open sets. Moreover, it is frequently possible to find a countable basis for the open sets of a metric space. For example, the open intervals with rational endpoints form a countable basis in R. In the Euclidean plane, a countable basis is given by the open spheres with rational radii, and with centers at points having both coordinates rational.

Theorem 3.1.10. A metric space M is separable if and only if there is a countable basis for the open sets of M.

Proof. Suppose that M has a countable basis $\{O_n\}$. Choose a point x_n from each O_n; we claim that $\overline{\{x_n\}} = M$. (Note that here $\{x_n\}$ is a countable set, and not a sequence.) To prove this, let x be any point in M not in $\{x_n\}$. Then any open sphere $S(x, \epsilon)$ contains some O_n and hence x_n. Therefore, x is a cluster point of $\{x_n\}$ and $\overline{\{x_n\}} = M$. Conversely, if $\{y_n\}$ is a countable dense subset of M and if $r_1, r_2, \ldots, r_k, \ldots$ is an enumeration of the positive rational numbers, then the set of open spheres

$$\{S(y_n: r_k): n, k = 1, 2, \ldots\}$$

is countable and is a basis for the open sets of M. For, if O is any open set and $x \in O$, then, for some $\epsilon > 0$, $S(x, \epsilon) \subseteq O$. The sphere $S(x, \epsilon/3)$ contains some y_n, and we have $x \in S(y_n, \eta) \subseteq S(x, \epsilon) \subseteq O$ if η is a rational number satisfying $\epsilon/3 < \eta < \epsilon/2$. This finishes the proof.

We give another application of bases.

Theorem 3.1.11. Let M be a metric space with a countable basis $\{A_n\}$, and let $\{O_\alpha\}$ be a collection of open sets of M such that $\bigcup_\alpha O_\alpha = M$. Then some countable subcollection of $\{O_\alpha\}$ covers M.

Proof. Any $x \in M$ is in some O_α, and there is an A_n such that $x \in A_n \subseteq O_\alpha$. That is, for every $x \in M$, there is an $A_{n(x)}$ containing x and contained in some O_α. Since $\{A_n\}$ is countable, so is $\{A_{n(x)}\}$; moreover $\bigcup_n A_n(x) = M$.

For every n choose an $O_{\alpha(n)}$ from $\{O_\alpha\}$ which contains $A_n(x)$. The collection $\{O_{\alpha(n)}\}$ is a countable subset of $\{O_\alpha\}$ and it covers M.

EXAMPLE 8. Let R^∞ be the set of all infinite sequences of real numbers, and define

$$d[\{x_n\}, \{y_n\}] = \sum_{n=1}^{\infty} k_n \frac{|x_n - y_n|}{1 + |x_n - y_n|}$$

where $\{k_n\}$ is a fixed sequence of *positive* real numbers such that $\sum_{n=1}^{\infty} k_n$ converges. Under this assumption, the series defining the distance function converges. Moreover, conditions (a) and (b) of definition 3.1.1 are obviously satisfied. To prove condition (c) we need the following inequality, which is valid for any two real numbers x and y.

(E)
$$\frac{|x + y|}{1 + |x + y|} \le \frac{|x|}{1 + |x|} + \frac{|y|}{1 + |y|}.$$

The inequality itself is obviously true if either $x = 0$ or $y = 0$. If x and y have the same sign, it suffices to consider them both positive. Then

$$\frac{|x + y|}{1 + |x + y|} = \frac{x + y}{1 + x + y} = \frac{x}{1 + x + y} + \frac{y}{1 + x + y} < \frac{x}{1 + x} + \frac{y}{1 + y}$$

$$= \frac{|x|}{1 + |x|} + \frac{|y|}{1 + |y|}.$$

If x and y have opposite signs, we can suppose $|x| \ge |y|$, in which case $|x + y| \le |x|$. Now for $t \ge 0$, the function $f(t) = t/(1 + t)$ is *increasing* since $f'(t) = (1 + t)^{-2} > 0$. Hence, $|x + y| \le |x|$ implies

$$\frac{|x + y|}{1 + |x + y|} \le \frac{|x|}{1 + |x|} < \frac{|x|}{1 + |x|} + \frac{|y|}{1 + |y|},$$

and the proof of (E) is finished. From (E) we see that

$$k_n \frac{|x_n - z_n|}{1 + |x_n - z_n|} = k_n \frac{|(x_n - y_n) + (y_n - z_n)|}{1 + |(x_n - y_n) + (y_n - z_n)|}$$

$$\le k_n \frac{|x_n - y_n|}{1 + |x_n - y_n|} + k_n \frac{|y_n - z_n|}{1 + |y_n - z_n|}$$

and we obtain the triangle inequality for the metric by summing over n.

In examples 2–5, it is evident that a sequence of points converges to a limit if and only if their coordinates converge to the coordinates of the limit point. For instance,

$$\lim_{n \to \infty} [(x_n - x_0)^2 + (y_n - y_0)^2]^{1/2} = 0$$

if and only if both $\lim\limits_{n \to \infty} x_n = x_0$ and $\lim\limits_{n \to \infty} y_n = y_0$. This also holds in R^∞; stated precisely, we have the following: A sequence of points

$$x_n = (x_{n1}, x_{n2}, \ldots, x_{nk}, \ldots), \qquad n = 1, 2, \ldots$$

in R^∞ converges to a limit

$$x_0 = (x_{01}, x_{02}, \ldots, x_{0k}, \ldots)$$

in the metric of R^∞ if and only if

$$\lim_{n \to \infty} x_{nk} = x_{0k}, \qquad \text{for all } k.$$

To prove this assertion we first assume that $d(x_n, x_0) \to 0$ as $n \to \infty$. Then, for any $\epsilon > 0$, there is an $N(\epsilon)$ such that

$$\sum_{j=1}^{\infty} k_j \frac{|x_{nj} - x_{0j}|}{1 + |x_{nj} - x_{0j}|} \leq \epsilon, \qquad \text{for } n \geq N(\epsilon).$$

In particular, for j fixed, we have

$$k_j \frac{|x_{nj} - x_{0j}|}{1 + |x_{nj} - x_{0j}|} \leq \epsilon \quad \text{or} \quad |x_{nj} - x_{0j}| \leq \frac{\epsilon}{k_j - \epsilon}, \qquad \text{for } n \geq N(\epsilon).$$

[We can assume $\epsilon < k_j$ without loss, so that the rearrangement of the inequality is legitimate.] Since $\epsilon/(k_j - \epsilon) \to 0$ as $\epsilon \to 0$, we have proved that

$$\lim_{n \to \infty} x_{nj} = x_{0j}, \qquad \text{for each } j.$$

To prove the converse, we assume that $\{x_n\}$ converges to x_0 in the coordinate-wise sense. Choose an $N(\epsilon)$ such that $\sum\limits_{j=N(\epsilon)}^{\infty} k_j < \epsilon/2$. Now

$$d(x_n, x_0) = \sum_{j=1}^{N(\epsilon)-1} k_j \frac{|x_{nj} - x_{0j}|}{1 + |x_{nj} - x_{0j}|} + \sum_{j=N(\epsilon)}^{\infty} k_j \frac{|x_{nj} - x_{0j}|}{1 + |x_{nj} - x_{0j}|}$$

$$\leq \sum_{j=1}^{N(\epsilon)-1} k_j \frac{|x_{nj} - x_{0j}|}{1 + |x_{nj} - x_{0j}|} + \frac{\epsilon}{2}, \qquad \text{and}$$

$$k_j \frac{|x_{nj} - x_{0j}|}{1 + |x_{nj} - x_{0j}|} \leq k_j |x_{nj} - x_{0j}|.$$

By the assumed coordinatewise convergence we can find, for any j, an $N_j(\epsilon)$ such that $k_j |x_{nj} - x_{0j}| < \epsilon/2[N(\epsilon) - 1]$ whenever $n \geq N_j(\epsilon)$. Then for $n \geq$

$$\max_{j=1}^{N(\epsilon)-1} \{N_j(\epsilon)\}$$ we have $d(x_n, x_0) \leq \epsilon$, as desired. This completes the proof.

We close this section with the following observations.

Any subset A of a metric space M with metric $d(x, y)$ is itself a metric space under this metric. The open sets and closed sets of A are characterized as follows.

(o) The set $O \subseteq A$ is open in A if and only if $O = O_1 \cap A$, where O_1 is open in M.

(c) The set $K \subseteq A$ is closed in A if and only if $K = K_1 \cap A$, where K_1 is closed in M.

The statement (o) becomes apparent if it is recognized that open spheres in A are the intersections with A of open spheres in M. Also, since the closure of a subset B of A, considered as a set in the metric space A, is the union of B with the set of cluster points of B which are in A, the second assertion is seen to be true.

EXERCISES

1.   Prove that the inequality
$$|d(x, y) - d(x, z)| \leq d(y, z)$$
holds in any metric space.

2.   Let M be any set and let δ be a real-valued function on $M \times M$ satisfying
   (i) $\delta(x, x) = 0$ for all $x \in M$, and
   (ii) $\delta(x, z) \leq \delta(y, x) + \delta(y, z)$
   a. Prove that δ also satisfies
   (iii) $\delta(x, y) \geq 0$ for all x and y in M, and
   (iv) $\delta(x, y) = \delta(y, x)$ for all x and y in M.
   b. Give an example to show that $\delta(x, y)$ may be zero when $y \neq x$.
   c. Define $x * y \Leftrightarrow \delta(x, y) = 0$. Show that $*$ is an equivalence relation, and that $x * x'$, $y * y'$, implies $\delta(x, y) = \delta(x', y')$.
   d. Show how the set of equivalence classes can be given a metric in a natural way.

3.   Show that, in any metric space M,
$$d(x_1, x_n) \leq d(x_1, x_2) + d(x_2, x_3) + \cdots + d(x_{n-1}, x_n).$$

4.   In examples 2, 3, and 4, prove that $(x_n, y_n) \rightarrow (x_0, y_0)$ in the metric

sense if and only if limit $x_n = x_0$ and limit $y_n = y_0$ in R. Prove the
$\underset{n \to \infty}{}$ $\underset{n \to \infty}{}$
analogous result for example 5.

5.   Consider the metric space M of example 7.
     a. Show that no subset of M has any cluster points.
     b. Show that a point $x \in M$ is an interior point of any set containing it.
     c. Show that all subsets of M are both open and closed.

6.   Prove statements (i)–(v) following definition 3.1.3.

7.   Prove lemma 3.1.4.

8.   Prove lemma 3.1.5.

9.   Prove lemma 3.1.6.

10. Prove lemma 3.1.7.

11. Prove that closed spheres are closed sets.

12. Show that the metric space R^∞ is separable.

13. Prove that a sequence of complex numbers $\{z_n\}$ converges to z_0 in the metric of example 6 if and only if both Re $(z_n) \to$ Re(z_0) and Im $(z_n) \to$ Im (z_0) in R.

14. Prove that a convergent sequence in a metric space is a bounded set.

15. Let d_1 and d_2 be two metrics on a set M.
     a. Show that $D_1(x, y) = d_1(x, y) + d_2(x, y)$ is a metric on M.
     d. Show that $D_2(x, y) = \max [d_1(x, y), d_2(x, y)]$ is a metric on M.

16. Let $M = R \times R$ and define $D[(x_1, y_1), (x_2, y_2)] = |x_1 - x_2| + f(y_1, y_2)$, where $f(y_1, y_2) = 1$ if $y_1 \neq y_2$, $f(y_1, y_2) = 0$ if $y_1 = y_2$. Show that D is a metric on M, and give a geometrical description of the convergence.

17. Let $\{x_n\}$ be a one-to-one sequence in a metric space M. Show that $\{x_n\}$, as a subset of M, may have a unique cluster point without, as a sequence, converging to that point.

18. Prove that if $x_n \to x_0$, $x_n \to y_0$, then $x_0 = y_0$.

19.   Show that, with obvious modifications, the definition and results on R^∞ carry over to C^∞, the set of all infinite sequences of complex numbers.

20.   Let M be any separable metric space. Show that the cardinality of M is not greater than that of R. (HINT: Let $\mathcal{B} = \{O_n\}$ be a countable basis for the open sets of M. To any x in M assign the subset of $\mathcal{B}$ consisting of those members of $\mathcal{B}$ not containing x. Show that this assignment is one-to-one. How many subsets can $\mathcal{B}$ have?)

Section 2   FUNCTIONS AND CONTINUITY

We consider a function f mapping the metric space M_1 with metric d_1 into the metric space M_2 with metric d_2.

Definition 3.2.1.

(a) f is said to have the limit $y_0 \in M_2$ at $x_0 \in M_1$ if, for any $\epsilon > 0$, there is a $\delta(x_0, \epsilon)$ such that $d_2[f(x), y_0] \le \epsilon$ whenever $0 < d_1(x, x_0) \le \delta(x_0, \epsilon)$. Symbolically, $\lim_{x \to x_0} f(x) = y_0$.

(b) f is said to be continuous at $x_0 \in M_1$ if f has the limit $f(x_0)$ at x_0. Symbolically, $\lim_{x \to x_0} f(x) = f(x_0)$.

The reader should note that the definition is an exact copy of the ordinary definition for real-valued functions of a real variable. We now give a lemma which exhibits the notion of continuity in other ways.

Lemma 3.2.2.   The following statements are equivalent.

(a) $f(x)$ is continuous at x_0.
(b) For every sequence $\{x_n\}$ such that $x_n \to x_0$, $\lim_{n \to \infty} f(x_n) = f(x_0)$.
(c) For any open set O_2 containing $f(x_0)$, there is an open set O_1 containing x_0 such that $f(O_1) \subseteq O_2$.

Proof. The proof of the equivalence of (a) and (c) is left as an exercise. We now prove (a) $\Leftrightarrow$ (b).

Let $x_n \to x_0$ and assume $f(x)$ is continuous at x_0. Then for any $\epsilon > 0$ there is a $\delta(x_0, \epsilon)$ such that $d_2[f(x), f(x_0)] \le \epsilon$ whenever $d(x, x_0) \le \delta(x_0, \epsilon)$. Now determine $N(\epsilon) = N[\delta(x_0, \epsilon)]$ such that $n \ge N(\epsilon)$ implies $d(x_n, x_0) \le$

$\delta(x_0, \epsilon)$. It follows that $d[f(x_n), f(x_0)] \leq \epsilon$ whenever $n \geq N(\epsilon)$, and $f(x_n) \to f(x_0)$ as desired. Hence, $(a) \Rightarrow (b)$.

Conversely, if $f(x)$ is not continuous at x_0, then there exists an $\epsilon_0 > 0$ such that $A_n = f[S(x_0, 1/n)]$ is not contained in $S[f(x_0), \epsilon_0]$ no matter how large n is taken. Thus for any n we can find an x_n such that $d(x_n, x_0) \leq 1/n$ but $d[f(x_n), f(x_0)] \geq \epsilon_0$. Hence (b) is false whenever (a) is false, i.e., $(b) \Rightarrow (a)$. This completes the proof.

Definition 3.2.3. The function f is said to be continuous from M_1 to M_2 if it is continuous at every point of M_1.

This can also be viewed in several ways.

Lemma 3.2.4. The following statements are equivalent for a function f from M_1 to M_2.

(a) f is continuous.
(b) $f^{-1}(O_2)$ is open for every open set O_2 in M_2.
(c) $f^{-1}(K_2)$ is closed for every closed set K_2 in M_2.
(d) $f(\bar{A}_1) \subseteq \overline{f(A_1)}$ for any $A_1 \subseteq M_1$.
(e) $\overline{f^{-1}(A_2)} \subseteq f^{-1}(\bar{A}_2)$ for any $A_2 \subseteq M_2$.

Proof. We show $a \Rightarrow d \Rightarrow e \Rightarrow c \Rightarrow b \Rightarrow a$.

$a \Rightarrow d$. Assume f is continuous, let $x_0 \in \bar{A}_1$, and choose a sequence $\{x_n\}$ of points of A_1 converging to x_0. By the *continuity* of f, $f(x_n) \to f(x_0)$, so that $f(x_0) \in \overline{f(A_1)}$. Since x_0 was an arbitrary member of $\bar{A}_1$, we have $f(\bar{A}_1) \subseteq \overline{f(A_1)}$.

$d \Rightarrow e$. Using the identities $f[f^{-1}(A_2)] \subseteq A_2$ and $f^{-1}[f(A_1)] \supseteq A_1$ with (d), we have $f[\overline{f^{-1}(A_2)}] \subseteq \overline{f[f^{-1}(A_2)]} \subseteq \bar{A}_2$, so that $\overline{f^{-1}(A_2)} \subseteq f^{-1}f[\overline{f^{-1}(A_2)}] \subseteq f^{-1}(\bar{A}_2)$, as desired.

$e \Rightarrow c$. Let K_2 be closed in M_2. Using (e), we have $f^{-1}(K_2) = f^{-1}(\bar{K}_2) \supseteq \overline{f^{-1}(K_2)} \supseteq f^{-1}(K_2)$, so that $f^{-1}(K_2) = \overline{f^{-1}(K_2)}$ and $f^{-1}(K_2)$ is closed.

$c \Rightarrow b$. We use (c) and the identity $f^{-1}(\tilde{A}_2) = \sim f^{-1}(A_2)$. Then, for O_2 open in M_2 and $K_2 = \tilde{O}_2$, closed, we have $f^{-1}(O_2) = f^{-1}(\tilde{K}_2) = \sim f^{-1}(K_2)$ which, as the complement of a closed set, is open.

$b \Rightarrow a$. If O_2 is an open set containing $f(x_0)$, then $f^{-1}(O_2)$ is open, and by lemma 3.2.2(c), f is continuous. This completes the entire proof.

We can also define the notion of *uniform continuity* for functions from one metric space to another.

Definition 3.2.5. A function f from metric space M_1 to metric space M_2 is *uniformly continuous* if for any $\epsilon > 0$ there is a $\delta(\epsilon)$ such that $d_2[f(x_1), f(x_2)] < \epsilon$ whenever $d_1(x_1, x_2) < \delta(\epsilon)$.

Clearly, uniformly continuous functions are also continuous. The converse is false, as shown by the function $f(t) = 1/t$ from $(0, \infty)$ to $(0, \infty)$.

Recall that the three metrics defined in the preceding section on $R^{(2)}$ had the same meaning for convergence of sequences; that is, $(x_n, y_n) \rightarrow (x_0, y_0)$ if and only if $x_n \rightarrow x_0$ and $y_n \rightarrow y_0$, so that these metrics should be considered equivalent in some sense. To make this precise, we introduce the following definition.

Definition 3.2.6. Let M be any set, let M_1 be the metric space obtained by using the metric $d_1(x, y)$ on M, and let M_2 be the metric space obtained by using the metric $d_2(x, y)$ on M. The metrics d_1 and d_2 are called *equivalent* on M if the identity mapping of M_1 onto M_2 is continuous in both directions. They are called *uniformly equivalent* if this mapping is uniformly continuous in both directions.

By lemma 3.2.2, d_1 and d_2 are equivalent if and only if $[d_1(x_n, x_0) \rightarrow 0 \Leftrightarrow d_2(x_n, x_0) \rightarrow 0]$. Moreover, if positive constants C_1 and C_2 exist such that

$$d_2(x, y) \leq C_1 d_1(x, y) \quad \text{and} \quad d_1(x, y) \leq C_2 d_2(x, y),$$

then d_1 and d_2 are uniformly equivalent. This criterion applies to the three metrics on $R^{(2)}$ mentioned before. (It can happen, however, that two metrics are uniformly equivalent without satisfying any such inequalities; see the exercises.)

EXAMPLE. Consider the following two metrics on C:

$$d_1(z_1, z_2) = |z_1 - z_2|$$

$$d_2(z_1, z_2) = \frac{|z_1 - z_2|}{(1 + |z_1|^2)^{1/2}(1 + |z_2|^2)^{1/2}}.$$

Since $d_2(z_1, z_2) \leq d_1(z_1, z_2)$ it is clear that convergence under d_1 implies convergence under d_2. The converse is also true. To see this we first show that $d_2(z_n, z_0) \rightarrow 0$ implies that $\{z_n\}$ is bounded in the d_1 metric.†

Suppose $\{z_n\}$ is not bounded in the d_1 metric: then a subsequence $\{z_{n(k)}\}$ satisfies $|z_{n(k)}| \rightarrow \infty$. Calling the subsequence $\{y_k\}$, we have

$$\frac{|y_k - z_0|}{(1 + |y_k|^2)^{1/2}(1 + |z_0|^2)^{1/2}} = \frac{|1 - z_0/y_k|}{(1 + 1/|y_k|^2)^{1/2}(1 + |z_0|^2)^{1/2}}$$

$$\geq \frac{|1 - |z_0|/|y_k||}{(1 + 1/|y_k|^2)^{1/2}(1 + |z_0|^2)^{1/2}} \rightarrow \frac{1}{(1 + |z_0|^2)^{1/2}} > 0.$$

†A sequence $\{x_n\}$ is said to be bounded if, considered as a subset, it is bounded.

Hence $\{y_k\}$, and therefore $\{z_n\}$, cannot converge.

Now, if $|z_n| \leq M$ for all n and $d_2(z_n, z_0) \to 0$, then

$$\frac{|z_n - z_0|}{(1 + |z_n|^2)^{1/2}(1 + |z_0|^2)^{1/2}} \geq \frac{|z_n - z_0|}{(1 + M^2)^{1/2}(1 + |z_0|^2)^{1/2}} \to 0$$

and $d_1(z_n, z_0) \to 0$.

It follows that d_1 is equivalent to d_2. Proof that these metrics are *not* uniformly equivalent is left for the exercises.

To further illustrate the notions of continuity and uniform continuity, we introduce the following definition.

Definition 3.2.7. Let A_1 and A_2 be subsets of a metric space M. Then define

$$d(x, A_1) = \inf\{d(x, x_1): x_1 \in A_1\} \quad \text{and}$$
$$d(A_1, A_2) = \inf\{d(x_1, x_2): x_1 \in A_1, x_2 \in A_2\}.$$

We note some properties of these functions.

1. For any x_0, the function from M to R given by $x \to d(x_0, x)$ is uniformly continuous. This is implied by the inequality $|d(x_0, y) - d(x_0, z)| \leq d(y, z)$.
2. $d(x, A) = 0$ if and only if $x \in \bar{A}$. (Why?)
3. For any A, the function from M to R given by $x \to d(x, A)$ is uniformly continuous. The proof goes as follows. For $\epsilon > 0$ and any x_1, there is an x_0 in A such that

$$d(x_1, x_0) \leq d(x_1, A) + \frac{\epsilon}{2}$$

Now

$$d(x_2, A) \leq d(x_2, x_0) \leq d(x_2, x_1) + d(x_1, x_0) \leq d(x_1, x_2)$$
$$+ \frac{\epsilon}{2} + d(x_1, A),$$

and so, if $d(x_1, x_2) \leq \epsilon/2$, we have

$$d(x_2, A) - d(x_1, A) \leq \epsilon.$$

Similarly, $d(x_1, x_2) \leq \epsilon/2$ implies $d(x_1, A) - d(x_2, A) \leq \epsilon$, so that $|d(x_1, A) - d(x_2, A)| \leq \epsilon$ whenever $d(x_1, x_2) \leq \epsilon/2$.
4. It can happen that $d(A_1, A_2) = 0$ even though $A_1 \cap A_2 = \emptyset$ and A_1 and A_2 are both closed. For example, in R let A_1 be the set $\{1, 2, 3, \ldots, n, \ldots\}$ and let A_2 be $\{0, 2 - \frac{1}{2}, 3 - \frac{1}{3}, \ldots, n - 1/n, \ldots\}$.

As a final application of these notions, we prove the following theorem.

Theorem 3.2.8. If K_1 and K_2 are disjoint closed sets in a metric space M, then there exist open sets O_1 and O_2 with $O_1 \supseteq K_1$, $O_2 \supseteq K_2$, and $O_1 \cap O_2 = \emptyset$.

Proof. For every $x_1 \in K_1$, consider the open sphere

$$\{x: d(x_1, x) < \tfrac{1}{3}d(x_1, K_2)\},$$

which has positive radius. The union of these spheres is an open set O_1 containing K_1. Similarly, obtain an open set O_2 containing K_2.

Now suppose $x_0 \in O_1 \cap O_2$; then for some $x_1 \in K_1$ and $x_2 \in K_2$, we have

$$x_0 \in \{x: d(x_1, x) < \tfrac{1}{3}d(x_1, K_2)\} \cap \{x: d(x_2, x) < \tfrac{1}{3}d(x_2, K_1)\}.$$

Hence

$$d(x_1, x_0) < \tfrac{1}{3}d(x_1, K_2), \ d(x_2, x_0) < \tfrac{1}{3}d(x_2, K_1),$$

and

$$0 < d(x_1, x_2) \leq d(x_1, x_0) + d(x_0, x_2) < \tfrac{1}{3}[d(x_1, K_2) + d(x_2, K_1)]$$
$$\leq \tfrac{1}{3}[d(x_1, x_2) + d(x_1, x_2)] = \tfrac{2}{3}d(x_1, x_2).$$

Since this is impossible, we conclude that there is no point x_0 in $O_1 \cap O_2$.

EXERCISES

1. Prove the equivalence of (a) and (c) in lemma 3.2.2.

2. a. Let d_1 and d_2 be two metrics on M such that $d_1(x, y) \leq C_2 d_2(x, y)$ and $d_2(x, y) \leq C_1 d_1(x, y)$ for all x and y. Show that d_1 and d_2 are uniformly equivalent.
    b. Show that the following three metrics on $C^{(n)}$ are uniformly equivalent.

$$d_1[(z_1, \ldots, z_n), (w_1, \ldots, w_n)] = \left[\sum_{k=1}^{n} |z_k - w_k|^2\right]^{1/2}$$

$$d_2[(z_1, \ldots, z_n), (w_1, \ldots, w_n)] = \max\left[|z_1 - w_1|, \ldots, |z_n - w_n|\right]$$

$$d_3[(z_1, \ldots, z_n), (w_1, \ldots, w_n)] = \sum_{k=1}^{n} |z_k - w_k|.$$

3. Show that the two metrics on C given in the example are not uniformly equivalent.

4. Let M be a metric space with metric $d_1(x, y)$. Define $d_2(x, y) = \dfrac{d_1(x, y)}{1 + d_1(x, y)}$.
   a. Show that $d_2(x, y)$ is a metric on M.
   b. Show that $d_2(x, y) \leq 1$ and $d_2(x, y) \leq d_1(x, y)$.
   c. Show that d_1 and d_2 are uniformly equivalent.
   d. Find an example where the inequality $d_1(x, y) \leq C_2 d_2(x, y)$ may not hold identically, no matter how the positive constant C_2 is chosen.

5. On R let $d_1(x, y) = |x - y|$ and $d_2(x, y) = |x^3 - y^3|$.

   a. Show that d_2 is a metric on R.
   b. Show that d_1 and d_2 are equivalent.
   c. Are they uniformly equivalent?

6. a. Show that equivalence of metrics on M is an equivalence relation.
   b. Show that uniform equivalence of metrics on M is an equivalence relation.

7. A function f from M_1 into M_2 is called *open* if $f(O)$ is open for every open set O in M_1. Show that, if f is a one-to-one open function of M_1 onto M_2, then the function f^{-1} is continuous.

8. Metric spaces M_1 and M_2 are called *homeomorphic* if there is a one-to-one function F from M_1 onto M_2 such that both f and f^{-1} are continuous. Define $M_1 * M_2$ to mean M_1 is homeomorphic to M_2; show that $*$ is an equivalence relation. Define the term "uniformly homeomorphic" and prove a similar result.

9. Show that the mapping $(x, y) \rightarrow x$ of $R \times R$ onto R is open (see exercise 7).

10. a. Let A be any subset of a metric space M and define $A_n = \{x : d(x, A) < 1/n\}$. Show that A_n is open. What is $\bigcap_{n=1}^{\infty} A_n$?
    b. Show that any closed set in M can be written as the intersection of countably many open sets in M.
    c. Show that any open set in M can be written as the union of countably many closed sets in M.

11. **Definition.** A collection of open sets $\{V_\alpha\}$ is called a fundamental system of neighborhoods of a set A if, for every open set $O \supseteq A$, there is a V_α such that $O \supseteq V_\alpha \supseteq A$.

    Does the family $\{A_n\}$ of exercise 10 always form a fundamental system of neighborhoods of A? [HINT: In $R \times R$ let $A = \{(x, y): y \leq 0\}$ and $O = \{(x, y): y < e^x\}$.]

12. Does a continuous function from M_1 onto M_2 have to be open? (See exercise 7.)

13. Let S be a dense subset of a metric space M_1, and let f be a continuous function from S into a metric space M_2. Show that f can be extended uniquely to a continuous function F from M_1 into M_2 if and only if, for every $x_0 \in M_1 - S$, $\lim_{x \to x_0} f(x)$ exists. (*F extends f* means F agrees with f on S.)

14. Show that the image of a bounded set under a uniformly continuous function may not be bounded. (See exercise 4.)

15. Let $f(x)$ be a real-valued function on the metric space M.

    Definition. a is a *value approached* by $f(x)$ at x_0 if there exists a sequence $\{x_n\}$ converging to x_0 such that $f(x_n) \to a$. Let $A(x_0)$ be the set of values approached by $f(x)$ at x_0, and define

    $$\overline{\lim_{x \to x_0}} f(x) = \sup A(x_0)$$

    $$\underline{\lim_{x \to x_0}} f(x) = \inf A(x_0).$$

    a. Show that $A(x_0)$ is a closed set.
    b. Define the functions

    $$f_U(x) = \overline{\lim_{y \to x}} f(y)$$

    $$f_L(x) = \underline{\lim_{y \to x}} f(y).$$

    (They are called the upper and lower envelopes, respectively, of $f(x)$.) Show that $f(x)$ is continuous at x_0 if and only if $f_U(x_0) = f_L(x_0)$.

16. Let $f(x) = f_1(x) + if_2(x)$ be a complex-valued function on a metric space M. Show that $f(x)$ is continuous if and only if both $f_1(x)$ and $f_2(x)$ are continuous. Use the metric $d(z_1, z_2) = |z_1 - z_2|$ on C. (f_1 and f_2 are real-valued.)

17. Let $f(x)$ and $g(x)$ be continuous complex-valued functions on the metric space M. Show that
   a. $f(x) + g(x)$ is continuous;
   b. $f(x)g(x)$ is continuous;
   c. $|f(x)|$ is continuous;
   d. $f(x)/g(x)$ is continuous whenever $g(x) \neq 0$.

Section 3 COMPLETENESS

For the metric space R, the real number system, we proved that a sequence converges if and only if it is a Cauchy sequence. We now define the notion of a Cauchy sequence in an arbitrary metric space.

Definition 3.3.1. A sequence $\{x_n\}$ in a metric space M is called a Cauchy sequence if for any $\epsilon > 0$ there is an $N(\epsilon)$ such that $d(x_m, x_n) \leq \epsilon$ whenever $m, n \geq N(\epsilon)$.

The following three results are quite elementary.

1. Cauchy sequences are bounded sets.
2. Convergent sequences are Cauchy sequences.
3. If a Cauchy sequence converges, then the limit is unique.

Since, for any $\epsilon > 0$, all but a finite number of terms in the Cauchy sequence $\{x_n\}$ are contained in the sphere $S(x_{N(\epsilon)}; \epsilon)$, it is clear that $\{x_n\}$ is bounded. This proves result 1. To prove result 2 we assume $x_n \to x_0$. Then

$$d(x_m, x_n) \leq d(x_m, x_0) + d(x_n, x_0) \leq \frac{\epsilon}{2} + \frac{\epsilon}{2} = \epsilon$$

for $m, n \geq N(\epsilon/2)$ and the convergent sequence $\{x_n\}$ is a Cauchy sequence. The third result is simply a restatement of the uniqueness of limits property mentioned in section 1.

The converse of result 2 is false in general. For example, in the rational numbers Q under the usual metric $d(r_1, r_2) = |r_1 - r_2|$, the sequence $1, 1.4, 1.41, 1.414, \ldots$ (the truncations of the infinite decimal representing $\sqrt{2}$) is a Cauchy sequence not converging to any member of Q. Since Q is a dense subset of R, a metric space in which all Cauchy sequences converge, we are led to the following definition.

Definition 3.3.2

(a) A metric space M is called *complete* if every Cauchy sequence in M converges to a limit in M.
(b) A metric space $\hat{M}$ is called a completion of the metric space M if $\hat{M}$ is complete, $\hat{M} \supseteq M$, and $\bar{M} = \hat{M}$.

By $\hat{M} \supseteq M$ we mean that not only is M a subset of $\hat{M}$ but also the metric $\hat{d}$ in $\hat{M}$ is an extension of the metric d in M; $\bar{M}$ is the closure of M as a subset of $\hat{M}$.

We will need the following lemma, whose proof is left as an exercise.

Lemma 3.3.3. Two Cauchy sequences $\{x_n\}$ and $\{y_n\}$ in a complete metric space M have the same limit if and only if $\lim_{n \to \infty} d(x_n, y_n) = 0$.

Our goal is to show that every metric space has a completion which is essentially unique. To make the uniqueness statement clear, we define metric spaces M_1 and M_2 to be *isometric* if there is a one-to-one function F mapping M_1 onto M_2 such that $d_1(x, y) = d_2[F(x), F(y)]$ for all x and y in M_1. We can now state our main theorem.

Theorem 3.3.4. Every metric space M has a completion $\hat{M}$, and any two completions of M are isometric.

Proof. We treat the uniqueness problem first. Let $\hat{M}_1$ and $\hat{M}_2$ be two completions of M. Then $\hat{M}_1 \supseteq M$ and $\hat{M}_2 \supseteq M$, and we define our isometry F to map M identically onto itself. We must now extend F to an isometry of $\hat{M}_1$ onto $\hat{M}_2$. Let $\hat{x}_1 \in \hat{M}_1 - M$, and choose a Cauchy sequence $\{x_n\}$ in M converging to $\hat{x}_1$ under d_1, the metric of $\hat{M}_1$. Then $\{x_n\} = \{F(x_n)\}$ is a Cauchy sequence under d_2, the metric of $\hat{M}_2$, since d_1 and d_2 coincide on M with the metric d of M. We define $F(\hat{x}_1) = \lim_{n \to \infty} F(x_n)$, but we must verify that, if $\{y_n\}$ is another Cauchy sequence converging to $\hat{x}_1$, then $\lim_{n \to \infty} F(y_n) = \lim_{n \to \infty} F(x_n)$. [Otherwise, $F(\hat{x}_1)$ would be dependent on the particular Cauchy sequence used.] However, $d_2[F(x_n), F(y_n)] = d_1[x_n, y_n] \to 0$ as $n \to \infty$, by the preceding lemma, and the Cauchy sequences $\{F(x_n)\}$ and $\{F(y_n)\}$ in $\hat{M}_2$ must have the same limit.

This completes the definition of the function F; it is easy to verify that F is one-to-one and maps $\hat{M}_1$ onto $\hat{M}_2$. This detail is left to the reader. We must now show that, for any $\hat{x}, \hat{y}$ in $\hat{M}_1$, $d_1(\hat{x}, \hat{y}) = d_2[F(\hat{x}), F(\hat{y})]$.

Let $x_n \to \hat{x}$ in $\hat{M}_1$; then $F(x_n) \to F(\hat{x})$ in $\hat{M}_2$ and, for any $\epsilon > 0$, there is an $N(\epsilon, \hat{x})$ such that $d_1(x_n, \hat{x}) \le \epsilon/4$ and $d_2[F(x_n), F(\hat{x})] \le \epsilon/4$ whenever $n \ge N(\epsilon, \hat{x})$. Similarly, $y_n \to \hat{y}$ implies $F(y_n) \to F(\hat{y})$, and there is an $N(\epsilon, \hat{y})$

such that $d_1(y_n, \hat{y}) \leq \epsilon/4$ and $d_2[F(y_n), F(\hat{y})] \leq \epsilon/4$ whenever $n > N(\epsilon, \hat{y})$. Hence, for $n \geq \max\{N(\epsilon, \hat{x}), N(\epsilon, \hat{y})\}$, we have

$$|d_1(\hat{x}, \hat{y}) - d(x_n, y_n)| = |d_1(\hat{x}, \hat{y}) - d_1(\hat{x}, y_n) + d_1(\hat{x}, y_n) - d(x_n, y_n)|$$
$$\leq |d_1(\hat{x}, \hat{y}) - d_1(\hat{x}, y_n)| + |d_1(\hat{x}, y_n) - d(x_n, y_n)|$$
$$\leq d_1(\hat{y}, y_n) + d_1(\hat{x}, x_n) \leq \frac{\epsilon}{4} + \frac{\epsilon}{4} = \frac{\epsilon}{2},$$

and similarly,

$$|d_2[F(\hat{x}), F(\hat{y})] - d(x_n, y_n)| \leq \frac{\epsilon}{2}.$$

From these two inequalities we find

$$|d_1(\hat{x}, \hat{y}) - d_2[F(\hat{x}), F(\hat{y})]|$$
$$\leq |d_1(\hat{x}, \hat{y}) - d(x_n, y_n)| + |d(x_n, y_n) - d_2[F(\hat{x}), F(\hat{y})]|$$
$$\leq \frac{\epsilon}{2} + \frac{\epsilon}{2} = \epsilon \qquad \text{for any } \epsilon > 0.$$

Hence $d_1(\hat{x}, \hat{y}) = d_2[F(\hat{x}), F(\hat{y})]$, and the uniqueness part of the theorem is proved.

Now let M be a metric space and let M_0 be the set of all Cauchy sequences in M. We define

$$\{x_n\} * \{y_n\} \Leftrightarrow \lim_{n \to \infty} d(x_n, y_n) = 0.$$

The relation $*$ is obviously symmetric and reflexive, and the inequality $d(x_n, z_n) \leq d(x_n, y_n) + d(y_n, z_n)$ yields its transitivity.

Let $\hat{M}$ be the set of equivalence classes of Cauchy sequences, and define

$$\hat{d}(|\{x_n\}|, |\{y_n\}|) = \lim_{n \to \infty} d(x_n, y_n).$$

This limit exists, for

$$|d(x_m, y_m) - d(x_n, y_n)| \leq |d(x_m, y_m) - d(x_m, y_n)| + |d(x_m, y_n) - d(x_n, y_n)|$$
$$\leq d(y_m, y_n) + d(x_m, x_n),$$

and so $\{d(x_n, y_n)\}$ is a Cauchy sequence in R. We must now show that the limit is independent of the choice of the representatives from the equivalence classes. Accordingly, let $\{x_n\} * \{x'_n\}$ and $\{y_n\} * \{y'_n\}$; we show $\lim_{n \to \infty} d(x_n, y_n)$ $= \lim_{n \to \infty} d(x'_n, y'_n)$. We have

$$|d(x_n, y_n) - d(x'_n, y'_n)| \leq |d(x_n, y_n) - d(x_n, y'_n)| + |d(x_n, y'_n) - d(x'_n, y'_n)|$$
$$\leq d(y_n, y'_n) + d(x_n, x'_n) \longrightarrow 0,$$

so that $\hat{d}$ is indeed well-defined. Verification that $\hat{d}$ is a metric on $\hat{M}$ is left to the reader.

If we identify each x in M with the equivalence class of Cauchy sequences which contains the Cauchy sequence $(x, x, \ldots, x, \ldots)$, then we can assert that $\hat{M} \supseteq M$ as desired. Note that $\hat{d}$ agrees with d on equivalence classes of this type.

Now let $\hat{x}$ be in $\hat{M} - M$ and let $\hat{x}$ be represented by the Cauchy sequence $\{x_n\}$. For any $\epsilon > 0$ there is an $N(\epsilon)$ such that $m, n \geq N(\epsilon)$ implies $d(x_m, x_n) \leq \epsilon$. Hence, for $m \geq N(\epsilon)$ we have $\lim_{n \to \infty} d(x_m, x_n) \leq \epsilon$. [Since $|d(x_m, x_n) - d(x_m, x_p)| \leq d(x_n, x_p) \longrightarrow 0$, we can be sure that this limit exists.] Letting x_m^* be the member of $\hat{M}$ represented by

$$(x_m, x_m, \ldots, x_m, \ldots), \text{ we have } \hat{d}(\hat{x}, x_m^*) \leq \epsilon.$$

This shows that $\hat{x}$ is the limit of a sequence of points of M, and thus $\bar{M} = \hat{M}$.

Finally, we show that $\hat{M}$ is complete. Let $\{\hat{x}_n\}$ be a Cauchy sequence of elements of $\hat{M}$. Since $\bar{M} = \hat{M}$ we can, for each n, find an x_n in M such that $\hat{d}(x_n, \hat{x}_n) \leq 1/n$. Then

$$d(x_m, x_n) \leq \hat{d}(x_m, \hat{x}_m) + \hat{d}(\hat{x}_m, \hat{x}_n) + \hat{d}(\hat{x}_n, x_n) \leq \frac{1}{m} + \frac{1}{n} + \hat{d}(\hat{x}_m, \hat{x}_n)$$

and $\{x_n\}$ is a Cauchy sequence in $\hat{M}$. It is equivalent to $\{\hat{x}_n\}$, and since $\{x_n\}$ has a limit in M, we see that M is complete. This finishes the proof of theorem 3.3.4.

At this point the only metric space we know to be complete is R, the real numbers, under $d(x, y) = |x - y|$. In fact, R is the completion of Q, the rationals, and may be constructed from Q by this procedure, as outlined in the exercises.

A simple result which is sometimes useful in determining completeness is as follows.

Lemma 3.3.5. A closed subset K of a complete metric space M is itself a complete metric space.

See the exercises for proof.

EXAMPLE 1. Consider $R^{(2)}$ under the Euclidean metric

$$d[(x_1, y_1), (x_2, y_2)] = [(x_1 - x_2)^2 + (y_1 - y_2)^2]^{1/2}.$$

If $\{(x_n, y_n)\}$ is a Cauchy sequence in R^2, then for any $\epsilon > 0$ there is an $N(\epsilon)$ such that $m, n \geq N(\epsilon)$ implies

$$\{(x_n - x_m)^2 + (y_n - y_m)^2\}^{1/2} \leq \epsilon.$$

It follows that $\{x_n\}$ and $\{y_n\}$ are Cauchy sequences in R, and hence $x_n \longrightarrow x_0$, $y_n \longrightarrow y_0$, so that $\{(x_n, y_n)\}$ converges coordinatewise to (x_0, y_0). But in $R^{(2)}$, coordinatewise convergence is equivalent to metric convergence, from which we have $d[(x_n, y_n), (x_0, y_0)] \longrightarrow 0$, and the completeness of $R^{(2)}$ is proved.

EXAMPLE 2. Example 1, applied to C, shows that C is a complete metric space under $d_1(z_1, z_2) = |z_1 - z_2|$, for this metric is exactly the Euclidean metric on $R^{(2)}$. We now consider C under the metric

$$d_2(z_1, z_2) = \frac{|z_1 - z_2|}{(1 + |z_1|^2)^{1/2}(1 + |z_2|^2)^{1/2}}.$$

Now, if $\{z_n\}$ is a Cauchy sequence under d_2 and if $|z_n| \leq M$ for all n, then

$$d_2(z_m, z_n) = \frac{|z_m - z_n|}{(1 + |z_m|^2)^{1/2}(1 + |z_n|^2)^{1/2}} \geq \frac{|z_m - z_n|}{1 + M^2}$$

and $\{z_n\}$ is Cauchy under d_1. Since C is complete under d_1 and since d_1 and d_2 are equivalent metrics, we see that $\{z_n\}$ converges under d_2.

We now show that any sequence $\{w_n\}$ such that $|w_n| \longrightarrow \infty$ is a Cauchy sequence under d_2. We have

$$d(w_m, w_n) = \frac{|w_m - w_n|}{(1 + |w_m|^2)^{1/2}(1 + |w_n|^2)^{1/2}} = \frac{|1/w_n - 1/w_m|}{(1 + 1/|w_n|^2)^{1/2}(1 + 1/|w_m|^2)^{1/2}}$$

$$\leq \frac{1}{|w_n|} + \frac{1}{|w_m|} \longrightarrow 0.$$

If $\{z_n\}$ is another sequence such that $|z_n| \longrightarrow \infty$, then

$$d(z_n, w_n) \leq \frac{1}{|z_n|} + \frac{1}{|w_n|} \longrightarrow 0$$

so that $\{z_n\}$ is equivalent to $\{w_n\}$.

It is left as an exercise to show that, if $\{z_n\}$ is Cauchy under d_2, then either (a) $|z_n| \leq M$ for all n, or (b) $|z_n| \longrightarrow \infty$. That is, the two types of Cauchy sequences just considered are the only possibilities. Once this is known, it is then clear that C is not complete under d_2 and that $\hat{C}$ consists of C plus a single additional point, the "point at infinity."

The preceding example shows that, even though two metrics on M may be equivalent, they do not necessarily lead to the same completion of M.

The difficulty is due to the fact that a sequence such as $1, 2, 3, \ldots, n, \ldots$ is a Cauchy sequence under d_2, but is not a Cauchy sequence under d_1. This illustrates the general proposition that the continuous image of a Cauchy sequence may not be a Cauchy sequence. In contrast to this we have the following lemma, whose proof is left as an exercise.

Lemma 3.3.6. The image of a Cauchy sequence under a uniformly continuous function is a Cauchy sequence.

Using this lemma, one can prove that uniformly equivalent metrics on a set M do lead to the same completion $\hat{M}$. (See exercise 5.) The lemma also enables us to prove the following extension theorem, which will be of great importance in later work.

Theorem 3.3.7. Let A be a dense subset of a metric space M_1, and let F be a uniformly continuous function from A into a complete metric space M_2. Then F can be uniquely extended to a continuous function $\hat{F}$ from M_1 into M_2, and this extension is uniformly continuous.

Proof. Let $x \in M_1 - A$ and let $\{x_n\}$ be a sequence in A converging to x. Then $\{x_n\}$ is a Cauchy sequence and, by the lemma, so is $\{F(x_n)\}$. If y is the limit of $\{F(x_n)\}$, then we define $\hat{F}(x) = y$. To be certain that this is unambiguous, let $\{x'_n\}$ be a second sequence converging to x. We must show that the Cauchy sequence $\{F(x'_n)\}$ also converges to y or, equivalently, that $d_2[F(x_n), F(x'_n)] \to 0$. By the uniform continuity of F on A we can, for any $\epsilon > 0$, find a $\delta(\epsilon)$ such that $d_1(x, y) \leq \delta(\epsilon)$ implies $d_2[F(x), F(y)] \leq \epsilon$. Since $\{x_n\}$ is equivalent to $\{x'_n\}$ we can find an $N[\delta(\epsilon)]$ such that $d_1(x_n, x'_n) \leq \delta(\epsilon)$ for $n \geq N[\delta(\epsilon)] = N(\epsilon)$ and, hence, $d_2[F(x_n), F(x'_n)] \leq \epsilon$ for $n \geq N(\epsilon)$. It follows that $\{F(x_n)\}$ is equivalent to $\{F(x'_n)\}$, and therefore, $\hat{F}$ is a well-defined function from M_1 into M_2. Moreover, $\hat{F}$ is continuous from M_1 into M_2, by lemma 3.2.2(b).

To show that $\hat{F}$ is uniformly continuous, let $\delta(\epsilon)$ be determined by the uniform continuity of F on A, and let x_0 and y_0 be any two points of M_1 for which $d_1(x_0, y_0) \leq \frac{1}{3}\delta(\epsilon)$. Let $\delta(x_0, \epsilon)$ and $\delta(y_0, \epsilon)$ be determined from the continuity of $\hat{F}$ on M_1, and choose points x and y in A such that

$$d_1(x_0, x) \leq \min\left[\tfrac{1}{3}\delta(\epsilon), \delta(x_0, \epsilon)\right] \quad \text{and} \quad d_1(y_0, y) \leq \min\left[\tfrac{1}{3}\delta(\epsilon), \delta(y_0, \epsilon)\right].$$

Then we have

$$d_2[\hat{F}(x_0), \hat{F}(x)] \leq \epsilon, \, d_2[\hat{F}(y_0), \hat{F}(y)] \leq \epsilon.$$

Also,

$$d_1(x, y) \leq d_1(x, x_0) + d_1(x_0, y_0) + d_1(y_0, y) \leq \delta(\epsilon),$$

so that $d_2[F(x), F(y)] \leq \epsilon$. Hence,

$$d_2[\hat{F}(x_0), \hat{F}(y_0)] \leq d_2[\hat{F}(x_0), F(x)] + d_2[F(x), F(y)] + d_2[F(y), \hat{F}(y_0)] \leq 3\epsilon,$$

and $\hat{F}$ is uniformly continuous.

Our last result of this section is a characterization of completeness analogous to the nested intervals theorem in R.

Theorem 3.3.8. A metric space M is complete if and only if, for every sequence of closed spheres $\{K_n\}$ such that $K_n \supseteq K_{n+1}$ for all n, and (diameter $K_n) \to 0$, we have $\bigcap\limits_{n=1}^{\infty} K_n = x_0$, a single point.

Proof. Assume that M is complete and that $\{K_n\}$ is a sequence of closed spheres $K_n = \{x: d(x, x_n) \leq a_n\}$, where $K_n \supseteq K_{n+1}$ and $a_n \to 0$. The sequence $\{x_n\}$ is a Cauchy sequence, for, if $n \geq m$, then $d(x_m, x_n) \leq a_m$. If x_0 is the limit of $\{x_n\}$, then x_0, as the limit of the sequence $x_m, x_{m+1}, \ldots, x_{m+p}, \ldots$, all of whose members are in K_m, is in K_m also (K_m is closed). Hence, $x_0 \in \bigcap\limits_{n=1}^{\infty} K_n$; it is left as an exercise to show that, in fact, $x_0 = \bigcap\limits_{n=1}^{\infty} K_n$.

We prove the converse contrapositively. Suppose that M is not complete; then there is a Cauchy sequence $\{x_n\}$ in M which has no limit. Now, for any k we can find an $N(k)$ such that $d(x_m, x_n) \leq 2^{-k}$ for $m, n \geq N(k)$. Let $\{y_k\}$ be the sequence $x_{N(1)}, x_{N(2)}, \ldots, x_{N(k)}, \ldots$, and let K_k be the closed sphere $\{x: d(y_k, x) \leq 2^{-k+1}\}$. Now, if $x \in K_{k+1}$, then

$$d(x, y_{k+1}) \leq 2^{-(k+1)+1} = 2^{-k}$$

and

$$d(x, y_k) \leq d(x, y_{k+1}) + d(y_k, y_{k+1}) \leq 2^{-k} + 2^{-k} = 2^{-k+1}.$$

Hence $K_{k+1} \subseteq K_k$, and the spheres are nested. Moreover, (diameter $K_k) \to 0$, and $\bigcap\limits_{k=1}^{\infty} K_k$ contains at most one point. If $y_0 = \bigcap\limits_{k=1}^{\infty} K_k$, then y_0 is the limit of $\{y_k\}$ and, hence, is the limit of $\{x_n\}$. Since $\{x_n\}$ has no limit, this is impossible. We conclude $\bigcap\limits_{k=1}^{\infty} K_k = \varnothing$. This finishes the proof.

EXAMPLE 3. We outline the steps needed to show that R^{∞} is complete.

(a) A sequence $\{x_n\}$ where $x_n = (a_{n1}, a_{n2}, \ldots, a_{nk}, \ldots)$ is Cauchy in R^{∞} if and only if, for every k, the sequence $(a_{1k}, a_{2k}, \ldots, a_{nk}, \ldots)$ is Cauchy.

(b) If $\{x_n\}$ is Cauchy in R^{∞}, then for every k, $\lim\limits_{n\to\infty} a_{nk} = a_k$ exists, and $\{x_n\}$ converges coordinatewise to $(a_1, a_2, \ldots, a_k, \ldots) = x_0$.

(c) Since convergence in R^{∞} is equivalent to coordinatewise convergence, R^{∞} is complete.

EXERCISES

1. Prove that two Cauchy sequences $\{x_n\}$ and $\{y_n\}$ have the same limit in a complete metric space M if and only if limit $\lim_{n \to \infty} d(x_n, y_n) = 0$.

2. Prove:
   a. A closed subset of a complete metric space is complete.
   b. A complete subset of a metric space is closed.

3. Fill in the details of example 2.

4. Prove that the image of a Cauchy sequence under a uniformly continuous mapping is a Cauchy sequence.

5. Prove that, if M_1 and M_2 are uniformly homeomorphic, then $\hat{M}_1$ and $\hat{M}_2$ are also uniformly homeomorphic. (See exercise 8 of section 2 for terminology.) Interpret this result for uniformly equivalent metrics on a set M.

6. Consider the real-number system R under the metric $d_2(x, y) = |x^3 - y^3|$. (See exercise 5 of section 2.) Is R complete under this metric?

7. Show that $C^{(n)}$ is complete under the Euclidean metric (See example 5 of section 1).

8. Let M be any set and consider M under the metric $d(x, y) = 1$ if $x \neq y$, $d(x, x) = 0$. Is M complete?

9. Give a full proof that R^∞ is complete. Does the same procedure apply to C^∞?

10. Let Q be the set of rational numbers under $d(x, y) = |x - y|$, and let $\hat{Q}$ be the set of equivalence classes of Cauchy sequences of rationals. Define:
$$|\{x_n\}| + |\{y_n\}| = |\{x_n + y_n\}|$$
$$|\{x_n\}| \cdot |\{y_n\}| = |\{x_n y_n\}|$$
$$|\{x_n\}| \in P \text{ if there exists a positive integer } N \text{ such that}$$
$$n \geq N \Rightarrow x_n \geq \frac{1}{N}.$$
   a. Show that these definitions are proper in that they are inde-

pendent of the representative sequences from the equivalence classes.

b. Show that $\hat{Q}$ is an ordered field.

c. Show that $\hat{Q}$ is a complete ordered field in the sense of chapter 2.

11. Prove that if $K_n = \{x: d(x, x_n) \leq a_n\}$, where $K_n \supseteq K_{n+1}$, $a_n \to 0$, then $\bigcap\limits_{n=1}^{\infty} K_n$ consists of at most one point.

Section 4 COMPACTNESS

The Archimedian property of the ordering in R guarantees that, if A is a bounded subset of R and if ϵ is any positive number, then A can be covered by a finite number of open intervals of length ϵ. The converse is equally clear, namely, if for any $\epsilon > 0$ a set A can be covered by finitely many intervals of length ϵ, then A is bounded. To generalize to metric spaces, we introduce the next definition.

Definition 3.4.1. A subset A of a metric space M is called *totally bounded* if, for any $\epsilon > 0$, A can be covered by a finite number of spheres of radius ϵ.

From our foregoing remarks, we see that for subsets of R boundedness is equivalent to total boundedness. The following lemma establishes some properties of totally bounded sets in metric spaces.

Lemma 3.4.2. A totally bounded subset A of a metric space M is bounded and separable.

Proof. Let A be totally bounded; then for any $\epsilon > 0$ we can find spheres $S(x_1; \epsilon), \ldots, S(x_N; \epsilon)$ such that $\bigcup\limits_{k=1}^{N} S(x_k; \epsilon) \supseteq A$. Let

$$c = \max_{i,j=1}^{N} d(x_i, x_j).$$

If x and y are any two points of A, then $x \in S(x_i; \epsilon)$ and $y \in S(x_j; \epsilon)$ for some i and j. Hence

$$d(x, y) \leq d(x, x_i) + d(x_i, x_j) + d(x_j, y) \leq 2\epsilon + c,$$

and A is bounded.

We now show that A is separable. For any integer n, A can be covered by finitely many spheres $S(x_{1n}; 1/n), \ldots, S(x_{N(n)n}; 1/n)$. The set

$$x_{11}, \ldots, x_{N(1)1}; x_{12}, \ldots, x_{N(2)2}; \ldots, x_{1n}, \ldots, x_{N(n)n}; \ldots$$

is countable. To see that it is dense, let $x \in A$ and note that for any n there is a $k(n)$ such that $x \in S(x_{k(n)n}; 1/n)$. The sequence

$$x_{k(1)1}, x_{k(2)2}, \ldots, x_{k(n)n} \ldots$$

then converges to x, and the proof is finished.

Boundedness does not, in general, imply total boundedness. For example, if M is any infinite set with the metric $d(x, y) = 1$, $x \neq y$, $d(x, x) = 0$, then M is bounded, but for $0 < \epsilon < 1$, no finite number of spheres of radius ϵ can cover M. Moreover, if M is denumerably infinite, then M is both bounded and separable, without being totally bounded, so that the converse of the preceding lemma is false.

Definition 3.4.3. A subset A of a metric space M is *compact* if it is totally bounded and complete.

It follows at once from our foregoing remarks and from exercise 2 of the preceding section that *a compact subset A of a metric space M is closed and bounded.* Equally immediate is the fact that *a closed subset B of a compact metric space M is compact.* In the spaces $C^{(n)}$ and $R^{(n)}$ boundedness is equivalent to total boundedness, so that *a subset of $C^{(n)}$ or $R^{(n)}$ is compact if and only if it is closed and bounded.*

Theorem 3.4.4. The following three statements about a metric space M are equivalent.

(a) M is compact.
(b) Every infinite sequence in M has a convergent subsequence.
(c) If $\{O_\alpha\}$ is any family of open subsets of M such that $\bigcup_\alpha O_\alpha = M$, then some finite subfamily also covers M.

Proof. (a) $\Rightarrow$ (c). Suppose that M is compact, that a family $\{O_\alpha\}$ of open sets covers M, but that no finite subfamily covers M. Let M be covered by finitely many spheres of radius $\frac{1}{2}$. Then at least one of them, $S(x_1; \frac{1}{2})$ is not covered by finitely many members of $\{O_\alpha\}$. Now let M be covered by finitely many spheres of radius 2^{-2}; then among those which have nonempty intersection with $S(x_1; \frac{1}{2})$, there must be at least one, $S(x_2; 2^{-2})$, which is not covered by finitely many O_α's. Continuing in this way, we obtain a sequence of spheres $S_n(x_n; 2^{-n})$, none of which is covered by finitely many O_α's. The sequence $\{x_n\}$ is Cauchy, for

$$d(x_n, x_{n+p}) \leq d(x_n, x_{n+1}) + d(x_{n+1}, x_{n+2}) + \cdots + d(x_{n+p-1}, x_{n+p})$$
$$\leq (2^{-n} + 2^{-(n+1)}) + (2^{-(n+1)} + 2^{-(n+2)}) + \cdots$$
$$+ (2^{-(n+p-1)} + 2^{-(n+p)})$$
$$< 2(2^{-n} + 2^{-(n+1)} + \cdots) = 2^{-n+1}(1 + 2^{-1} + 2^{-2} + \cdots) = 2^{-n+2}.$$

Let x_0 be the limit of $\{x_n\}$; then $x_0 \in O_\alpha$ for some α, and hence $x_0 \in S(x_0; \epsilon) \subseteq O_\alpha$ for some ϵ. Choosing n so that $2^{-n} < \epsilon/2$ and $d(x_n, x_0) < \epsilon/2$, we see that $S(x_n; 2^{-n}) \subseteq S(x_0; \epsilon) \subseteq O_\alpha$, in contradiction to the construction of the sequence $\{S(x_n; 2^{-n})\}$. This finishes the proof that (a) $\Rightarrow$ (c).

(c) $\Rightarrow$ (b). Let $\{x_n\}$ be an infinite sequence in M and let K_n be the closure of the set $\{x_n, x_{n+1}, \ldots, x_{n+p}, \ldots\}$. We claim that $\bigcap_{n=1}^{\infty} K_n$ is not empty, for otherwise $M = \bigcup_{n=1}^{\infty} \tilde{K}_n$, and by (c), $\tilde{K}_{n(1)} \cup \cdots \cup \tilde{K}_{n(k)} = M$. But this implies $K_{n(1)} \cap \cdots \cap K_{n(k)} = \varnothing$ which is clearly not so. Let $x_0 \in \bigcap_{n=1}^{\infty} K_n$; then for every $\epsilon > 0$ and n the set $S(x_0; \epsilon) \cap \{x_n, x_{n+1}, \ldots\}$ is nonempty. Using this fact, we can easily select a subsequence of $\{x_n\}$ converging to x_0.

(b) $\Rightarrow$ (a). Clearly, (b) implies that M is complete. We assume M is not totally bounded. Then there exists $\epsilon_0 > 0$ such that no finite set of spheres of radius ϵ_0 cover M. Let x_1 be arbitrary and choose x_2 in $\sim[S(x_1; \epsilon_0)]$; then choose x_3 in $\sim[S(x_1; \epsilon_0) \cup S(x_2; \epsilon_0)]$ and continue the process. (x_n is in $\sim[S(x_1; \epsilon_0) \cup \ldots \cup S(x_{n-1}; \epsilon_0)]$.) Since $d(x_m, x_n) > \epsilon_0$ for all m and n, the sequence $\{x_n\}$ can have no convergent subsequence. Hence (b) implies that M is totally bounded as well as complete. This finishes the entire proof.

The following lemma will be useful in exploring some ramifications of the notion of compactness.

Lemma 3.4.5. The closure $\bar{A}$ of a totally bounded subset A of a metric space M is totally bounded.

Proof. For any $\epsilon > 0$, a finite number of spheres $S(x_1; \epsilon/2), \ldots, S(x_n; \epsilon/2)$ cover A. We show that $\bar{A}$ is covered by $S(x_1; \epsilon), \ldots, S(x_n; \epsilon)$. Let $x_0 \in \bar{A} - A$; from

$$\bigcup_{k=1}^{n} \overline{S\left(x_k; \frac{\epsilon}{2}\right)} = \overline{\bigcup_{k=1}^{n} S\left(x_k; \frac{\epsilon}{2}\right)} \supseteq \bar{A},$$

we see that $x_0 \in \overline{S(x_j; \epsilon/2)} \subseteq S(x_j; \epsilon)$ for some j. This finishes the proof.

Now, if A is a totally bounded subset of a metric space M, then the closure of A in $\hat{M}$, the completion of M, is a compact subset of $\hat{M}$ (closed subsets of complete spaces are complete). In particular, if M is complete, then $\bar{A}$ is compact in M. This situation arises often enough to merit special terminology. A subset A of a metric space M is called *relatively compact* if the closure $\bar{A}$ of A in M is compact. We can then say that *totally bounded subsets of complete metric spaces are relatively compact.*

We now give two theorems which illustrate the use of compactness.

Theorem 3.4.6. If F is a continuous function from a compact metric space M_1 into a metric space M_2, then F is uniformly continuous.

Proof. By the continuity of F, there is, for $\epsilon > 0$ and $x_0 \in M_1$, a $\delta(\epsilon, x_0)$ such that, if $x \in S[x_0; \delta(\epsilon, x_0)]$, then $d[F(x), F(x_0)] \leq \epsilon$. M_1 is then covered by the spheres $\{S[x, \frac{1}{2}\delta(\epsilon, x)]\}$ and, by the compactness of M_1, we have $M_1 = \bigcup_{k=1}^{n} S[x_k; \frac{1}{2}\delta(\epsilon, x_k)]$. Let

$$\delta(\epsilon) = \min_{k=1}^{n} \{\tfrac{1}{2}\delta(\epsilon, x_k)\},$$

and let x and x' be any two points in M_1 such that $d(x, x') \leq \delta(\epsilon)$. We show that $d[F(x), F(x')] \leq 2\epsilon$. Now, x is in some $S[x_i; \frac{1}{2}\delta(x_i, \epsilon)]$, and

$$d(x_i, x') \leq d(x_i, x) + d(x, x') \leq \tfrac{1}{2}\delta(x_i, \epsilon) + \delta(\epsilon) \leq \delta(x_i, \epsilon).$$

Hence we have

$$d[F(x), F(x')] \leq d[F(x), F(x_i)] + d[F(x_i), F(x')] \leq \epsilon + \epsilon = 2\epsilon,$$

and the proof is finished.

Theorem 3.4.7. If F is a continuous function from metric space M_1 into M_2 and if K is a compact subset of M_1, then $F(K)$ is compact in M_2.

Proof. If $\{O_\alpha\}$ is a family of open sets in M_2 covering $F(K)$, then the family $\{F^{-1}(O_\alpha)\}$ consists of open sets in M_1 and it covers K. Hence a finite subfamily $F^{-1}(O_1), \ldots, F^{-1}(O_n)$ covers K, and from

$$F(K) \subseteq F[F^{-1}(O_1) \cup \cdots \cup F^{-1}(O_n)]$$
$$= F[F^{-1}(O_1)] \cup \cdots \cup F[F^{-1}(O_n)] \subseteq O_1 \cup \cdots \cup O_n,$$

we see that $F(K)$ is compact.

As an application of theorem 3.4.7, let F be a continuous function from a compact metric space M into the real numbers R. Then $F(M)$ is compact, i.e., closed and bounded, in R, and $F(M)$ has a maximum c_2 and a minimum c_1 (both *in* $F(M)$). Hence there exist x_1 and x_2 in M such that $F(x_1) = c_1$ and

$F(x_2) = c_2$. In summary, *a continuous, real-valued function on a compact metric space assumes maximum and minimum values*. In the exercises, we indicate an analogous result for complex-valued, continuous functions on a compact metric space.

The following lemma gives further applications of the notion of compactness.

Lemma 3.4.8. Let M be a metric space, x_0 a point of M, A an arbitrary subset of M, and C, C_1, and C_2 compact subsets of M. Then

(a) $d(x_0, C) = d(x_0, x_1)$ for some $x_1 \in C$.
(b) $d(A, C) = d(A, x_1)$ for some $x_1 \in C$.
(c) $d(C_1, C_2) = d(x_1, x_2)$ for some $x_1 \in C_1$ and $x_2 \in C_2$.
(d) $d(A, C) = 0 \Leftrightarrow \bar{A} \cap C \neq \varnothing$.

We prove (a). The function $d(x_0, x)$ for $x \in C$ is a continuous function on a compact metric space and, therefore, assumes a minimum at some $x_1 \in C$. But then $d(x_0, x_1) = d(x_0, C)$. The proofs of (b), (c), and (d) are left as exercises.

EXERCISES

1. Let $x_i = (a_{i1}, a_{i2}, \ldots, a_{in}, \ldots)$, $i = 1, 2, \ldots, N$, be N points in R^∞. Define

$$x_0 = (b_1, b_2, \ldots) \quad \text{by} \quad b_n = \overset{N}{\underset{i=1}{\max}} |a_{in}| + 1.$$

   a. Show that $d(x_i, x_0) \geq \frac{1}{2} \overset{\infty}{\underset{n=1}{\sum}} k_n$ for $i = 1, 2, \ldots, N$.
   b. Show that R^∞ is not totally bounded.

2. Show that boundedness and total boundedness are equivalent for subsets of $R^{(n)}$ and $C^{(n)}$.

3. Let $x(t)$ be a continuous, complex-valued function on a metric space M. If M is compact and $a = \sup \{|x(t)|: t \in M\}$ show that, for some $t_0 \in M, |x(t_0)| = a$.

4. Prove parts (b), (c), and (d) of lemma 3.4.8.

5. A metric space M is called *locally compact* if, for every $x \in M$, there is an open set O containing x such that $\bar{O}$ is compact.

    a. Show that M compact $\Rightarrow$ M locally compact.
    b. Show that $R^{(n)}$ and $C^{(n)}$ are locally compact.

6. Show that, if an infinite sequence in a compact metric space has a unique cluster point, then it converges to that point.

7. Let M be a compact metric space and $\{O_\alpha\}$ a family of open sets whose union is M. Show that there exists an $\epsilon > 0$ such that every sphere of radius ϵ is contained in some O_α.

8. A family $\{K_\alpha\}$ of closed sets is said to have the finite interesction property if every finite subfamily $K_{\alpha(1)}, \ldots, K_{\alpha(n)}$ satisfies $\bigcap\limits_{i=1}^{n} K_{\alpha(i)}$ $\neq \varnothing$. Show that M is compact if and only if, for every family $\{K_\alpha\}$ of closed sets having the finite intersection property, $\bigcap\limits_{\alpha} K_\alpha \neq \varnothing$.

9. (Refer to exercises 10 and 11 of section 2). If K is a compact subset of M, show that the sets $K_n = \{x : d(x, K) < 1/n\}$ constitute a fundamental system of neighborhoods of K.

10. a. Write a precise statement of "$f(x)$ is not uniformly continuous on M."
    b. Prove theorem 3.4.6 by the contrapositive method.

11. Let F be a *continuous one-to-one* function from a compact metric space M_1 *onto* a metric space M_2. Show that $F(O)$ is open if O is open and, hence, that F^{-1} is continuous.

12. Show that if M is a compact metric space under metric d_1, and if d_2 is equivalent to d_1, then M is compact under d_2, and d_1 is uniformly equivalent to d_2.

Section 5 CONNECTEDNESS

In this section we formalize our intuitive notion that certain metric spaces seem to be composed naturally of pieces which are separate or disconnected from each other.

Definition 3.5.1. A metric space M is connected if M and $\varnothing$ are the only subsets of M which are both open and closed.

Equivalently, M is connected if it is impossible to write $M = O_1 \cup O_2$ where O_1 and O_2 are nonempty disjoint open sets, for in this case, O_1 and O_2 are also closed. A subset A of M is connected if, as a metric space, it is connected.

Lemma 3.5.2. If A is a connected subset of M, and if $A \subseteq A_1 \subseteq \bar{A}$, then A_1 is connected.

Proof. Suppose A_1 is not connected. Then $A_1 = O_1 \cup O_2$, where O_1 and O_2 are nonempty disjoint open subsets of A_1. We then have $A = (O_1 \cap A) \cup (O_2 \cap A)$, where $(O_1 \cap A)$ and $(O_2 \cap A)$ are nonempty disjoint open subsets of A. Hence, if A_1 is disconnected, it follows that A is also disconnected, and the lemma is proved.

Another fundamental property of connectedness is given by the following theorem.

Theorem 3.5.3. If F is a continuous function from M_1 to M_2 and if A is connected in M_1, then $F(A)$ is connected in M_2.

Proof. Suppose $F(A)$ is not connected. Then $F(A) = O_1 \cup O_2$, where O_1 and O_2 are nonempty disjoint subsets of $F(A)$. It follows that

$$A = [A \cap F^{-1}(O_1)] \cup [A \cap F^{-1}(O_2)],$$

where $A \cap F^{-1}(O_1)$ and $A \cap F^{-1}(O_2)$ are nonempty disjoint open subsets of A. Hence A is not connected if $F(A)$ is not connected, and the proof is finished.

EXAMPLE. *The real line, R, is connected*, because any proper open subset of R is a countable union of disjoint open intervals. Since such a set is not closed, there are no proper subsets of R which are simultaneously open and closed. Further, for any open interval (a, b) there is a one-to-one continuous function mapping R onto (a, b), and hence, (a, b) is connected, by the preceding theorem. Finally, the lemma shows that the intervals of type $[a, b)$ and $[a, b]$ are connected.

This example contains half of the proof of the following theorem.

Theorem 3.5.4. A subset of R is connected if and only if it is an interval.

Proof. Let A be a connected subset of R. If A consists of a single point, then A is an interval; so we assume now that A contains at least two points. We assert that, if a and b are in A, and if $a < c < b$, then $c \in A$. For if $c \notin A$, we then have $A = [A \cap (-\infty, c)] \cup [A \cap (c, \infty)]$, and A is not connected. Now let $a_0 = \inf \{a : a \in A\}$, and $b_0 = \sup \{b : b \in A\}$. By our earlier remarks, it is clear that $A \supseteq (a_0, b_0)$; but we also have $A \cap \{\sim [a_0, b_0]\}$

$= \varnothing$ by the definition of a_0 and b_0. Therefore, A must be one of the sets (a_0, b_0), $[a_0, b_0)$, $(a_0, b_0]$, or $[a_0, b_0]$. It may happen that $a_0 = -\infty$ or $b_0 = +\infty$ (or both). This completes the proof.

A corollary to theorems 3.5.3 and 3.5.4 is the intermediate-value theorem of calculus. Namely, if $x(t)$ is a continuous, real-valued function on an interval, and if $f(t_1) < c < f(t_2)$ where $t_1 < t_2$, then there exists a t_0 such that $t_1 < t_0 < t_2$ and $f(t_0) = c$. The proof is left as an exercise.

We now characterize connected open subsets of $R^{(n)}$.

Definition 3.5.5. A subset A of $R^{(n)}$ is *polygonally connected* if any two points of A can be joined by a broken line lying entirely in A.

We understand that a broken line consists of a finite number of straight-line segments joined end to end.

Theorem 3.5.6. An *open* subset of $R^{(n)}$ is connected if and only if it is polygonally connected.

Proof. Let the open set O be connected, and let x be any point of O. If O_1 is defined as the set of all points in O which can be joined to x by a polygonal line in O and if $O_2 = O - O_1$, then clearly $O = O_1 \cup O_2$, $O_1 \cap O_2 = \varnothing$. To see that O_1 is open, let $y \in O_1$; then some sphere $S(y: \epsilon)$ is contained in O, and any z in $S(y; \epsilon)$ can be joined to x by a polygonal line if the polygonal line from x to y is extended radially from y to z. Hence O_1 is open. Similarly, if $y \in O_2$, then some sphere $S(y; \epsilon)$ is contained in O. This sphere cannot meet O_1, for otherwise we could conclude $y \in O_1$. Therefore O_2 is also open. Finally, O_1 is nonempty because $x \in O_1$. To avoid contradicting the connectedness of O, we must conclude $O_2 = \varnothing$; hence $O = O_1$, and O is polygonally connected.

Conversely, let O be a polygonally connected open set which is not connected. Then $O = O_1 \cup O_2$, where O_1 and O_2 are nonempty disjoint subsets of O. Let $x_1 \in O_1$, $x_2 \in O_2$, and join x_1 to x_2 by a broken line. By examining this line, we can find a segment $I = [y_1, y_2]$ with $y_1 \in O_1$, $y_2 \in O_2$. Then $I = (O_1 \cap I) \cup (O_2 \cap I)$, where $O_1 \cap I$ and $O_2 \cap I$ are nonempty disjoint subsets of I. Hence I is not connected. This contradicts theorem 3.5.4, and thus the polygonally connected open set O must also be connected.

We finally proceed to carry out the discussion promised at the beginning of this section. For this we need the following lemma.

Lemma 3.5.7. Let $\{C_\alpha\}$ be a family of connected subsets of M such that $\bigcap_\alpha C_\alpha \neq \varnothing$. Then $\bigcup_\alpha C_\alpha$ is connected.

Proof. Suppose that $\bigcup_\alpha C_\alpha = O_1 \cup O_2$, where O_1 and O_2 are nonempty disjoint open subsets of $\bigcup_{\alpha} C_\alpha$. Let $x_0 \in \bigcap_{\alpha} C_\alpha$ and suppose that $x_0 \in O_1$. Now, for some β, $O_2 \cap C_\beta \neq \emptyset$ and $x_0 \in C_\beta \cap O_1$. Then

$$C_\beta = (O_1 \cap C_\beta) \cup (O_2 \cap C_\beta),$$

where $O_1 \cap C_\beta$ and $O_2 \cap C_\beta$ are nonempty disjoint open subsets of C_β. This contradicts the assumed connectedness of C_β, and the proof is complete.

Now let M be any metric space and define $x \sim y$ to mean that there exists a connected subset of M containing x and y. The relation $\sim$ is obviously reflexive and symmetric, and an application of the lemma 3.5.7 shows that it is also transitive. M is thus partitioned into disjoint subsets A_α which, again by the lemma, are connected. Moreover, by lemma 3.5.2, the A_α are closed. They are called the *components* of M.

EXERCISES

1. Prove the intermediate-value theorem of elementary calculus by the method of this section.

2. Does theorem 3.5.6 generalize to $C^{(n)}$?

3. Show in detail that the components of a metric space are connected and closed.

4. Show that if M has only a finite number of components, then these components are open subsets of M.

5. Show that M may have components which are not open subsets of M.

6. Show that if $C_1, \ldots, C_n$ is a finite sequence of connected subsets of M and if $C_k \cap C_{k+1} \neq \emptyset$, then $\bigcup_{k=1}^{n} C_k$ is connected.

7. Show that any connected subset of M is contained entirely in one of the components of M.

8. Does exercise 6 generalize to an arbitrary infinite collection of connected sets?

9. A metric space M is *locally connected* if, for every $x \in M$ and open

set O containing x, there is an open set $O_1 \subseteq O$ such that O_1 is connected and $x \in O_1$.

a. Show that $R^{(n)}$ and $C^{(n)}$ are locally connected.

b. Give an example to show that M may be connected but not locally connected.

Section 6 PRODUCT SPACES

Let M_1 and M_2 be two metric spaces. We can define a metric on the Cartesian product $M_1 \times M_2$ in various ways, as follows:

$$d[(x_1, x_2), (y_1, y_2)] = \max [d_1(x_1, y_1), d_2(x_2, y_2)];$$
$$d_s[(x_1, x_2), (y_1, y_2)] = d_1(x_1, y_1) + d_2(x_2, y_2);$$
$$d_e[(x_1, x_2), (y_1, y_2)] = [d_1(x_1, y_1)^2 + d_2(x_2, y_2)^2]^{1/2}.$$

We leave to the exercises the proofs that d, d_s, and d_e are legitimate metrics and that they are uniformly equivalent. *Henceforth we use the metric d on $M_1 \times M_2$.*

Lemma 3.6.1.

$$S[(x_1, x_2); \epsilon] = S_1[x_1; \epsilon] \times S_2[x_2; \epsilon]$$

The proof is left as an exercise. A corollary to the lemma is that sets of the form $S_1 \times S_2$, where S_1 is a sphere in M_1 and S_2 is a sphere in M_2 form a basis for the open sets of $M_1 \times M_2$.

Lemma 3.6.2. The functions $P_1: (x_1, x_2) \to x_1$ and $P_2: (x_1, x_2) \to x_2$ which project $M_1 \times M_2$ onto M_1 and M_2, respectively, are both continuous and open. Moreover, they are uniformly continuous.

Proof. It suffices to consider P_1. For an open set O_1 in M_1, we have $P^{-1}(O_1) = O_1 \times M_2$ which is open in $M_1 \times M_2$; hence P_1 is continuous. Now let O be open in $M_1 \times M_2$, and let $x_1 \in P_1(O)$. Then for some $x_2 \in M_2$, and $\epsilon > 0$, we have $(x_1, x_2) \in O$, and some sphere $S_1(x_1; \epsilon) \times S_2(x_2; \epsilon)$ is contained in O. It follows that $S_1(x_1; \epsilon) \subseteq P_1(O)$ and $P_1(O)$ is open. This completes the proof, except for the uniformity, which is left as an exercise.

We come now to the principal result on product spaces.

Theorem 3.6.3

(a) $(x_{1n}, x_{2n}) \longrightarrow (x_1, x_2) \Leftrightarrow x_{1n} \longrightarrow x_1$ and $x_{2n} \longrightarrow x_2$.

(b) $\{(x_{1n}, x_{2n})\}$ is Cauchy in $M_1 \times M_2$ if and only if $\{x_{1n}\}$ is Cauchy in M_1 and $\{x_{2n}\}$ is Cauchy in M_2.

(c) $M_1 \times M_2$ is complete $\Leftrightarrow$ M_1 and M_2 are complete.

(d) $M_1 \times M_2$ is totally bounded $\Leftrightarrow$ M_1 and M_2 are totally bounded.

(e) $M_1 \times M_2$ is compact $\Leftrightarrow$ M_1 and M_2 are compact.

(f) $M_1 \times M_2$ is separable $\Leftrightarrow$ M_1 and M_2 are separable.

(g) $M_1 \times M_2$ is connected $\Leftrightarrow$ M_1 and M_2 are connected.

Proof of (d). If M_1 and M_2 are totally bounded and $\epsilon > 0$, then there are spheres $S_1(x_{11}; \epsilon), \ldots, S_1(x_{1n}; \epsilon)$ covering M_1 and spheres $S_2(x_{21}; \epsilon), \ldots,$ $S(x_{2m}; \epsilon)$ covering M_2. Hence the mn spheres $S_1(x_{1i}; \epsilon) \times S_2(x_{2j}; \epsilon)$ cover $M_1 \times M_2$, and $M_1 \times M_2$ is totally bounded. Conversely, if the spheres $S_1(x_{1i}; \epsilon) \times S_2(x_{2i}; \epsilon), i = 1, \ldots, p$ cover $M_1 \times M_2$, then $M_1 = \bigcup\limits_{i=1}^{p} S_1(x_{1i}; \epsilon)$ and $M_2 = \bigcup\limits_{i=1}^{p} S_2(x_{2i}; \epsilon)$ and M_1 and M_2 are totally bounded.

Proof of (g). If $M_1 \times M_2$ is connected, then M_1, as the continuous image of $M_1 \times M_2$ under P_1, is connected; similarly, M_2 is connected. Now assume that M_1 and M_2 are connected. Let (x_1, x_2) and (y_1, y_2) be any points in $M_1 \times M_2$. The sets $\{x_1\} \times M_2$ and $M_1 \times \{y_2\}$ are connected in $M_1 \times M_2$, for they are continuous images of M_2 and M_1, respectively, and the point (x_1, y_2) is in both of them. Hence their union is connected (see lemma 3.5.7), and it contains both (x_1, x_2) and (y_1, y_2). We have proved that, if M_1 and M_2 are connected, then any two points of $M_1 \times M_2$ are contained in a connected subset of $M_1 \times M_2$. It follows that $M_1 \times M_2$ is connected, for it has only one connected component.

The proofs of the other parts of the theorem are left as exercises.

EXERCISES

1. Prove that the three metrics defined on $M_1 \times M_2$ are legitimate and uniformly equivalent.

2. Prove lemma 3.6.1.

3. Show that for $A_1 \subseteq M_1$, $A_2 \subseteq M_2$ the set $A_1 \times A_2$ in $M_1 \times M_2$ is closed if and only if A_1 is closed in M_1, and A_2 is closed in M_2.

4. Prove parts (a,) (b), (c), (e), and (f) of theorem 3.6.3.

5. Show that $M_1 \times M_2$ is bounded if and only if both M_1 and M_2 are bounded.

6. Let $f: t \rightarrow [f_1(t), f_2(t)]$ be a function from a metric space M into the metric space $M_1 \times M_2$. Show that f is (uniformly) continuous if and only if both f_1 and f_2 are (uniformly) continuous.

7. Extend the definitions and results of this section to a product $M_1 \times M_2 \times \cdots \times M_n$ of n metric spaces.

8. Let $\{M_n\}$ be an infinite sequence of metric spaces, and let M be the set of all sequences $(x_1, \ldots, x_n, \ldots)$ where $x_n \in M_n$. Define the function $d: M \times M \rightarrow R$ by

$$d[\{x_n\}, \{y_n\}] = \sum_1^\infty k_n \frac{d_n(x_n, y_n)}{1 + d_n(x_n, y_n)}, \qquad \text{where } \sum_{n=1}^\infty k_n < \infty,$$
$$k_n > 0.$$

a. Show that M is a metric space under d.
b. Show that convergence in M is equivalent to coordinatewise covergence.
c. Let

$$S(x_1, \ldots, x_n; \epsilon) = S_1(x_1; \epsilon) \times \cdots \times S_n(x_n; \epsilon)$$
$$\times M_{n+1} \times M_{n+2} \times \cdots \times M_{n+k} \times \cdots.$$

Show that the collection of all subsets of M of this type form a basis for the open sets of M.

9. If M_1 and M_2 are (a) locally compact, or (b) locally connected, show that $M_1 \times M_2$ is also. (See exercise 5, section 4, and exercise 9, section 5.)

4 Normed Vector Spaces

In this chapter we consider metric spaces which are equipped with algebraic structure. Some of the basic function spaces of analysis will be introduced here. Section 1 treats that part of the algebraic theory of vector spaces which is relevant to future developments. The reader who is already familiar with this material may omit this section.

Section 1 VECTOR SPACES

We begin with the definition and some examples.

Definition 4.1.1. A vector space V is a set of objects called vectors. Two binary operations, one from $V \times V$ to V and the other from $F \times V$ to V, are given, where F is a field of scalars, either R or C.† These operations are addition of vectors and multiplication of vectors by scalars, and they are assumed to satisfy the following conditions.

1. There is a vector 0 in V such that, for all x in V,

$$x + 0 = x.$$

2. For every x in V there is a vector $-x$ in V such that

$$x + (-x) = 0.$$

---

†The usual definition actually requires only that F be a field. We will always use either R or C.

3. $x + (y + z) = (x + y) + z$.
4. $x + y = y + x$.
5. $\alpha(x + y) = \alpha x + \alpha y$.
6. $(\alpha + \beta)x = \alpha x + \beta x$.
7. $(\alpha\beta)x = \alpha(\beta x)$.
8. $1x = x$.

EXAMPLE 1. $F^{(n)}$, the set of all n-tuples of elements of F, is a vector space under the operations

$$(\alpha_1, \ldots, \alpha_n) + (\beta_1, \ldots, \beta_n) = (\alpha_1 + \beta_1, \ldots, \alpha_n + \beta_n), \quad \text{and}$$

$$\alpha(\alpha_1, \ldots, \alpha_n) = (\alpha\alpha_1, \ldots, \alpha\alpha_n).$$

EXAMPLE 2. Let A be any set, and let $\mathscr{F}(A)$ be set of all functions from A into F. Then $\mathscr{F}(A)$ is a vector space if we define

$$(x + y)(t) = x(t) + y(t) \quad \text{and} \quad (\alpha x)(t) = \alpha[x(t)].$$

EXAMPLE 3. A sequence $(\alpha_1, \alpha_2, \ldots, \alpha_n, \ldots)$ of scalars is *bounded* if $|\alpha_n| \leq M$ for all n. Let (m) designate the set of all bounded sequences, and define

$$\{\alpha_n\} + \{\beta_n\} = \{\alpha_n + \beta_n\} \quad \text{and} \quad \alpha\{\alpha_n\} = \{\alpha\alpha_n\}.$$

In most examples, the verification of conditions 1–8 is trivial and will accordingly be omitted. In many cases it is less obvious that the operations are well-defined. While this is clear in examples 1 and 2, a short argument is needed for example 3, as follows.

Let

$$M_1 = \sup_n |\alpha_n| \quad \text{and} \quad M_2 = \sup_n |\beta_n|.$$

Then

$$|\alpha_n + \beta_n| \leq |\alpha_n| + |\beta_n| \leq M_1 + M_2,$$

so that $\{\alpha_n + \beta_n\}$ is bounded, and

$$\sup_n |\alpha_n + \beta_n| \leq \sup_n |\alpha_n| + \sup_n |\beta_n|.$$

Also,

$$\sup_n |\alpha\alpha_n| = |\alpha| \sup_n |\alpha_n|,$$

so that $\{\alpha\alpha_n\}$ is bounded.

EXAMPLE 4. Let $l^{(p)}$ be the set of all sequences $\{\alpha_n\}$ of scalars such that $\sum_{n=1}^{\infty} |\alpha_n|^p$ converges, where p is a fixed, positive, real number. Clearly,

$$\sum_{n=1}^{\infty} |\alpha\alpha_n|^p = |\alpha|^p \sum_{n=1}^{\infty} |\alpha_n|^p,$$

so that $l^{(p)}$ is closed under the multiplication by scalars given by

$$\alpha\{\alpha_n\} = \{\alpha\alpha_n\}.$$

To see that $l^{(p)}$ is closed under the addition

$$\{\alpha_n\} + \{\beta_n\} = \{\alpha_n + \beta_n\},$$

we must show that $\sum_{n=1}^{\infty} |\alpha_n + \beta_n|^p$ converges wherever $\sum_{n=1}^{\infty} |\alpha_n|^p$ and $\sum_{n=1}^{\infty} |\beta_n|^p$ converge.

We have

$$|\alpha_n + \beta_n| \leq |\alpha_n| + |\beta_n| \leq 2 \max \{|\alpha_n|, |\beta_n|\},$$

and so

$$|\alpha_n + \beta_n|^p \leq 2^p \max \{|\alpha_n|^p, |\beta_n|^p\}$$
$$\leq 2^p[|\alpha_n|^p + |\beta_n|^p].$$

Therefore,

$$\sum_{n=1}^{\infty} |\alpha_n + \beta_n|^p \leq 2^p \left[\sum_{n=1}^{\infty} |\alpha_n|^p + \sum_{n=1}^{\infty} |\beta_n|^p \right]$$

and the sequence $\{\alpha_n + \beta_n\}$ is indeed in $l^{(p)}$.

EXAMPLE 5. Let $C_C(M)$ be the set of all continuous functions from a compact metric space M into the complex numbers C. Since the sum of two continuous functions is continuous and a scalar multiple of a continuous function is continuous, $C_C(M)$ is a vector space. (See exercise 16, section 2, chapter 3.) Similarly, $C_R(M)$, the set of all real-valued continuous functions on M is a vector space.

There are some elementary consequences of definition 4.1.1 which we collect in the following lemma.

Lemma 4.1.2. In any vector space V, we have

(a) There is only one zero element 0 with the property $0 + x = x$ for all x.
(b) $\alpha 0 = 0$ and $0x = 0$ for all α and x.
(c) For any x, $-x = (-1)x$ so that additive inverses are unique.
(d) $\alpha x = 0$ implies either $\alpha = 0$ or $x = 0$.

See the exercises for proof. Henceforth we will write $x - y$ for $x + (-y)$.

Definition 4.1.3. A subset S of a vector space V is called a subspace if it is itself a vector space under the operations of V.

Note that *S is a subspace if it is closed under addition and multiplication by scalars.* In this case, $x \in S \Rightarrow (-1)x = -x \in S$ and $x + (-x) = 0 \in S$, so that S satisfies conditions 1 and 2 of definition 4.1.1, while the other conditions are automatically met.

Lemma 4.1.4. If S_1 and S_2 are subspaces of V, then $S_1 \cap S_2$ and $S_1 + S_2 = \{s_1 + s_2 : s_1 \in S_1, s_2 \in S_2\}$ are subspaces of V.

The proof is left as an exercise. If $S_1 \cap S_2 = \{0\}$, then the *sum* $S_1 + S_2$ is called *direct* and is written $S_1 \oplus S_2$.

Lemma 4.1.5. The sum $S_1 + S_2$ is direct if and only if every vector x in $S_1 + S_2$ has a unique representation $x = t_1 + t_2$ with $t_1 \in S_1$, $t_2 \in S_2$.

Proof. Assume $S_1 \cap S_2 = \{0\}$ and that $x \in (S_1 + S_2)$ can be written $x = s_1 + s_2 = t_1 + t_2$. Then $s_1 - t_1 = t_2 - s_2$ is a vector in both S_1 and S_2 and is therefore the zero vector. Hence $s_1 = t_1$, $s_2 = t_2$, and the representation of x is unique. Conversely, if some nonzero vector x is in $S_1 \cap S_2$, then we have $t_1 + t_2 = (t_1 + x) + (t_2 - x)$, and the representation of $y = t_1 + t_2$ is not unique. This completes the proof.

EXAMPLE 6. Let V be the vector space of all complex-valued functions on a real interval from $-a$ to a. Let

$$E = \{x \in V : x(t) = x(-t) \quad \text{for all } t\}$$

$$D = \{x \in V : x(t) = -x(-t) \quad \text{for all } t\}.$$

Clearly, E and D are subspaces of V and $E \cap D = \{0\}$. Moreover, $V = E \oplus D$ since, for any x in V, we can write

$$x(t) = \tfrac{1}{2}[x(t) + x(-t)] + \tfrac{1}{2}[x(t) - x(-t)].$$

Definition 4.1.6

(a) A *finite* sum $\alpha_1 x_1 + \cdots + \alpha_n x_n$ is called a *linear combination* of the vectors $x_1, \ldots, x_n$.

(b) Vectors $x_1, \ldots, x_n$ are called *linearly dependent* if, for some $\alpha_1, \ldots, \alpha_n$ *not all zero*, $\alpha_1 x_1 + \ldots + \alpha_n x_n = 0$.

(c) Vectors $x_1, \ldots, x_n$ are called *linearly independent* if they are not linearly dependent.

The notion of linear independence is sufficiently important to warrant explicit formulation. Vectors $x_1, \ldots, x_n$ are linearly independent if and only if

$$\alpha_1 x_1 + \cdots + \alpha_n x_n = 0 \quad \text{implies} \quad \alpha_1 = \alpha_2 = \cdots = \alpha_n = 0.$$

An *infinite set*, A, of vectors is said to be linearly independent if every finite subset is independent, and similarly, A is linearly dependent if some finite subset is linearly dependent. Clearly, any subset of an independent set is independent, and any set which contains a dependent subset is itself dependent.

Lemma 4.1.7. A set of vectors is dependent if and only if one of them can be written as a linear combination of other members of the set.

The proof is left as an exercise.

Let A be any subset of a vector space V, and let $S(A)$ be the set of all linear combinations of vectors of A. Clearly, $S(A)$ is a subspace of V; it is called the subspace *spanned* by A.

Definition 4.1.8. A subset B of a vector space V is a *basis* for V if $V = S(B)$ and if B is independent.

Lemma 4.1.9. If $\{x_\alpha\}$ is a basis for V, then every vector in V has a unique representation

$$x = \beta_1 x_{\alpha(1)} + \cdots + \beta_n x_{\alpha(n)} \qquad (n \text{ is not fixed}).$$

Proof. Suppose that a vector x has two such representations. By using some zero coefficients, we then have

$$x = \sum_{k=1}^{n} \beta_k x_{\alpha(k)} = \sum_{k=1}^{n} \beta_k' x_{\alpha(k)}$$

or

$$\sum_{k=1}^{n} (\beta_k - \beta_k') x_{\alpha(k)} = 0.$$

The independence of $\{x_\alpha\}$ then implies that $\beta_k = \beta'_k$ for $k = 1, \ldots, n$, and the two representations are identical. This proves the uniqueness part of the lemma; the existence of at least one such representation is guaranteed by the definition of a basis.

We now wish to define the notion of *dimension* in a vector space. To do this we need the following lemma.

Lemma 4.1.10. If $\{x_1, \ldots, x_n\}$ is a basis in V, then any $n + 1$ vectors in V are linearly dependent.

Proof. We use induction; if $n = 1$, then any two vectors in V are of the form αx_1 and βx_1 and are obviously dependent. Now assume the theorem is true for vector spaces having bases consisting of $n - 1$ or fewer vectors, and let V have basis $\{x_1, \ldots, x_n\}$. We must show that, if $y_1, \ldots, y_n, y_{n+1}$ are any $n + 1$ vectors in V, then they are dependent.

Clearly we can write

$$y_1 = \alpha_{11}x_1 + \alpha_{12}x_2 + \cdots + \alpha_{1n}x_n$$
$$y_2 = \alpha_{21}x_1 + \alpha_{22}x_2 + \cdots + \alpha_{2n}x_n$$
$$\vdots \qquad\qquad \vdots$$
$$y_n = \alpha_{n1}x_1 + \alpha_{n2}x_2 + \cdots + \alpha_{nn}x_n$$
$$y_{n+1} = \alpha_{n+1,1}x_1 + \alpha_{n+1,2}x_2 + \cdots + \alpha_{n+1,n}x_n.$$

We distinguish two cases. If

$$\alpha_{11} = \alpha_{21} = \cdots = \alpha_{n1} = \alpha_{n+1,1} = 0,$$

then the vectors $y_1, y_2, \ldots, y_n, y_{n+1}$ lie in the vector space W which has for a basis the set $\{x_2, \ldots, x_n\}$. They are then dependent by the induction hypothesis.

If the complementary case holds, we can assume $\alpha_{11} \neq 0$ (renumber the y's if necessary). Consider the n vectors

$$\alpha_{11}y_2 - \alpha_{21}y_1 = (\alpha_{11}\alpha_{22} - \alpha_{21}\alpha_{12})x_2 + \cdots + (\alpha_{11}\alpha_{2n} - \alpha_{21}\alpha_{1n})x_n$$
$$\vdots \qquad\qquad\qquad \vdots$$
$$\alpha_{11}y_{n+1} - \alpha_{n+1,1}y_1 = (\alpha_{11}\alpha_{n+1,2} - \alpha_{n+1,1}\alpha_{12})x_2 + \cdots$$
$$+ (\alpha_{11}\alpha_{n+1,n} - \alpha_{n+1,1}\alpha_{1n})x_n.$$

By the induction hypothesis, there exist scalars $\beta_2, \ldots, \beta_{n+1}$, not all zero, such that

$$\beta_2(\alpha_{11} y_2 - \alpha_{21} y_1) + \cdots + \beta_{n+1}(\alpha_{11} y_{n+1} - \alpha_{n+1,1} y_1) = 0$$

or

$$(\beta_2 \alpha_{11}) y_2 + \cdots + (\beta_{n+1} \alpha_{11}) y_{n+1} - (\beta_2 \alpha_{21} + \cdots + \beta_{n+1} \alpha_{n+1,1}) y_1 = 0.$$

Since the coefficients are not all zero, the set $\{y_1, \ldots, y_n, y_{n+1}\}$ is dependent, and the proof is complete.

Corollary. If $\{x_1, \ldots, x_n\}$ and $\{x_1', \ldots, x_m'\}$ are bases for V, then $m = n$.

To prove the corollary, we note that if $m > n$ then, by the lemma, the set $\{x_1', \ldots, x_m'\}$ is dependent, in contradiction to the fact that it is a basis. Similarly $n > m$ is impossible, and we conclude that $m = n$.

Definition 4.1.11. Let V be any vector space.

dim $V = \infty$ if there is no finite set A which is a basis of V.
dim $\{0\} = 0$.
dim $V = n$, a positive integer, if V has a basis of n vectors.

The definition is justified by the preceding corollary, which ensures that dimension is unambiguously determined.

Lemma 4.1.12. If vectors $x_1, \ldots, x_k, k < n$, are independent in a vector space V of dimension n, then there exist vectors $x_{k+1}, \ldots, x_n$ such that the set $\{x_1, \ldots, x_k; x_{k+1}, \ldots, x_n\}$ is a basis of V.

Proof. Since V has dimension n, we can find a basis $\{y_1, \ldots, y_n\}$ for V. Consider now the list of vectors $x_1, \ldots, x_k; y_1, \ldots, y_n$. If y_1 can be expressed as a linear combination of $x_1, \ldots, x_k$, then delete it from the list; if not, then retain it. Now consider y_2 and delete or retain it, depending on whether it is or is not a linear combination of the vectors which precede it in the revised list. Continuing in this way to the end of the list, we obtain a new set $\{x_1, \ldots, x_k; x_{k+1}, \ldots, x_m\}$. Since the original set $\{x_1, \ldots, x_k; y_1, \ldots, y_n\}$ spanned V, we know that the new set also spans V, for the only vectors which we deleted are obtainable as linear combinations of those remaining. The new set is also independent, for if

$$\alpha_1 x_1 + \cdots + \alpha_k x_k + \alpha_{k+1} x_{k+1} + \cdots + \alpha_m x_m = 0$$

then, by the independence of $x_1, \ldots, x_k$, not all of the scalars $\alpha_{k+1}, \ldots, \alpha_m$ are zero. Let α_j be the last nonzero scalar; then $x_j (k < j \le m)$ can be written as a linear combination of the preceding vectors, in contradiction to our construction. Finally, from the fact that $\{x_1, \ldots, x_k; x_{k+1}, \ldots, x_m\}$ spans

V and is independent, we conclude that it is a basis for V and that $m = n$. The proof is complete.

Every n-dimensional vector space is essentially the same as $F^{(n)}$ (see example 1). To make this precise, we need the following definition and subsequent lemma.

Definition 4.1.13. A linear transformation T is a function from a vector space V into a vector space W such that

1. $T(\alpha x + \beta y) = \alpha T(x) + \beta T(y)$ for all α, β, x, and y.

Condition 1 is clearly equivalent to the pair of conditions:

1(a). $T(x + y) = T(x) + T(y)$, and

1(b). $T(\alpha x) = \alpha T(x)$.

We note from 1(a) that $T(0) = T(0 + 0) = T(0) + T(0)$; hence $T(0) = 0$ for any linear transformation T.

Lemma 4.1.14. If T is a one-to-one linear transformation of vector space V onto vector space W, then T^{-1} is also linear.

Proof. Since T is one-to-one, the transformation T^{-1} is defined. We must show that $T^{-1}(\alpha_1 y_1 + \alpha_2 y_2) = \alpha_1 T^{-1}(y_1) + \alpha_2 T^{-1}(y_2)$. Now, there are unique vectors x_1 and x_2 such that $T(x_1) = y_1$ and $T(x_2) = y_2$. Then

$$T^{-1}(\alpha_1 y_1 + \alpha_2 y_2) = T^{-1}[\alpha_1 T(x_1) + \alpha_2 T(x_2)] = T^{-1}[T(\alpha_1 x_1 + \alpha_2 x_2)]$$
$$= \alpha_1 x_1 + \alpha_2 x_2 = \alpha_1 T^{-1}(y_1) + \alpha_2 T^{-1}(y_2),$$

and T^{-1} is linear.

We now define V to be isomorphic to W if there is a one-to-one linear transformation of V onto W. It is left as an exercise to show that the relation $V * W \Leftrightarrow (V$ is isomorphic to $W)$ is an equivalence relation (see exercise 11). We can now state the principal result about finite dimensional vector spaces.

Theorem 4.1.15. If V has dimension n, then V is isomorphic to $F^{(n)}$.

Proof. Let $x_1, \ldots, x_n$ be a basis for V, and define

$$T: (\alpha_1, \ldots, \alpha_n) \longrightarrow \sum_{k=1}^{n} \alpha_k x_k.$$

It is easy to show that T is linear and onto. Further, if $T[(\alpha_1, \ldots, \alpha_n)] = T[(\beta_1, \ldots, \beta_n)]$ then

$$\sum_{k=1}^{n} \alpha_k x_k = \sum_{k=1}^{n} \beta_k x_k \quad \text{and} \quad \sum_{k=1}^{n} (\alpha_k - \beta_k) x_k = 0.$$

The independence of $x_1, \ldots, x_n$ now yields $\alpha_1 = \beta_1, \ldots, \alpha_n = \beta_n$, and T is one-to-one. This completes the proof.

EXERCISES

1. By expanding $(1 + 1)(x + y)$ in two ways, show that the commutativity of addition is a consequence of the other vector space axioms.

2. Let P be the set of all positive real numbers, and define

$$x \circ y = xy \qquad x, y \in P$$
$$\alpha \cdot x = x^{\alpha} \qquad x \in P, \quad \alpha \in R.$$

Show that P is a vector space and is isomorphic to $R^{(1)}$. Is P a subspace of $R^{(1)}$?

3. a. Let V and W be vector spaces over the field R, and let T be a function from V into W satisfying $T(x + y) = T(x) + T(y)$. Show that
   (i) $T(mx) = mT(x)$ for any positive integer m.
   (ii) $T\left(\dfrac{1}{n}x\right) = \dfrac{1}{n}T(x)$ for any positive integer n.
   (iii) $T(-x) = -T(x)$.
   (iv) $T\left(\dfrac{p}{q}x\right) = \dfrac{p}{q}T(x)$ for any rational number $\dfrac{p}{q}$.

   b. Let V and W be vector spaces over C, and let T be a function from V into W satisfying $T(x + y) = T(x) + T(y)$ and $T(ix) = iT(x)$. Show that

$$T\left(\left[\frac{p}{q} + \frac{r}{s}i\right]x\right) = \left[\frac{p}{q} + \frac{r}{s}i\right]T(x)$$

   for all rational numbers p/q and r/s.

4. Prove lemma 4.1.2.

5. Prove lemma 4.1.4.

6. Prove lemma 4.1.7.

7. Is the set $\{0\}$ dependent or independent?

8. Let V be the set of all polynomial functions from C to C.
a. Show that V is an infinite dimensional vector space.
b. Show that a polynomial is an even (odd) function if and only if it contains only even (odd) powers of the variable.

9. Let V be the set of all differentiable real-valued functions on $[-a, a]$.
a. Show that the derivative of an even (odd) function is odd (even).
b. Let T be the mapping of V into V which sends $x(t)$ into $x'(t)$. Show that T is a linear transformation which maps $E \cap V$ into D, and $D \cap V$ into E (see example 6).

10. Let V be the set of all real-valued functions on $[a, b]$ whose graphs consist of a finite number of (nonvertical) straight-line segments connected end to end, i.e., V is the set of all "broken-line" functions.
a. Given a broken-line function, find an explicit formula for it (i.e., show what form such a formula must have, and give directions for evaluating any constants therein).
b. Show that V is an infinite dimensional vector space.
c. If the functions in V are multiplied in the usual way, is the product in V?

11. a. If T_1 is a linear transformation from V_1 into V_2, and if T_2 is a linear transformation from V_2 into V_3, show that the function T_2T_1, defined by
$$(T_2T_1)(x) = T_2[T_1(x)],$$
is a linear transformation from V_1 into V_3.
b. Show that the relation $V * W$, defined in the text, is an equivalence relation.

12. Show that $0 < \text{dimension } V < \infty$ if and only if V is spanned by a finite subset of V.

13. Let W be a subspace of V. Define $x * y \Leftrightarrow x - y \in W$.
a. Show that $*$ is an equivalence relation and that
$$(x' * x, y' * y) \Rightarrow [(x' + y') * (x + y), \alpha x' * \alpha x].$$
b. Show that the equivalence classes form a vector space under
$$\lfloor x \rfloor + \lfloor y \rfloor = \lfloor x + y \rfloor \quad \text{and} \quad \alpha \lfloor x \rfloor = \lfloor \alpha x \rfloor.$$
This space is called the quotient space of V by W.

14. Show that, if T is a linear transformation from V into W, then the range of T is a subspace of W.

15. a. Show that a linear transformation T is one-to-one if and only if
$$\{x: Tx = 0\} = \{0\}.$$

   b. Show that, for any linear transformation T, the null space of T, $\{x: Tx = 0\}$, is a subspace of the domain of T.

Section 2 NORMED VECTOR SPACES

We now consider sets which are both metric spaces and vector spaces. Without requiring some cooperation between the metric and vector space structures, there would be nothing new to say about such an object. Hence we introduce the following definition.

Definition 4.2.1. A metric vector space is a set V which is a vector space with a metric $d(x, y)$ such that the algebraic operations

$$(\alpha, x) \longrightarrow \alpha x$$
$$(x, y) \longrightarrow x + y$$

are continuous functions from $F \times V \longrightarrow V$ and $V \times V \longrightarrow V$, respectively.

EXAMPLE 1. The complex number system C under

$$d(z_1, z_2) = \frac{|z_1 - z_2|}{(1 + |z_1|^2)^{1/2}(1 + |z_2|^2)^{1/2}}$$

is a metric vector space. Verification of the continuity of addition and multiplication by scalars is left as an exercise.

EXAMPLE 2. The space R^∞ of all sequences of real numbers under the metric

$$d[\{\alpha_n\}, \{\beta_n\}] = \sum_{n=1}^{\infty} k_n \frac{|\alpha_n - \beta_n|}{1 + |\alpha_n - \beta_n|}, \qquad k_n > 0, \qquad \sum_{n=1}^{\infty} k_n < \infty$$

and operations

$$\{\alpha_n\} + \{\beta_n\} = \{\alpha_n + \beta_n\}$$
$$\alpha\{\alpha_n\} = \{\alpha\alpha_n\}$$

is a metric vector space. It is left to the reader to verify the continuity of the operations.

In many metric vector spaces, the metric and the algebraic operations are connected by stronger ties than mere continuity. We say that a metric on a vector space is *translation invariant* if

$$d(x + z, y + z) = d(x, y)$$

for all x, y, and z. We say that it is *homogeneous* if

$$d(\alpha x, \alpha y) = |\alpha| d(x, y)$$

for all α, x, and y. The metric in example 1 has neither of these properties, while that in example 2 is translation invariant without being homogeneous.

Note that if d is translation invariant, then

$$d(x, y) = d(x - y, 0),$$

so that distances from 0 determine all other distances.

Lemma 4.2.2. Let V be a vector space with a homogeneous, translation-invariant metric, and define $\|x\| = d(x, 0)$. Then the function $x \longrightarrow \|x\|$ from V onto R satisfies

(a) $\|x\| \geq 0$; $\|x\| = 0 \Leftrightarrow x = 0$;
(b) $\|\alpha x\| = |\alpha| \|x\|$;
(c) $\|x + y\| \leq \|x\| + \|y\|$.

Moreover, V is a metric vector space.

Proof. (a) and (b) are obvious, while (c) is proved as follows:

$$\|x + y\| = d(x + y, 0) = d(x, -y) \leq d(x, 0) + d(0, -y)$$
$$= d(x, 0) + d(y, 0) = \|x\| + \|y\|.$$

Finally, let $d(x_n, x_0) \longrightarrow 0$ and $d(y_n, y_0) \longrightarrow 0$. Then

$$d(x_n + y_n, x_0 + y_0) = d([x_n - x_0] + [y_n - y_0], 0)$$
$$= \|(x_n - x_0) + (y_n - y_0)\|$$
$$\leq \|x_n - x_0\| + \|y_n - y_0\|$$
$$= d(x_n - x_0, 0) + d(y_n - y_0, 0)$$
$$= d(x_n, x_0) + d(y_n, y_0),$$

so that $d(x_n + y_n, x_0 + y_0) \longrightarrow 0$, and addition is (uniformly) continuous. Also, if $|\alpha_n - \alpha_0| \longrightarrow 0$, we have

$$d(\alpha_n x_n, \alpha_0 x_0) = d(\alpha_n x_n - \alpha_0 x_0, 0) = \|\alpha_n x_n - \alpha_0 x_0\|$$
$$= \|\alpha_n x_n - \alpha_n x_0 + \alpha_n x_0 - \alpha_0 x_0\|$$
$$\leq |\alpha_n| \|x_n - x_0\| + |\alpha_n - \alpha_0| \|x_0\|.$$

But $|\alpha_n| \leq M$ for all n, so that

$$d(\alpha_n x_n, \alpha_0 x_0) \leq M\, d(x_n, x_0) + \|x_0\| |\alpha_n - \alpha_0|,$$

and multiplication by scalars is a continuous operation. Thus V is a metric vector space.

Corollary. For any x and y in V, $\big|\|x\| - \|y\|\big| \leq \|x \pm y\|$, and hence $d(x_n, x_0) \to 0$ implies $\|x_n\| \to \|x_0\|$.

The proof of the corollary is left for the exercises.

Definition 4.2.3. A *normed vector space* is a vector space with a real-valued function $x \to \|x\|$ satisfying the conditions

(a) $\|x\| \geq 0$; $\|x\| = 0 \Leftrightarrow x = 0$;
(b) $\|\alpha x\| = |\alpha| \|x\|$;
(c) $\|x + y\| \leq \|x\| + \|y\|$.

It should be clear that, if in a normed vector space V we define

$$d(x, y) = \|x - y\|,$$

then V becomes a metric vector space in which the metric is both translation invariant and homogeneous.

Most of the metric vector spaces we will consider will be normed. Here are some examples.

EXAMPLE 3. The vector space $C^{(n)}$ can be normed in various ways:

(a) $\|(\alpha_1, \ldots, \alpha_n)\|_1 = \max [|\alpha_1|, \ldots, |\alpha_n|]$;
(b) $\|(\alpha_1, \ldots, \alpha_n)\|_2 = |\alpha_1| + \cdots + |\alpha_n|$;
(c) $\|(\alpha_1, \ldots, \alpha_n)\| = [|\alpha_1|^2 + \cdots + |\alpha_n|^2]^{1/2}$.

The legitimacy of these norms and their equivalence (i.e., the equivalence of the induced metrics) was essentially established in section 1 of chapter 3.

EXAMPLE 4. In the preceding section (see example 3 there) we considered the vector space (m) of all bounded sequences of scalars. If we define $\|\{\alpha_n\}\| = \sup_n |\alpha_n|$, then we see that (m) is a normed vector space.

We now prove that (m) is *complete*. Let $\{x_n\}$ be Cauchy sequence in (m), where

$$x_n = (\alpha_{n1}, \alpha_{n2}, \ldots, \alpha_{nk}, \ldots).$$

Then

$$|\alpha_{mk} - \alpha_{nk}| \le \sup_k |\alpha_{mk} - \alpha_{nk}| = \|x_m - x_n\|$$

so that for each k, the sequence $\{\alpha_{nk}\}$ is a Cauchy sequence in R or C, whichever we are using. Hence there exists a sequence $\{\alpha_k\}$ such that $\alpha_{nk} \to \alpha_k$ for every k, and the sequence $\{x_n\}$ converges coordinatewise to $x_0 = (\alpha_1, \alpha_2, \ldots, \alpha_k, \ldots)$. We must now show that x_0 is in (m) and that $\|x_n - x_0\| \to 0$.

Now, since Cauchy sequences are bounded we have $\|x_n\| = \sup_k |\alpha_{nk}| \le M$ for all n, and so $|\alpha_{nk}| \le M$ for all n and k. Hence, $\lim\limits_{n \to \infty} |\alpha_{nk}| = |\alpha_k| \le M$ for all k, $\{\alpha_k\} = x_0$ satisfies $\|x_0\| \le M$, and $x_0 \in (m)$.

Finally, since $\{x_n\}$ is Cauchy, we have, for any $\epsilon > 0$, an $N(\epsilon)$ such that $\|x_n - x_m\| \le \epsilon$ for $m, n \ge N(\epsilon)$. That is,

$$\sup_k |\alpha_{nk} - \alpha_{mk}| \le \epsilon \quad \text{or} \quad |\alpha_{nk} - \alpha_{mk}| \le \epsilon \qquad \text{for all } k.$$

Letting $m \to \infty$, we have $|\alpha_{nk} - \alpha_k| \le \epsilon$ for $n \ge N(\epsilon)$, or $\|x_n - x_0\| \le \epsilon$ for $n \ge N(\epsilon)$. This finishes the proof of the completeness of (m); a proof that (m) is not separable will be sketched in the exercises.

EXAMPLE 5. Let M be any compact metric space, and let $C_C(M)$ be the set of all continuous complex-valued functions on M, with $\|x\| = \sup\limits_{t \in M} |x(t)|$. As indicated earlier, $C_C(M)$ is a vector space; moreover, $\|x\|$ is finite by the compactness of M. The norm obviously satisfies (a) $\|x\| \ge 0$, $\|x\| = 0 \Leftrightarrow x = 0$, and (b) $\|\alpha x\| = |\alpha| \|x\|$. To prove the triangle inequality (c), we write

$$|x(t) + y(t)| \le |x(t)| + |y(t)| \le \|x\| + \|y\|,$$

and hence $\|x + y\| \le \|x\| + \|y\|$.

In the next section the properties of $C_C(M)$ and $C_R(M)$ will be investigated in detail. Convergence in $C_C(M)$ and $C_R(M)$ will be called *uniform* convergence, for $\|x_n - x_0\| \to 0$ is equivalent to the assertion that, for any $\epsilon > 0$, there is an $N(\epsilon)$ such that $n \ge N(\epsilon)$ implies $|x_n(t) - x_0(t)| \le \epsilon$ for *all t* in M.

Definition 4.2.4. A normed vector space which is complete as a metric space is called a *Banach space*.

We know that a normed vector space V which is not complete can be imbedded in a complete metric space $\hat{V}$, but operations of addition and multiplication by scalars are not defined on members of $\hat{V}$ which are not in V. Once the problem is recognized, it is clear that we must define

$$\alpha \hat{x} = \operatorname*{limit}_{n \to \infty} \alpha x_n$$

$$\hat{x} + \hat{y} = \operatorname*{limit}_{n \to \infty} (x_n + y_n),$$

where $\{x_n\}$ and $\{y_n\}$ are Cauchy sequences in V which converge to $\hat{x}$ and $\hat{y}$, respectively. After checking that these definitions are unambiguous, we can verify that the metric $\hat{d}$ on $\hat{V}$ is homogeneous and translation invariant, so that $\hat{V}$ is a Banach space whose norm is an extension of the norm on V. Details are left to the exercises.

A criterion for the completeness of a normed vector space is given in the following theorem. In preparation, we consider series $\sum\limits_{n=1}^{\infty} x_n$ in a normed vector space V, where *convergence is equated to norm convergence of the sequence of partial sums.* If $\sum\limits_{n=1}^{\infty} x_n$ does converge, then its partial sums form a Cauchy sequence, and in particular,

$$\left\| \sum_{k=1}^{n} x_k - \sum_{m=1}^{n-1} x_k \right\| = \| x_n \|$$

approaches zero as $n \to \infty$. A series $\sum\limits_{n=1}^{\infty} x_n$ is called *absolutely convergent* if $\sum\limits_{n=1}^{\infty} \| x_n \|$ converges. We can now ask if, as in the case of R or C, absolute convergence implies convergence. We have

$$\left\| \sum_{k=1}^{n} x_k - \sum_{k=1}^{n+p} x_k \right\| = \left\| \sum_{k=n+1}^{n+p} x_k \right\| \leq \sum_{k=n+1}^{n+p} \| x_k \|,$$

so that if $\sum\limits_{n=1}^{\infty} x_n$ is absolutely convergent, then the partial sums of $\sum\limits_{n=1}^{\infty} x_n$ form a Cauchy sequence. Thus, if V is complete, we see that absolutely convergent series must converge. We have proved half of the following theorem.

Theorem 4.2.5. A normed vector space V is complete if and only if every absolutely convergent series in V is convergent.

The proof will be finished if we can show that, in an incomplete normed vector space V, there is an absolutely convergent series which does not converge. Now, by assumption, there is a Cauchy sequence $\{x_n\}$ in V which has no limit. For every k there is an $N(k)$ such that $n \geq N(k)$ implies $\|x_n - x_{n+p}\| \leq 2^{-k}$ for all positive integers p. Letting $y_k = x_{N(k)}$, we have $\|y_k - y_{k+1}\| \leq 2^{-k}$ for all k so that the infinite series

$$y_1 + (y_2 - y_1) + (y_3 - y_2) + \cdots + (y_k - y_{k-1}) + \cdots$$

converges absolutely, for

$$\|y_1\| + \sum_{k-1}^{\infty} \|y_{k+1} - y_k\| \leq \|y_1\| + \sum_{k=1}^{\infty} 2^{-k}.$$

However, this series cannot converge because its sequence of partial sums is $\{y_k\}$ which is a subsequence of $\{x_n\}$, and convergence of $\{y_k\}$ would imply convergence of $\{x_n\}$. (If a subsequence of a Cauchy sequence converges, then so does the Cauchy sequence.) This finishes the proof of the theorem.

EXAMPLE 6. We now give a simple proof of the M-test of Weierstrass for uniform convergence of a series of functions in $C_c(M)$. This criterion states that, if $\sum_{n=1}^{\infty} x_n(t)$ is such a series and if $|x_n(t)| \leq M_n$ where $\sum_{n=1}^{\infty} M_n < \infty$, then the series converges uniformly. Now, the assumption $\sum_{n=1}^{\infty} M_n < \infty$ implies that $\sum_{n=1}^{\infty} \|x_n(t)\|$ converges. In section 4 we will prove that $C_c(M)$ is complete; this fact, together with theorem 4.2.5 now imply the uniform convergence of $\sum_{n=1}^{\infty} x_n(t)$ [i.e., convergence in the norm of $C_c(M)$].

We close this section with a general theorem on rearrangements of series, which again shows the importance of the notion of absolute convergence.

If f is a one-to-one mapping of the positive integers I^+ onto I^+, then the sequence

$$y_1 = x_{f(1)}, y = x_{f(2)}, \ldots, y_n = x_{f(n)}, \ldots$$

is said to be a *rearrangement* of the sequence

$$x_1, x_2, \ldots, x_n, \ldots,$$

and the series $\sum_{n=1}^{\infty} y_n$ is a rearrangement of the series $\sum_{n=1}^{\infty} x_n$.

Theorem 4.2.6. Let V be a Banach space and let $\sum_{n=1}^{\infty} x_n$ be an absolutely convergent series in V. If $\sum_{n=1}^{\infty} y_n$ is any rearrangement of $\sum_{n=1}^{\infty} x_n$, then $\sum_{n=1}^{\infty} y_n$ is absolutely convergent, and $\sum_{n=1}^{\infty} x_n = \sum_{n=1}^{\infty} y_n$.

Proof. To show that $\sum_{n=1}^{\infty} ||y_n||$ converges, it is only necessary to show that its partial sums are bounded. Now, for any positive integer n, we let $M(n)$ be the largest of the integers $f(1), f(2), \ldots, f(n)$. Then

$$\sum_{k=1}^{n} ||y_k|| \leq \sum_{k=1}^{M(n)} ||x_k||,$$

because every term of the first sum occurs in the second. Since the partial sums of $\sum_{k=1}^{\infty} ||x_k||$ are bounded (it converges), so are those of $\sum_{k=1}^{\infty} ||y_k||$, and our first assertion is proved.

To show that $\sum_{n=1}^{\infty} x_n = \sum_{n=1}^{\infty} y_n$, we show that the Cauchy sequences of partial sums are equivalent. (The completeness of V, together with the convergence of $\sum_{n=1}^{\infty} ||y_n||$, ensures that $\sum_{n=1}^{\infty} y_n$ exists.) By the absolute convergence of $\sum_{n=1}^{\infty} x_n$ we can, for any $\epsilon > 0$, find an $N(\epsilon)$ such that $\sum_{k=n+1}^{n+p} ||x_k|| \leq \epsilon$ for $n \geq N(\epsilon)$ and any p. Let $M(\epsilon)$ be the largest of the integers $f^{-1}(1)$, $f^{-1}(2), \ldots, f^{-1}[N(\epsilon)]$. Then $\sum_{k=n}^{n+p} ||y_k|| \leq \epsilon$ for $n \geq M(\epsilon)$ and any p. Finally, for $n \geq \max[N(\epsilon), M(\epsilon)]$ we have

$$\left\| \sum_{k=1}^{n} y_k - \sum_{k=1}^{n} x_k \right\| \leq \left\| \sum_{k=1}^{n} y_k - \sum_{k=1}^{M(\epsilon)} y_k \right\| + \left\| \sum_{k=1}^{M(\epsilon)} y_k - \sum_{k=1}^{N(\epsilon)} x_k \right\|$$
$$+ \left\| \sum_{k=1}^{N(\epsilon)} x_k - \sum_{k=1}^{n} x_k \right\| \leq 3\epsilon,$$

and the proof is finished.

The theorem applies, of course, to series of real or complex numbers. In the exercises, the reader will meet the problem of rearranging nonabsolutely convergent series.

EXERCISES

1. Prove that addition and multiplication by scalars are continuous operations in examples 1 and 2.

2. Is multiplication by scalars in a normed vector space a *uniformly* continuous operation?

3. Let V be the set of all bounded sequences of vectors in a normed vector space W. That is, for any $(x_1, \ldots, x_n, \ldots) \in V$, there is

an M such that $||x_n|| \leq M$ for all n (the value of M may be different for different sequences). Show that, if operations are defined coordinatewise and if $||(x_1, \ldots, x_n, \ldots)||$ is taken as $\sup\limits_n ||x_n||$, then V is a normed vector space. Show that V is complete if W is complete.

4. Let V be the set of all continuous functions from a compact metric space M into a normed vector space W. Show that V is a normed vector space if the operations and norm are defined in the obvious way.

5. Write out the full proof that a normed vector space V can be imbedded in a Banach space $\hat{V}$ in a natural way.

6. Let $A = \{x_\alpha\}$ be any *countable* set of vectors in a Banach space V. Then A is said to be *absolutely summable* if, for any one-to-one mapping f of I^+ onto A, the series $\sum\limits_{n=1}^{\infty} f(n)$ is absolutely convergent. Show:

a. If A is absolutely summable, then a sum $\sum\limits_A x_\alpha$ can be unambiguously ascribed to A.

b. A is absolutely summable if and only if the set of all *finite* sums of norms of members of A is bounded.

c. Any subset of an absolutely summable set is absolutely summable.

d. If A is absolutely summable, then for any $\epsilon > 0$, there is a finite subset $F(\epsilon)$ of A such that, whenever F_1 and F_2 are finite subsets of A satisfying $F_1 \cap F(\epsilon) = \phi$, $F_2 \supseteq F(\epsilon)$, it follows that

$$\sum_{F_1} ||x_\alpha|| \leq \epsilon \quad \text{and} \quad ||\sum_A x_\alpha - \sum_{F_2} x_\alpha|| \leq 2\epsilon.$$

e. Let $A = \{x_\alpha\}$ be a countable subset of a Banach space V, and suppose A is partitioned by a *finite* number of subsets $A_1, \ldots, A_n$. Show that A is absolutely summable if and only if all of the sets $A_1, \ldots, A_n$ are absolutely summable. Show that, in this case,

$$\sum_A x_\alpha = \sum_{A_1} x_\alpha + \cdots + \sum_{A_n} x_\alpha.$$

f. Let A be absolutely summable and let $\{B_n\}$ be a partition of A into a denumerable number of denumerable sets.

(i) Show that B_n is absolutely summable and let $Z_n = \sum\limits_{B_n} x_\alpha$.

(ii) Show that $\{Z_n\}$ is absolutely summable and $\sum\limits_n Z_n = \sum\limits_A x_\alpha$.

7. Let A be any countable set of real numbers, and let P and N be the subsets of A consisting of the positive and negative members of A, respectively (we can assume $0 \notin A$). Show that A is absolutely summable if and only if both P and N are absolutely summable (see exercise 6). Show that, in this case, $\sum_A x_\alpha = \sum_P x_p + \sum_N x_n$.

8. Are the following series absolutely convergent?

a. $1 - \dfrac{1}{2^2} + \dfrac{1}{3} - \dfrac{1}{4^2} + \dfrac{1}{5} - \dfrac{1}{6^2} + \dfrac{1}{7} - \cdots$

b. $1 - \dfrac{1}{2^2} + \dfrac{1}{3^3} - \dfrac{1}{4^2} - \dfrac{1}{5^3} + \dfrac{1}{6^2} - \dfrac{1}{7^3} + \cdots$

(See exercise 7.)

9. a. Let (c) be the set of all *convergent* sequences of complex numbers. Show that (c) is a closed subspace of (m).

b. Let (c_0) be the set of all sequences of complex numbers which converge to zero. Show that (c_0) is a closed subspace of (c).

10. Let (cs) be the set of all sequences of complex numbers for which $\sum\limits_{n=1}^{\infty} \alpha_n$ converges.

a. Show that (cs), under coordinatewise operations and

$$\|\{\alpha_n\}\| = \sup_n \left| \sum_{k=1}^{n} \alpha_k \right|$$

is a Banach space.

b. Show that (c) and (cs) are connected by a norm-preserving isomorphism.

11. Let (bs) be the set of all sequences of complex numbers for which $\sup\limits_{n} \left| \sum\limits_{k=1}^{n} \alpha_k \right|$ is finite.

a. Show that (bs) is a Banach space.

b. Is there any connection between (bs) and (m)?

12. A subset A of a normed vector space V is *convex* if $x, y \in A \Rightarrow [\alpha x + (1 - \alpha)y] \in A$ for all real $\alpha \in [0, 1]$. Show that all spheres are convex.

13. Let A be the subset of (m) consisting of all sequences of zeros and ones.

a. Show that A is uncountable.

b. Calculate $\|x - y\|$ for two distinct members of A.

c. Show that A, and hence (m), cannot be separable.

14. a. Let A be the set of all sequences of complex numbers $\alpha + i\beta$, where α and β are rational. Let B be the subset of A consisting of those members of A which are ultimately constant (i.e., $\{\alpha_n + i\beta_n\} \in B$ if, for some N, $\alpha_n + i\beta_n = \alpha_N + i\beta_N$ wherever $n \geq N$; the N may be different for different members of B.) Show that B is countable.

b. Show that (c) is separable (see exercise 9).

15. Let V be the subspace of $C_R[0, 1]$ consisting of all broken-line functions (see exercise 10 of section 1).

a. Show that V is dense in $C_R[0, 1]$.

b. Find a countable subset of V which is also dense in $C_R[0, 1]$, thus establishing that $C_R[0, 1]$ is separable.

16. Show that the closure of a subspace of a normed vector space is itself a subspace.

17. Let W be a subspace of the normed vector space V, and consider the quotient space V/W (see section 1, exercise 13).

a. Define: $|||x||| = \inf \{||z|| : z \in \lfloor x \rfloor\}$.

Show: $|||x||| \geq 0$

$$||\alpha\lfloor x \rfloor||| = |\alpha| \, |||x|||$$

$$|||\lfloor x \rfloor + \lfloor y \rfloor||| \leq |||x||| + |||y|||$$

b. Show that if W is *closed* in V then $(|||x||| = 0) \Leftrightarrow (\lfloor x \rfloor = \lfloor 0 \rfloor)$, so that in this case V/W is a normed vector space.

c. Show that V/W is complete if V is complete and W is closed. [HINT: Let $\{X_n\}$ be a Cauchy sequence in V/W; choose a subsequence $\{Y_k\}$ such that $||Y_k - Y_{k+1}|| \leq 2^{-k}$. Show that $\{Y_k\}$ converges.]

Section 3 EQUIVALENCE OF NORMS AND FINITE DIMENSIONAL NORMED VECTOR SPACES

Two norms $||x||_1$ and $||x||_2$ are called equivalent on a vector space V if the corresponding metrics are equivalent. This means, we recall, that the identity mapping of V onto V is continuous in both directions. Since this mapping is obviously linear, the following theorem is useful here as well as in many other situations.

Theorem 4.3.1. Let T be a linear transformation from a normed vector space V into a normed vector space W. Then

(a) T is everywhere continuous if and only if T is continuous at $x = 0$.
(b) T is continuous at 0 if and only if there exists an M such that

$$||Tx|| \leq M\,||x||$$

for all x in V.

Thus, T is continuous if and only if it sends bounded sets into bounded sets.

Proof

(a) Let $||x_n - x_0|| \to 0$. The continuity of T at 0 implies

$$||Tx_n - Tx_0|| = ||T(x_n - x_0)|| \leq M\,||x_n - x_0||.$$

Hence $Tx_n \to Tx_0$, and T is continuous everywhere.

(b) If the inequality holds, then for $||x_n - 0|| \to 0$, we have

$$||Tx_n - T0|| = ||Tx_n|| \leq M\,||x_n|| = M\,||x_n - 0||$$

and T is continuous at 0. If there is no M for which the inequality holds, then there exists a sequence $\{x_n\}$ such that $||Tx_n|| \geq n\,||x_n||$. Letting $y_n = \dfrac{1}{n}\dfrac{x_n}{||x_n||}$, we see that

$$||y_n - 0|| \to 0 \quad \text{but} \quad ||Ty_n - T0|| = \frac{1}{n\,||x_n||}||Tx_n|| \geq 1 \qquad \text{for all } n.$$

Thus T is not continuous at $x = 0$.

Corollary. A continuous linear transformation is uniformly continuous. The corollary is an immediate consequence of the inequality

$$||T(x_n - x_0)|| = ||Tx_n - Tx_0|| \leq M\,||x_n - x_0||$$

and of the definition of uniform continuity. The inequality leads to the following characterization of equivalence of norms.

Lemma 4.3.2. If $||x||_1$ and $||x||_2$ are equivalent on V, then they are uniformly equivalent. Moreover, they are equivalent if and only if there exist positive constants C_1 and C_2 such that

$$||x||_1 \leq C_2\,||x||_2 \quad \text{and} \quad ||x||_2 \leq C_1\,||x||_1$$

for all x in V.

Proof. The first part follows from the linearity of the identity mapping and the foregoing corollary, while the second part is a consequence of the theorem itself.

Thus the equivalence of norms on a vector space is a simpler matter than that of equivalence of metrics on a vector space. In the finite dimensional case, we have the following result.

Theorem 4.3.3. Any two norms on a finite dimensional vector space V are equivalent.

Proof. Let vectors $x_1, x_2, \ldots, x_n$ be a basis in the n-dimensional vector space V. Then any vector x in V has a unique representation $x = \sum_{k=1}^{n} \alpha_k x_k$. Define

$$\|x\|_1 = \sum_{k=1}^{n} |\alpha_k|.$$

This is easily verified to be a norm on V. Let $\|x\|_2$ be any other norm; it suffices to show that $\|x\|_2$ is equivalent to $\|x\|_1$. We have

$$\|x\|_2 = \left\| \sum_{k=1}^{n} \alpha_k x_k \right\|_2 \le \sum_{k=1}^{n} |\alpha_k| \|x_k\|_2.$$

Let $M = \max(\|x_1\|_2, \|x_2\|_2, \ldots, \|x_n\|_2)$. Then

$$\|x\|_2 \le M \sum_{k=1}^{n} |\alpha_k| = M \|x\|_1,$$

and the proof is half done.

Now consider the function F from V_1 onto $R(V_1$ is V under the first norm), defined by

$$F: x \longrightarrow \|x\|_2.$$

It is continuous, as shown by the following calculation:

$$\left| \|x\|_2 - \|y\|_2 \right| \le \|x - y\|_2 = \left\| \sum_{k=1}^{n} (\alpha_k - \beta_k) x_k \right\|_2$$
$$\le M \sum_{k=1}^{n} |\alpha_k - \beta_k| = M \|x - y\|_1.$$

Let $S = \{x: \|x\|_1 = 1\}$; then S is compact in V_1. (Convergence in V_1 is equivalent to coordinatewise convergence. Since $x = \sum_{k=1}^{n} \alpha_k x_k \in S$ implies $|\alpha_k| \le 1$ for $k = 1, \ldots, n$, the Bolzano-Weierstrass theorem can be applied coordinatewise to any infinite subset of S to show that this subset must contain a convergent sequence.) Now, 0 is not in $F(S)$, for otherwise $0 \in S$, which is not the case. It follows that $m = \inf \{F(x): x \in S\} > 0$, for $F(S)$, as a compact subset of $(0, \infty)$, contains its infimum. Hence we have

$$\|x\|_2 \ge m > 0$$

for $\|x\|_1 = 1$. Then

$$\left\|\frac{x}{\|x\|_1}\right\|_2 \geq m$$

for all nonzero $x \in V$, or

$$\|x\|_1 \leq \frac{1}{m}\|x\|_2,$$

which holds for $x = 0$ also.

The two inequalities

$$\|x\|_2 \leq M\|x\|_1 \quad \text{and} \quad \|x\|_1 \leq \frac{1}{m}\|x\|_2$$

yield the equivalence of the two norms.

Corollary. Every finite dimensional normed vector space is a Banach space.

Corollary. Every finite dimensional subspace W of a normed vector space V is a closed subspace of V.

The proofs are left as exercises.

We now give a characterization of finite dimensional normed linear spaces. We need the following lemma, due to F. Riesz.

Lemma 4.3.4. Let V be a normed vector space, W a closed proper subspace of V, and $U = \{x \in V: \|x\| = 1\}$. Then for any $\epsilon > 0$, there exists a vector $x \in U$ such that $d(x, W) \geq 1 - \epsilon$.

Proof. Note that for $x \notin W$, $d(x, W) = \inf_{z \in W}\|x - z\| > 0$, since W is a closed proper subspace of V. Also $d(y, W) \leq 1$ for every $y \in U$, since $0 \in W$.

If y_1 is any vector not in W, and y_0 is any vector in W, then let

$$x = \frac{y_1 - y_0}{\|y_1 - y_0\|}.$$

Now we have

$$d(x, W) = \inf_{z \in W}\|x - z\| = \inf_{z \in W}\left\|\frac{y_1 - y_0}{\|y_1 - y_0\|} - z\right\|$$

$$= \frac{1}{\|y_1 - y_0\|}\inf_{z \in W}\|(y_1 - y_0 - z\|y_1 - y_0\|)\|$$

$$= \frac{1}{\|y_1 - y_0\|}\inf_{w \in W}\|y_1 - w\| = \frac{d(y_1, W)}{\|y_1 - y_0\|} \leq 1.$$

However, by suitably choosing y_0 in W the quantity

$$\frac{d(y_1, W)}{\|y_1 - y_0\|}$$

can clearly be greater than $1 - \epsilon$ for any preassigned $\epsilon > 0$. The corresponding x then satisfies the required conditions of the lemma, and the proof is finished.

We now apply the lemma to obtain the promised characterization of finite dimensional normed vector spaces.

Theorem 4.3.5. A normed vector space V is finite dimensional if and only if the unit sphere $S = \{x: \ \|x\| \leq 1\}$ is compact in V.

Proof. Since a finite dimensional normed vector space is equivalent to $R^{(n)}$ (or $C^{(n)}$) and the unit sphere in that space is compact, we are half done.

Now let V be a normed vector space in which the unit sphere S is compact, and choose $\eta < 1$. Since S is totally bounded, we can find spheres $S(x_k, \eta)$, $k = 1, \ldots, n$ such that $\bigcup_{k=1}^{n} S(x_k, \eta) \supseteq S$. Let W be the subspace of V spanned by the vectors $x_1, \ldots, x_n$. We claim that $W = V$, for otherwise there is a vector z in S such that $d(z, W) > \eta$, by the lemma. This contradicts the fact that every z in S is within distance η of some x_k. Hence $W = V$, and V is finite dimensional.

Examination of lemma 4.3.4 reveals that it is proved entirely from the fact that, if W is a closed proper subspace of V, and if $x \notin W$, then for any $\epsilon > 0$ there is a vector $z \in W$ such that $d(x, z) \leq d(x, W) + \epsilon$. This fact is obvious from the definition $d(x, W) = \inf_{z \in W} d(x, z)$. Now, in approximation theory one is frequently confronted with a vector x outside of a closed subspace W of a normed vector space V, and one desires to find the vector in W closest to x, i.e., the vector in W which best approximates x. The problem may have no solution, and even if it does have a solution, that solution may not be unique. For example, consider $R^{(2)}$ under

$$\|(\alpha_1, \alpha_2)\| = \max\,(|\alpha_1|, |\alpha_2|),$$

let $x = (2, 1)$, and let W be the set of vectors of the form $(\alpha, 0)$. We have

$$d[(2, 1), (\alpha, 0)] = \max\,[|\alpha - 2|, |1 - 0|],$$

which obviously takes on a minimum value of 1 for all vectors $(\beta, 0)$, where $1 \leq \beta \leq 3$.

The example shows that the uniqueness question may have a different answer if the norm is replaced by an equivalent norm, for if the Euclidean metric is used in $R^{(2)}$, then there is in any subspace a unique closest vector to any given vector outside that subspace.

The existence question is related to the Riesz lemma as follows. If for some $x_0 \notin W$ there is a $w_0 \in W$ such that $d(x, W) = \|x_0 - w_0\|$, then there also exists a vector y, with $\|y\| = 1$ and $d(y, W) = 1$. The reader can easily show that it suffices to take $y = \dfrac{x - w_0}{\|x - w_0\|}$. The possible nonexistence of a best approximation, in the closed proper subspace W, to a vector $x_0 \notin W$, can then be demonstrated by giving an example in which the indicated strengthening of the Riesz lemma is impossible.

EXAMPLE 1. Let V be the normed vector space $C_R[0, 1]$ consisting of all continuous real-valued functions on $[0, 1]$ with $\|x\| = \sup_t |x(t)|$. Let W be the subset of V consisting of all functions x for which

$$\int_0^{1/2} x(t)\, dt = \int_{1/2}^1 x(t)\, dt.$$

It is clear that W is a proper subspace of V; the fact that W is closed is a consequence of the advanced calculus theorem which states that, if $x_n \to x$ uniformly, then

$$\int_a^b x_n(t) \to \int_a^b x(t)\, dt.$$

Now suppose that z is in V but not in W, $\|z\| = 1$, and $d(z, W) = 1$. Let y be any vector in V but not in W. A short calculation shows that $w = (z - \alpha y) \in W$ if we let

$$\alpha = \frac{\displaystyle\int_0^{1/2} z(t)\, dt - \int_{1/2}^1 z(t)\, dt}{\displaystyle\int_0^{1/2} y(t)\, dt - \int_{1/2}^1 y(t)\, dt}.$$

We also see that $1 \le d(w, z) = |\alpha|\,\|y\|$, or

$$\left| \int_0^{1/2} y(t)\, dt - \int_{1/2}^1 y(t)\, dt \right| \le \left| \int_0^{1/2} z(t)\, dt - \int_{1/2}^1 z(t)\, dt \right| \|y\|$$

for all y not in W. For $n \ge 2$, let

$$y_n(t) = \begin{cases} 1; & 0 \le t \le \dfrac{1}{2} - \dfrac{1}{n} \\[2mm] n\left(\dfrac{1}{2} - t\right); & \dfrac{1}{2} - \dfrac{1}{n} \le t \le \dfrac{1}{2} + \dfrac{1}{n} \\[2mm] -1; & \dfrac{1}{2} + \dfrac{1}{n} \le t \le 1. \end{cases}$$

Then

$$1 - \frac{1}{n} \leq \left| \int_0^{1/2} z(t)\,dt - \int_{1/2}^1 z(t)\,dt \right| \qquad \text{for all } n$$

and

$$\left| \int_0^{1/2} z(t)\,dt - \int_{1/2}^1 z(t)\,dt \right| \geq 1.$$

However, no z in V of norm one can satisfy this inequality. Hence there can be no z such that $\|z\| = 1$ and $d(z, W) = 1$.

We can prove, however, that in the case where W is finite dimensional, there always exist best approximations.

Lemma 4.3.6. If W is a finite dimensional subspace of a normed vector space V, and if $x_0 \notin W$, then there exists a $w_0 \in W$ such that $\|x_0 - w_0\| = d(x_0, W)$.

Proof. If W has dimension n, then the subspace W_1 of V spanned by x_0 and W has dimension $n + 1$, and we can safely consider that $V = W_1$ for the purpose at hand. Choose a sequence $\{w_n\}$ in W such that $d(x_0, W) = \lim_{n \to \infty} \|x_0 - w_n\|$. The sequence $\{w_n\}$ is a bounded infinite set in a *finite dimensional* normed vector space. The Bolzano-Weierstrass principle applies in finite dimensional spaces, and hence, some subsequence $\{w_{n(k)}\}$ of $\{w_n\}$ converges; let w_0 be the limit. Then

$$\|w_{n(k)} - x_0\| \longrightarrow d(x_0, W),$$

but from the inequality

$$\left| \|x_0 - w_{n(k)}\| - \|x_0 - w_0\| \right| \leq \|w_{n(k)} - w_0\|,$$

we see that it also converges to $\|x_0 - w_0\|$. Hence $d(x_0, W) = \|x_0 - w_0\|$ by the uniqueness of limits. This finishes the proof.

To illustrate the lemma, we consider the following example.

EXAMPLE 2. Let $V = C_R[a, b]$ and let W_n be the set of all polynomial functions on $[a, b]$ of degree less than or equal to n. Then W_n is a closed proper $(n + 1)$ dimensional subspace of V. Let x_0 be any fixed member of $C_R[a, b]$ not in W_n. The lemma guarantees the existence of a polynomial P_n in W_n such that $d[x_0, W_n] = \|x_0 - P_n\|$. P.L. Tchebycheff has shown that the polynomial P_n is *unique*, and he has described its properties.

The Weierstrass approximation theorem, which we will prove later, states that for any $\epsilon > 0$ and any $x_0 \in C_R[a, b]$ there is a polynomial P such that $|x_0(t) - P(t)| < \epsilon$ for all $t \in [a, b]$. A consequence of this theorem is the fact that $\lim_{n \to \infty} d[x_0, W_n] = 0$.

EXERCISES

1. a. Prove the first corollary following theorem 4.3.3.
    b. Prove the second corollary following theorem 4.3.3.

2. Let W be the subspace of $R^{(2)}$ consisting of all vectors of the form $(\alpha, 2\alpha)$, and let x_0 be the vector $(1, 1)$.
    a. If $\|(\alpha, \beta)\|_1 = \max [|\alpha|, |\beta|]$, find $d_1(x_0, W)$ and the unique vector $w_1 \in W$ such that $d_1(x_0, W) = \|x_0 - w_1\|_1$.
    b. If $\|(\alpha, \beta)\|_2 = |\alpha| + |\beta|$, find $d_2(x_0, W)$ and the unique vector $w_2 \in W$ such that $\|x_0 - w_2\| = d_2(x_0, W)$.

3. Let T be a continuous function from a normed vector space V into a normed vector space W satisfying $T(x + y) = T(x) + T(y)$.
    a. If R is the field of scalars in V and W, show that $T(\alpha x) = \alpha T(x)$.
    b. If C is the field of scalars in V and W and if $T(ix) = iT(x)$, show that $T(\alpha x) = \alpha T(x)$.
    (See exercise 3, section 1, chapter 4.)

4. Let W be the subset of (m), the Banach space of all bounded sequences under $\|\{\alpha_n\}\| = \sup_n |\alpha_n|$, consisting of all sequences $\{\beta_n\}$ such that $\beta_{2k} = 0$ for all k.
    a. Show that W is a closed subspace of (m) of infinite dimension.
    b. Show that there exist vectors $x = \{\alpha_n\}$ in (m) such that $\|x\| = 1$ and $d(x, W) = 1$. Find the form of the most general such vector.
    c. Let A be the set of all vectors found in part b, and let z_0 be any vector not in W. Show that there exists a vector y_0 in A and a scalar λ such that $z_0 + \lambda y_0$ is in W.
    d. Show that for any z_0 not in W there is a vector w_0 in W such that $d(z_0, W) = \|z_0 - w_0\|$. Find the most general such vector.

5. Let V be the subset of $C_R[0, 1]$ consisting of all functions satisfying $x(0) = 0$. Let W be the subset of V consisting of all members of V satisfying $\int_0^1 x(t)\, dt = 0$.

   a. Show that V and W are closed subspaces of $C_R[0, 1]$.
   b. Assume there exists z_0 such that $\|z_0\| = 1$ and $d(z_0, W) = 1$.
   Let y be any function in V but not in W. Show that for a suitable
   α, the function $w = z_0 - \alpha y$ is in W.
   c. Continue the argument along the lines of example 1 to show
   that z_0 cannot exist.

6.   a. A norm on a vector space V is called *strictly convex* if

$$[x \neq y, \ \|x\| = \|y\| = 1] \Rightarrow \left[\left\|\frac{x+y}{2}\right\| < 1\right].$$

   Show that a norm is strictly convex if and only if

$$[x \neq y, \ \|x\| = \|y\| = 1] \Rightarrow [\|\alpha x + (1 - \alpha)y\| < 1$$
   for all real α satisfying $0 < \alpha < 1].$

   b. A norm on a vector space V is called *strong* if

$$[x \neq 0, y \neq 0, \|x + y\| = \|x\| + \|y\|]$$
$$\Rightarrow [x = \alpha y \quad \text{for some real positive } \alpha].$$

   Show that a norm is strong if and only if it is strictly convex.

7.   Let W be a closed subspace of a normed vector space V, let $z_0 \notin W$,
   and let $A = \{x \in W: \ d(z_0, W) = \|z_0 - x\|\}$. Assume that A
   is not empty.
   a. Show that A is closed in W.
   b. Show that $x_1, x_2 \in A \Rightarrow \frac{1}{2}(x_1 + x_2) \in A$.
   c. Show that if α_1 and α_2 are positive integers such that $\alpha_1 + \alpha_2 = 2^n$, then $2^{-n}(\alpha_1 x_1 + \alpha_2 x_2) \in A$.
   d. Show that A is convex (see exercise 12 of the preceding section).

8.   Let W be a closed subspace of a strongly normed vector space V,
   and let $z_0 \notin W$. Show that if $d(z_0, W) = d(z_0, x_1) = d(z_0, x_2)$
   for x_1 and x_2 in W, then $x_1 = x_2$ (see exercises 6, 7).

9.   Suppose for x and y in $C_R[a, b]$ it happens that $\|x + y\| = \|x\| + \|y\|$. What can be said about x and y? Is $C_R[a, b]$ strongly normed (see exercise 6)?

10. Let $x_1, \ldots, x_n$ be any n linearly independent vectors in a normed
   vector space V, and let F be the mapping

$$(\alpha_1, \ldots, \alpha_n) \longrightarrow \sum_{k=1}^{n} \alpha_k x_k$$

   of $C^{(n)}$ into V. Show that F is continuous (Use any norm on $C^{(n)}$.)

11. Let $x_1, \ldots, x_n$ be a basis for the n-dimensional normed vector space V. Show that the function F_j which sends

$$x = \sum_{k=1}^{n} \alpha_k x_k \longrightarrow \alpha_j$$

    is continuous from V to C.

12. Let W_n be the subspace of $C_R[a, b]$ consisting of all polynomials of degree less than or equal to n. Show that B is a bounded subset of W_n if and only if the set of all coefficients occurring in members of B is bounded in R.

13. Let $x_1, \ldots, x_n$ be any n linearly independent members of $C_R[a, b]$, and let W_n be the subspace spanned by them. Show that the result in exercise 12 holds here.

14. Let A and B be any two subsets of a normed vector space V, and let $A + B = \{x + y : \ x \in A, y \in B\}$. Show that $\bar{A} + \bar{B} \subseteq \overline{A + B}$. Give an example to show that the inclusion can be proper. (It suffices to take $V = R$.)

15. Let x be a function from a metric space M into a finite dimensional normed vector space V, and let $y_1, \ldots, y_n$ be a basis for V. Then

    $$x(t) = x_1(t)y_1 + \cdots + x_n(t)y_n, \qquad \text{for all } t \in M.$$

    Show that x is continuous if and only if x_k is continuous for $k = 1, \ldots, n$.

Section 4   THE VECTOR SPACE OF CONTINUOUS FUNCTIONS ON A COMPACT METRIC SPACE.

In example 5 of section 2 we introduced $C_C(M)$, the vector space of all continuous functions from a compact metric space M into C, the complex numbers, and we showed that $C_C(M)$ is normed by

$$\|x\| = \sup_{t \in M} |x(t)|.$$

Convergence in $C_C(M)$ was called uniform convergence. The weaker statement, $x_n \longrightarrow x_0$ *pointwise*, simply means $|x_n(t) - x_0(t)| \longrightarrow 0$ for every $t \in M$. Unless the word *pointwise* is used, convergence will mean convergence in the norm sense.

Lemma 4.4.1. $C_C(M)$ is complete.

Proof. Let $\{x_n\}$ be a Cauchy sequence in $C_C(M)$. Then for any $\epsilon > 0$, there is an $N(\epsilon)$ such that for $m, n \geq N(\epsilon)$ we have

$$|x_n(t_0) - x_m(t_0)| \leq \sup_{t \in M} |x_m(t) - x_n(t)| = \|x_m - x_n\| \leq \epsilon,$$

and so $\{x_n(t_0)\}$ is a Cauchy sequence in C for any $t_0 \in M$. Since C is complete, $\lim_{n \to \infty} x_n(t_0)$ exists for all $t_0 \in M$, and the sequence $\{x_n\}$ converges pointwise to a function which we call x_0.

To see that x_0 is continuous, we notice that, since

$$|x_n(t_0) - x_m(t_0)| \leq \epsilon, \qquad \text{for} \quad m, n \geq N(\epsilon),$$

we can let $m \longrightarrow \infty$ without disturbing the inequality, thus obtaining

$$|x_n(t_0) - x_0(t_0)| \leq \epsilon, \qquad \text{for} \quad n \geq N(\epsilon) \text{ and all } t_0.$$

Now, for any fixed $n \geq N(\epsilon)$ we have

$$|x_0(t_1) - x_0(t_2)| \leq |x_0(t_1) - x_n(t_1)| + |x_n(t_1) - x_n(t_2)|$$
$$+ |x_n(t_2) - x_0(t_2)| \leq 2\epsilon + |x_n(t_1) - x_n(t_2)|.$$

Since x_n is uniformly continuous on M, there is a $\delta(\epsilon)$ such that $d(t_1, t_2) \leq \delta(\epsilon)$ implies

$$|x_n(t_1) - x_n(t_2)| \leq \epsilon.$$

Hence

$$|x_0(t_1) - x_0(t_2)| \leq 3\epsilon, \qquad \text{for} \quad d(t_1, t_2) \leq \delta(\epsilon),$$

and x_0 is continuous (uniformly) on M.

Finally, the inequality $|x_n(t_0) - x_0(t_0)| \leq \epsilon$ for all $t_0 \in M$ and $n \geq N(\epsilon)$ implies $\sup_{t \in M} |x_n(t) - x_0(t)| \leq \epsilon$, or $\|x_n - x_0\| \leq \epsilon$, for $n \geq N(\epsilon)$, and therefore $\{x_n\}$ converges to x_0 in the norm sense.

Corollary. A uniformly convergent sequence of continuous complex-valued functions on a metric space M has a continuous limit function.

The corollary follows from the fact that the compactness of M was not essential in the pertinent part of the foregoing proof.

We shall now restrict our attention to $C_R(M)$, where M is compact. In the exercises, some aspects of the cases where the functions are complex-valued or even Banach-space-valued will be indicated.

Definition 4.4.2. A family of functions $\{x_\alpha\}$ in $C_R(M)$ is *equicontinuous at t_0* if for any $\epsilon > 0$ there is a $\delta(\epsilon, t_0)$ such that, for all α, $|x_\alpha(t_0) - x_\alpha(t)| \leq \epsilon$ whenever $d(t_0, t) \leq \delta(\epsilon, t_0)$.

The family $\{x_\alpha\}$ is *uniformly equicontinuous* if for any $\epsilon > 0$ there is a $\delta(\epsilon)$ such that $d(t_1, t_2) \leq \delta(\epsilon)$ implies $|x_\alpha(t_1) - x_\alpha(t_2)| \leq \epsilon$ for all α.

The argument used to prove the theorem which states that a continuous function on a compact metric space is uniformly continuous can be repeated almost verbatim to show that, if the family $\{x_\alpha\}$ in $C_R(M)$ is equicontinuous at every point, when it is uniformly equicontinuous.

Equicontinuity plays a crucial role in the characterization of compact subsets of $C_R(M)$. The following lemmas indicate its importance.

Lemma 4.4.3. Let $\{x_n\}$ be a convergent sequence in $C_R(M)$. Then
(a) $\{x_n\}$ is a bounded subset of $C_R(M)$, and
(b) $\{x_n\}$ is a uniformly equicontinuous subset of $C_R(M)$.

Proof. (a) follows from the fact that a convergent sequence in any metric space is bounded. To prove (b), we let x_0 be the (continuous) limit function of $\{x_n\}$ and then write

$$|x_n(t_1) - x_n(t_2)| \leq |x_n(t_1) - x_0(t_1)| + |x_0(t_1) - x_0(t_2)| + |x_0(t_2) - x_n(t_2)|.$$

For $n \geq N(\epsilon)$, the first and last terms of the right member of this inequality are less than ϵ, and for $d(t_1, t_2) \leq \delta_0(\epsilon)$ the middle term is too. The functions $x_1, \ldots, x_{N(\epsilon)-1}$ determine numbers $\delta_1(\epsilon), \ldots, \delta_{N(\epsilon)-1}(\epsilon)$ by their uniform continuity. Then if we let

$$\delta(\epsilon) = \min [\delta_0(\epsilon), \delta_1(\epsilon), \ldots, \delta_{N(\epsilon)-1}(\epsilon)]$$

we have $|x_n(t_1) - x_n(t_2)| \leq 3\epsilon$ whenever $d(t_1, t_2) \leq \delta(\epsilon)$, for all n. This finishes the proof.

If the reader has not already met sequences of continuous functions in $C_R[a, b]$ converging pointwise but not uniformly to a continuous limit function, he will do so in the exercises. The existence of such examples adds interest to the following lemma, which complements the preceding one.

Lemma 4.4.4. Let $\{x_n\}$ be a uniformly equicontinuous sequence in $C_R(M)$ converging pointwise to a function x_0. Then the convergence is uniform, $x_0 \in C_R(M)$, and the sequence is bounded in $C_R(M)$.

Proof. By the uniform equicontinuity of $\{x_n\}$, we can, for any $\epsilon > 0$, find a $\delta(\epsilon)$ such that $|x_n(t_1) - x_n(t_2)| \leq \epsilon$ whenever $d(t_1, t_2) \leq \delta(\epsilon)$, for all n. Letting $n \to \infty$ and applying the pointwise convergence, we see that $|x_0(t_1) - x_0(t_2)| \leq \epsilon$ whenever $d(t_1, t_2) \leq \delta(\epsilon)$, so that x_0 is (uniformly) continuous on M.

Now, cover M with spheres $S[t, \delta(\epsilon)]$ about each point, and extract a finite covering $S[t_1; \delta(\epsilon)], \ldots, S[t_k; \delta(\epsilon)]$. Let t be any point in M and suppose, for definiteness, $t \in S[t_i; \delta(\epsilon)]$. We have

$$|x_0(t) - x_n(t)| \leq |x_0(t) - x_0(t_i)| + |x_0(t_i) - x_n(t_i)| + |x_n(t_i) - x_n(t)|$$
$$\leq \epsilon + |x_0(t_i) - x_n(t_i)| + \epsilon.$$

By the pointwise convergence, there exist integers $N_1(\epsilon), \ldots, N_k(\epsilon)$ such that $|x_0(t_i) - x_n(t_i)| \leq \epsilon$ whenever $n \geq N_i(\epsilon)$. Let $N(\epsilon) = \max_{i=1}^{n} N_i(\epsilon)$; then $|x_0(t_i) - x_n(t_i)| \leq \epsilon$ for $n \geq N(\epsilon)$ and any i. Thus $|x_0(t) - x_n(t)| \leq 3\epsilon$, provided that $n \geq N(\epsilon)$, and $\{x_n\}$ converges uniformly to x_0. From this we also know that it is bounded, and the proof is finished.

We come now to our main theorem.

Theorem 4.4.5. A subset K of $C_R(M)$ is compact if and only if it is closed, bounded, and equicontinuous.

Proof. Since compact subsets of any metric space are closed and bounded, we need to show that K compact implies K equicontinuous. By the total boundedness of K we can, for any $\epsilon > 0$, cover K by a finite number of spheres $S(x_1, \epsilon), \ldots, S(x_n, \epsilon)$. Corresponding to each x_j, $j = 1, \ldots, n$, there is a modulus of continuity $\delta_j(\epsilon)$. Letting $\delta(\epsilon) = \min[\delta_1(\epsilon), \ldots, \delta_n(\epsilon)]$, we have $d(t_1, t_2) \leq \delta(\epsilon)$ implies $|x_j(t_1) - x_j(t_2)| \leq \epsilon$ for $j = 1, \ldots, n$. Now, any x in K is in some $S(x_j, \epsilon)$, so that

$$|x(t_1) - x(t_2)| \leq |x(t_1) - x_j(t_1)| + |x_j(t_1) - x_j(t_2)| + |x_j(t_2) - x(t_2)|.$$

The first and last terms of the right member are less than or equal to ϵ since $x \in S(x_j, \epsilon)$, while for $d(t_1, t_2) \leq \delta(\epsilon)$, the middle term is also less than or equal to ϵ. This proves the equicontinuity of K.

Now, let K be a closed, bounded, equicontinuous subset of $C_R(M)$. Then K is complete, and we need only show that it is totally bounded. By the equicontinuity of K we can, for any $\epsilon > 0$, find a $\delta(\epsilon)$ such that $d(t, t') \leq \delta(\epsilon)$ implies $|x(t) - x(t')| \leq \epsilon$ for all $x \in K$. The compactness of M enables us to find finitely many spheres,

$$S[t_1, \delta(\epsilon)], \ldots, S[t_n, \delta(\epsilon)]$$

covering M. By the boundedness of K, we can find a number B such that $|x(t)| \leq B$ for all $t \in M$, $x \in K$. Now choose an N such that $N\epsilon \geq 2B$, and let the points $-B = b_0, b_1, \ldots, b_{N-1}, b_N = B$ divide $[-B, B]$ into N equal parts of length $2B/N \leq \epsilon$.

Consider the collection of all mappings of the set $\{t_1, \ldots, t_n\}$ into the set $\{b_0, b_1, \ldots, b_N\}$; there are only finitely many of these, and each is specified by n pairs $(t_1, b_{j(1)}), \ldots, (t_n, b_{j(n)})$. For every such mapping, choose a function x in K such that

$$|x(t_i) - b_{j(i)}| \leq \epsilon, \qquad \text{for} \quad i = 1, \ldots, n$$

if such a function exists in K. We now have finitely many functions in K, say $x_1, \ldots, x_m$, which are to serve as the centers of spheres in $C_R(M)$ covering K. To see how this goes, let y be any member of K. Then choose $b_{j(1)}, \ldots, b_{j(n)}$ such that

$$|y(t_i) - b_{j(i)}| \leq \epsilon, \qquad \text{for} \quad i = 1, \ldots, n.$$

Let x_k be the function from the set $x_1, \ldots, x_m$ for which

$$|x_k(t_i) - b_{j(i)}| \leq \epsilon, \qquad i = 1, \ldots, n.$$

Then

$$|y(t_i) - x_k(t_i)| \leq 2\epsilon, \qquad \text{for} \quad i = 1, \ldots, n.$$

Finally, let t be any point in M, and suppose it lies in the sphere $S[t_j, \delta(\epsilon)]$. Then

$$|y(t) - x_k(t)| \leq |y(t) - y(t_j)| + |y(t_j) - x_k(t_j)| + |x_k(t_j) - x_k(t)|$$
$$\leq \epsilon + 2\epsilon + \epsilon = 4\epsilon,$$

where the first and last terms are less than or equal to ϵ due to the choice of $\delta(\epsilon)$. Hence $\|y - x_k\| \leq 4\epsilon$, the spheres $S(x_k, 4\epsilon)$, $k = 1, \ldots, m$ cover K, and K is totally bounded. This completes the proof.

We now give an example to illustrate the importance of the preceding theorem.

EXAMPLE. It is desired to find a function x_0 which minimizes the value of the integral

$$\int_a^b F[t, x(t)]\, dt.$$

We assume that $F(t, s)$ is continuous in the t-s plane and that F satisfies the Lipschitz condition

$$|F(t, s_1) - F(t, s_2)| \leq c|s_1 - s_2|.$$

(If $\partial F/\partial s$ is continuous and bounded, this condition will certainly be satisfied.) If we consider the function Φ from $C_R[a, b]$ into R defined by

$$x \longrightarrow \int_a^b F[t, x(t)]\, dt,$$

we can prove that Φ is continuous as follows:

$$|\Phi(x) - \Phi(y)| = \left| \int_a^b F[t, x(t)]\, dt - \int_a^b F[t, y(t)]\, dt \right|$$

$$\leq \int_a^b |F[t, x(t)] - F[t, y(t)]|\, dt$$

$$\leq c \int_a^b |x(t) - y(t)|\, dt$$

$$\leq c(b - a)\|x - y\|.$$

Now, if the domain of Φ is restricted to a compact subset of $C_R[0, 1]$, then we can be certain that the desired minimum *exists*.

In the exercises, the reader will find several concrete examples of equicontinuous sets of functions.

EXERCISES

1. Show that if the family $\{x_a\}$ of real-valued continuous functions on a compact metric space M is equicontinuous at every point of M, then it is uniformly equicontinuous on M.

2. Consider the sequence $\{x_n\}$ in $C_R[0, 2]$ defined by

$$x_n(t) = \begin{cases} n^2 t; & 0 \leq t \leq \dfrac{1}{n} \\[2mm] n^2\left(\dfrac{2}{n} - t\right); & \dfrac{1}{n} \leq t \leq \dfrac{2}{n} \\[2mm] 0; & \dfrac{2}{n} \leq t \leq 2 \end{cases}$$

   a. Show that $\{x_n\}$ converges pointwise to the zero function, but not uniformly.
   b. Modify this sequence to obtain a sequence in $C_R[0, 2]$ which is bounded and pointwise convergent to the zero function, but not uniformly convergent.
   c. Show that $\{x_n\}$ is equicontinuous at every point of $(0, 2]$, but is not equicontinuous at $t = 0$.

3. Show that the closure of an equicontinuous set in $C_R(M)$ is equicontinuous.

4. For each of the following sequences of functions, determine the pointwise limit if it exists, and ascertain whether or not the limit function is continuous, and whether the convergence is uniform. Also find the points where the sequences are equicontinuous and the points where they are not equicontinuous.

a. $x_n(t) = t^n$, $[-1, 1]$
b. $x_n(t) = t^{1/n}$, $[0, 1]$
c. $x_n(t) = nte^{-nt}$, $[0, 1]$
d. $x_n(t) = nt^n(1 - t)$, $[0, 1]$
e. $x_n(t) = \dfrac{nt}{1 + n^2 t^2}$, $[-1, 1]$

5. Let $\{x_n\}$ be a sequence in $C_R(M)$ which converges uniformly to $x_0 \in C_R(M)$ on a subset A of M. Show that the convergence is uniform on $\bar{A}$.

6. Let c be a fixed positive number and let D be the subset of $C_R[0, 1]$ consisting of all x for which x' exists, is continuous, and satisfies $|x'(t)| \le c$ for $0 \le t \le 1$. Show that D is equicontinuous.

7. Let c and α be fixed positive numbers and let S be a subset of $C_R(M)$ consisting of functions x satisfying $|x(t_1) - x(t_2)| \le cd(t_1, t_2)^\alpha$. Show that S is equicontinuous.

8. Let $C(M_1, M_2)$ be the set of all continuous functions from a compact metric space M_1 into a compact metric space M_2. Define $d(x, y) = \sup_{t \in M_1} \{d_2[x(t), y(t)]\}$.
    a. Show that $C(M_1, M_2)$ is a metric space under d.
    b. Show that a subset S of $C(M_1, M_2)$ is compact if and only if it is closed and equicontinuous.

9. Let M be a compact metric space, V a normed vector space, and $C_V(M)$ the set of all continuous functions from M into V.
    a. Show that $C_V(M)$ is a normed linear space under $\|x\| = \sup_{t \in M} \|x(t)\|$. Show that $C_V(M)$ is complete if V is complete.
    b. Let V be finite dimensional. Show that a subset K of $C_V(M)$ is compact if and only if it is bounded, closed, and equicontinuous.
    c. Let V be a Banach space, not necessarily finite dimensional. For each t in M define the functions F_t from $C_V(M)$ into V by

    $$F_t: \quad x \longrightarrow x(t).$$

    Show that F_t is continuous.

d. With V as in part c, let K be a compact subset of $C_V(M)$. Show that K is closed, bounded, and equicontinuous, and show also that $F_t(K)$ is compact for each $t \in M$.

e. Prove the converse of part d.

10. Prove that, if $\{x_n\}$ is a Cauchy sequence in $C_R[0, 1]$, then

$$\int_0^1 [x_m(t) - x_n(t)]^2 \, dt \longrightarrow 0 \qquad \text{as} \quad m, n \longrightarrow \infty.$$

b. Let $\{x_n\}$ be an equicontinuous sequence in $C_R[0, 1]$ such that

$$\int_0^1 [x_m(t) - x_n(t)]^2 \, dt \longrightarrow 0 \qquad \text{as} \quad m, n \longrightarrow \infty.$$

Show that $\{x_n\}$ must be pointwise convergent, and hence that it is a Cauchy sequence in $C_R[0, 1]$.

c. Show that part b may be false if the requirement of equicontinuity is dropped.

11. Let D_C be the set of differentiable functions on $[0, 1]$ such that $|x'(t)| \leq C$ for all t. Show that $\bar{D}_C$, the closure of D_C in $C_R[0, 1]$, is the set of functions satisfying the Lipschitz condition.

$$\text{Lip} \, (C): \quad |x(t_2) - x(t_1)| \leq C|t_2 - t_1|.$$

[HINT: For $x \in \text{Lip} \, (C)$, consider the functions

$$y_n(t) = \int_{t-1/n}^{t+1/n} x(s) \, ds,$$

extending x horizontally for $t < 0$ and $t > 1$ so that y_n is well-defined.]

Section 5   SOME SPECIAL RESULTS

Consider the two spaces $C_R[a, b]$ and $C_R[0, 1]$. It is easily verified that the mapping

$$F: \quad x(t) \longrightarrow y(s) = x\left[\frac{s - a}{b - a}\right]$$

is a one-to-one linear isometry of $C_R[0, 1]$ onto $C_R[a, b]$; therefore we can restrict our attention to $C_R[0, 1]$. We now give a proof, due to S. Bernstein, that the subspace W of $C_R[0, 1]$ consisting of all polynomial functions is dense in $C_R[0, 1]$. This is the Weierstrass approximation theorem.

Theorem 4.5.1. For any $x \in C_R[0, 1]$ and any $\epsilon > 0$ there exists a polynomial p such that $\|x - p\| \leq \epsilon$.

Proof. Let

$$p_n(t) = \sum_{k=0}^{n} x\left(\frac{k}{n}\right)\left[\binom{n}{k} t^k (1 - t)^{n-k}\right],$$

where

$$\binom{n}{k} = \frac{n!}{k!(n - k)!},$$

and notice that

$$1 = 1^n = [t + (1 - t)]^n = \sum_{k=0}^{n} \binom{n}{k} t^k (1 - t)^{n-k}.$$

Then

$$|x(t) - p_n(t)| = \left| \sum_{k=0}^{n} \left[x(t) - x\left(\frac{k}{n}\right) \right] \binom{n}{k} t^k (1 - t)^{n-k} \right|$$

$$\leq \sum_{k=0}^{n} \left| x(t) - x\left(\frac{k}{n}\right) \right| \binom{n}{k} t^k (1 - t)^{n-k}.$$

Now let $\delta(\epsilon)$ be the modulus of uniform continuity of x, so that $|t - t'| \leq \delta(\epsilon)$ implies $|x(t) - x(t')| \leq \epsilon$. Divide the set $\{1, 2, \ldots, n\}$ into two disjoint parts

$$A_1 = \left\{ k: \left| \frac{k}{n} - t_0 \right| \leq \delta(\epsilon) \right\}$$

$$A_2 = \left\{ k: \left| \frac{k}{n} - t_0 \right| > \delta(\epsilon) \right\}$$

where t_0 is a fixed point in $[0, 1]$. Then, using an obvious abbreviation, we have

$$|x(t_0) - p_n(t_0)| \leq \sum_{A_1} + \sum_{A_2} \leq \epsilon \sum_{A_1} \binom{n}{k} t_0^k (1 - t_0)^{n-k}$$

$$+ 2\|x\| \sum_{A_2} \binom{n}{k} t_0^k (1 - t_0)^{n-k}$$

$$\leq \epsilon + 2\|x\| \sum_{A_2} \binom{n}{k} t_0^k (1 - t_0)^{n-k}.$$

Using the identity

$$\sum_{k=0}^{n} \binom{n}{k}(k - nt_0)^2 t_0^k(1 - t_0)^{n-k} = nt_0(1 - t_0)$$

(see the exercises for a derivation), we have

$$nt_0(1 - t_0) = n^2 \sum_{k=0}^{n} \binom{n}{k}\left(\frac{k}{n} - t_0\right)^2 t_0^k(1 - t_0)^{n-k}$$

$$\geq n^2 \sum_{A_2} \left(\frac{k}{n} - t_0\right)^2 \binom{n}{k} t_0^k(1 - t_0)^{n-k}$$

$$\geq n^2 \delta(\epsilon)^2 \sum_{A_2} \binom{n}{k} t_0^k(1 - t_0)^{n-k},$$

and so

$$\sum_{A_2} \binom{n}{k} t_0^k(1 - t_0)^{n-k} \leq \frac{t_0(1 - t_0)}{n\delta(\epsilon)^2} \leq \frac{1}{4n\delta(\epsilon)^2},$$

since $t_0(1 - t_0) \leq \frac{1}{4}$ for $t_0 \in [0, 1]$. Finally,

$$|x(t_0) - p_n(t_0)| \leq \epsilon + \frac{\|x\|}{2n\delta(\epsilon)^2}$$

for all $t_0 \in [0, 1]$ and all n. Hence, for sufficiently large n, we have $\|x - p_n\| \leq 2\epsilon$, and the proof is finished.

The motivation for the foregoing theorem and proof is as follows. If a coin is weighted so that it comes up *head* with probability t and *tail* with probability $1 - t$, then the probability of exactly k heads in n tosses is $\binom{n}{k}t^k (1 - t)^{n-k}$. If the function x determines a payoff of amount $x(k/n)$ when exactly k heads occur in n tosses, then the expected payoff in n tosses is

$$\sum_{k=1}^{n} x\left(\frac{k}{n}\right)\binom{n}{k}t^k(1 - t)^{n-k}.$$

Intuitively, we feel that for very large n this amount should be close to $x(t)$, and the theorem indicates that our intuition is correct.

The separability of $C_R[0, 1]$ can now be shown easily. Let D be the set of all polynomials with rational coefficients; the reader can verify that D is countable. By the theorem we can, for any $x \in C_R[0, 1]$, choose a polynomial p such that $\|x - p\| \leq \epsilon/2$. Now if $p(t) = \sum_{k=0}^{n} \alpha_k t^k$, we choose rationals $\rho_0, \rho_1, \ldots, \rho_n$ such that $|\rho_k - \alpha_k| \leq \dfrac{\epsilon}{(n+1)2}$ for $k = 0, \ldots, n$.

Then, letting $q(t) = \sum\limits_{k=0}^{n} \rho_k t^k$, we have

$$\|p - q\| = \sup_{t \in [0,1]} \left| \sum_{k=0}^{n} (\alpha_k - \rho_k) t^k \right| \leq \sum_{k=0}^{n} |\alpha_k - \rho_k| \leq \frac{\epsilon}{2}$$

and $\|x - q\| \leq \|x - p\| + \|p - q\| \leq \epsilon$. We have proved the following theorem.

Theorem 4.5.2. $C_R[0, 1]$ is separable, and the polynomials with rational coefficients are a countable dense set in $C_R[0, 1]$.

In the next section we will show that $C_R(M)$ is separable for any compact metric space M.

A sequence of functions $\{x_n\}$ is said to be *monotone nonincreasing* if $x_n(t) \geq x_{n+1}(t)$ for all t. We now give a refinement of theorem 4.5.1 as follows.

Lemma 4.5.3. If $x \in C_R[0, 1]$, then there exists a monotone nonincreasing sequence of polynomials $\{p_n\}$ converging uniformly to $x(t)$.

Proof. Let $\{\epsilon_n\}$ be any sequence of positive numbers such that $\sum\limits_{n=1}^{\infty} \epsilon_n = 1$. By theorem 4.5.1 we can find polynomials $p_1, p_2, \ldots, p_n, \ldots$ such that for all $t \in [0, 1]$

$$|x(t) + (1 - \tfrac{1}{2}\epsilon_1) - p_1(t)| < \tfrac{1}{2}\epsilon_1$$
$$|x(t) + (1 - \epsilon_1 - \tfrac{1}{2}\epsilon_2) - p_2(t)| < \tfrac{1}{2}\epsilon_2$$
$$\vdots$$
$$|x(t) + (1 - \epsilon_1 - \cdots - \epsilon_{n-1} - \tfrac{1}{2}\epsilon_n) - p_n(t)| < \tfrac{1}{2}\epsilon_n.$$

Letting $x_0(t) = x(t) + 1$, and $x_n(t) = x(t) + \left(1 - \sum\limits_{k=1}^{n} \epsilon_k\right)$ for $n \geq 1$, we have

$$x_0(t) \geq p_1(t) \geq x_1(t) \geq p_2(t) \geq \cdots \geq x_n(t) \geq p_n(t) \geq \cdots.$$

Since $x_n \longrightarrow x$ uniformly, the lemma is proved.

Clearly, we could replace "monotone nonincreasing" by "monotone nondecreasing" and prove the corresponding modified statement. Another result on monotone convergence, in its original form due to Dini, is as follows.

Theorem 4.5.4. If $\{x_n\}$ is a monotone nonincreasing sequence in $C_R(M)$, where M is compact, converging *pointwise* to the *continuous* function x_0, then the convergence is uniform.

Proof. By the monotone pointwise convergence we can, for any $\epsilon > 0$ and $t_0 \in M$, find an $N(\epsilon, t_0)$ such that $0 \le [x_n(t_0) - x_0(t_0)] \le \epsilon/2$ for $n \ge N(\epsilon, t_0)$. Since the function $x_{N(\epsilon, t_0)} - x_0$ is continuous and the convergence is monotone, there is a sphere $S[t_0, \delta(\epsilon, t_0)]$ such that $t' \in S[t_0, \delta(\epsilon, t_0)]$ implies $0 \le [x_{N(\epsilon, t_0)}(t') - x_0(t')] \le \epsilon$; again using the monotonicity, we have

$$0 \le [x_n(t') - x_0(t')] \le \epsilon$$

for $n \ge N(\epsilon, t_0)$ and $t' \in S[t_0, \delta(\epsilon, t_0)]$. If such a sphere is associated with every $t_0 \in M$, then by compactness, there is a finite covering,

$$S[t_1, \delta(\epsilon, t_1)], \ldots, S[t_m, \delta(\epsilon, t_m)],$$

of M. Let

$$N(\epsilon) = \max_{k=1}^{m} N(\epsilon, t_k).$$

Then for $n \ge N(\epsilon)$ and any $t \in M$, we have $0 \le [x_n(t) - x_0(t)] \le \epsilon$, since t must be in one of these spheres. This gives the uniform convergence, and the proof is complete.

For real-valued functions there js a weakening of the notion of continuity which is sometimes useful.

Definition 4.5.5. A real-valued function x on a metric space M is called upper semicontinuous at t_0 if for any $\epsilon > 0$ there is a $\delta(\epsilon, t_0)$ such that $x(t) \le x(t_0) + \epsilon$ whenever $d(t, t_0) \le \delta(\epsilon, t_0)$. If x is upper semicontinuous at every point of M, it is said to be upper semicontinuous on M. Lower semicontinuity is analogously defined.

Clearly, x is continuous if and only if it is both upper semicontinuous and lower semicontinuous.

Lemma 4.5.6. If x is upper semicontinuous on the compact metric space M, then x is bounded above. Moreover, if $A = \sup_{t \in M} x(t)$, then $x(p_0) = A$ for some $p_0 \in M$.

Proof. By the upper semicontinuity, there is a sphere $S[t, \delta(\epsilon, t)]$ around every point $t \in M$ such that $t' \in S[t, \delta(\epsilon, t)]$ implies $x(t') \le x(t) + \epsilon$. The compactness of M gives a finite number of these spheres $S_1, \ldots, S_n$ covering

M, and the number max $[x(t_1), \dots, x(t_n)] + \epsilon$ is then an upper bound for x. Now let $\{t_n\}$ be a sequence in M such that limit $x(t_n) = A$. By the compact-

ness of M there is a convergent subsequence $\{p_k\} = \{t_{n(k)}\}$ of $\{t_n\}$; let p_0 be the limit of this subsequence. If $x(p_0) < A$ then, letting $\epsilon = \frac{1}{2}[A - x(p_0)]$, there is a $\delta(\epsilon, p_0)$ such that

$$x(p) \leq x(p_0) + \epsilon = \tfrac{1}{2}[A + x(p_0)] \qquad \text{for} \quad d(p, p_0) \leq \delta(\epsilon, p_0).$$

But this is in contradiction to the fact that limit $x(p_k) = A$, and we conclude

that $x(p_0) < A$ is impossible. Hence $x(p_0) = A$, and the proof is finished.

Corollary. If x is lower semicontinuous on the compact metric space M, then x is bounded below and $x(p_0) = \inf_{t \in M} x(t)$ for some $p_0 \in M$.

EXAMPLE. A real-valued function x on $[0, 1]$ is called a step function if there is a finite set of points $0 = t_0, t_1, \dots, t_{n-1}, t_n = 1$ on $[0, 1]$ such that x is constant on each of the open intervals (t_{k-1}, t_k), $k = 1, 2, \dots, n$. It is easy to verify that the set $S[0, 1]$ of all step functions on $[0, 1]$ is a normed vector space under $\|x\| = \sup_{t \in [0, 1]} |x(t)|$ (see the exercises). A step function x is upper semicontinuous if and only if

$$x(t_k) \geq \max [x(t_k^-), x(t_k^+)] \qquad \text{for} \quad k = 0, \dots, n,$$

where $x(t_k^-)$ and $x(t_k^+)$ are the left and right limits of x at t_k.

It is clear that we cannot, in general, uniformly approximate an upper semicontinuous function x by continuous functions, for in that case x would be continuous. Hence we can only hope to approximate x in the pointwise sense.

Theorem 4.5.7. Let x be an upper semicontinuous function on a compact metric space M. Then there exists a sequence $\{x_n\}$ of continuous functions on M such that $x_{n+1}(t) \leq x_n(t)$ and limit $x_n(t) = x(t)$ for all t.

Proof. By lemma 4.5.6, x is bounded above so that the function $x_n(t) = \sup_{t_1 \in M} \{x(t_1) - nd(t, t_1)\}$ is finite valued. To see that it is continuous, we note that

$$x_n(t) \geq x(t_1) - nd(t_1, t) \geq x(t_1) - n[d(t_1, t') + d(t', t)];$$

therefore

$$x_n(t) \geq \sup_{t_1 \in M} [x(t_1) - nd(t_1, t')] - nd(t', t)$$

or

$$x_n(t) \geq x_n(t') - nd(t', t).$$

Hence $x_n(t') - x_n(t) \leq nd(t', t)$, and since we can interchange t' and t, we obtain

$$|x_n(t') - x_n(t)| \leq nd(t', t),$$

which establishes the continuity of x_n. The inequality

$$x(t_1) - nd(t, t_1) \geq x(t_1) - (n + 1)d(t, t_1)$$

yields the monotonicity of the sequence $\{x_n\}$, i.e., $x_1(t) \geq \cdots \geq x_n(t) \geq \cdots$, while $x_n(t) \geq x(t) - nd(t, t) = x(t)$ gives $\lim_{n \to \infty} x_n(t) \geq x(t)$.

Now for a given t_0 we determine $\delta(\epsilon, t_0)$ so that $d(t, t_0) \leq \delta(\epsilon, t_0)$ implies $x(t) \leq x(t_0) + \epsilon$; hence $x(t) - nd(t, t_0) \leq x(t_0) + \epsilon$ for such values of t, and all n. For $d(t, t_0) > \delta(\epsilon, t_0)$, we have

$$x(t) - nd(t, t_0) \leq x(t) - n\delta(\epsilon, t_0) \leq A - n\delta(\epsilon, t_0)$$

where $A = \sup_{t \in M} x(t)$. It follows that, for all sufficiently large n, we have

$$x(t) - nd(t, t_0) \leq x(t_0) + \epsilon \quad \text{and} \quad x_n(t_0) \leq x(t_0) + \epsilon.$$

Hence $\lim_{n \to \infty} x_n(t_0) \leq x(t_0) + \epsilon$ for all ϵ, and therefore $\lim_{n \to \infty} x_n(t_0) \leq x(t_0)$. Combining this inequality with the one previously derived, we obtain $\lim_{n \to \infty} x_n(t_0) = x(t_0)$ for all $t_0 \in M$, and the proof is finished.

The reader is advised to consider some examples for the case $M = [0, 1]$ to see the motivation for the definition of the sequence $\{x_n\}$ in the foregoing proof. Of course, a similar theorem holds for lower semicontinuous functions. By combining the various results of this section and by using some of the techniques employed, the reader can prove the following corollary.

Corollary. If x is an upper semicontinuous function on $[0, 1]$, then there exists a sequence of polynomials p_n such that $p_{n+1}(t) \leq p_n(t)$ and $\lim_{n \to \infty} p_n(t) = x(t)$ for all $t \in [0, 1]$.

EXERCISES

1.   Verify in detail that the mapping $x(t) \longrightarrow y(s) = x\left(\dfrac{s - a}{b - a}\right)$ is a one-to-one linear isometric correspondence of $C_R[0, 1]$ onto $C_R[a, b]$.

2. a. Show that x is upper semicontinuous if and only if $x(t_0) \geq \overline{\lim_{t \to t_0}} x(t)$ for all t_0. (See exercise 15, section 2, chapter 3)

   b. Formulate and prove the analog of part a for lower semicontinuity.

3. a. Show that $S[0, 1]$, the set of all step functions on $[0, 1]$, is a normed vector space under the norm $\|x\| = \sup_{t \in [0, 1]} |x(t)|$.

   b. Show that it is not complete.

   c. Show that the step function

   $$x(t) = \alpha_k, \ t_{k-1} < t < t_k, \qquad k = 1, \ldots, n$$

   is upper semicontinuous if and only if

   $$x(t_k) \geq \max [x(t_k^-), x(t_k^+)], \qquad k = 1, 2, \ldots, n.$$

4. Let x be an upper semicontinuous function on a metric space M, not necessarily compact.

   a. Show that the conclusion of theorem 4.5.7 still holds if x is bounded above.

   b. Show that the conclusion of theorem 4.5.7 still holds if there is a continuous function x_0 such that $x_0(t) \geq x(t)$ for all $t \in M$. [HINT: $x - x_0$ is upper semicontinuous (why?) and bounded above.]

5. Let

   $$x(t) = \begin{cases} 2, & 0 \leq t - \frac{1}{2} \\ 3, & \frac{1}{2} = t \\ 1, & \frac{1}{2} < t \leq 1 \end{cases}$$

   Find explicitly the functions x_n which approximate x in the sense of theorem 4.5.7.

6. Prove the corollary to theorem 4.5.7.

7. Prove the converse of theorem 4.5.7: If x is a real-valued function on a metric space M, and if there is a sequence of continuous functions $\{x_n\}$ such that $x_{n+1}(t) \leq x_n(t)$ and $\lim_{n \to \infty} x_n(t) = x(t)$ pointwise, then x is upper semicontinuous.

8. a. Show that, if x and y are upper semicontinuous, then $x + y$ is upper semicontinuous.

   b. Show that if $\alpha \geq 0$, and x is upper semicontinuous, then αx is upper semicontinuous.

   c. Show that x is upper semicontinuous if and only if $-x$ is lower semicontinuous.

d. Show that parts a and b above remain valid if "upper" is replaced by "lower."

9. Let $\{x_n\}$ be a sequence in $C_R(M)$, M compact, with $x_n(t) \geq 0$ for all t and n. Show that if $\sum_{n=1}^{\infty} x_n$ converges pointwise to $x \in C_R(M)$, then the convergence is uniform.

10. By two differentiations of the identity
$$1 = [1 + (1 - t)]^n = \sum_{k=0}^{n} \binom{n}{k} t^k (1 - t)^{n-k},$$
prove

a. $\displaystyle\sum_{k=0}^{n} k \binom{n}{k} t^k (1 - t)^{n-k} = nt,$

b. $\displaystyle\sum_{k=0}^{n} k(k-1) \binom{n}{k} t^k (1 - t)^{n-k} = n(n - 1)t^2,$ and

c. $\displaystyle\sum_{k=0}^{n} (k - nt)^2 \binom{n}{k} t^k (1 - t)^{n-k} = nt(1 - t)$

11. Let $C^{(1)}[0, 1]$ be the set of all real-valued functions on $[0, 1]$ having a continuous derivative.
a. Show that $C^{(1)}[0, 1]$ is a normed vector space under $\|x\| = \sup_{t \in [0, 1]} |x(t)|$, but is not complete.
b. Show that $C^{(1)}[0, 1]$ is a Banach space under $\|x\|_2 = \sup_{t \in [0, 1]} |x|(t)| + \sup_{t \in [0, 1]} |x'(t)|$.

12. Let V be the vector space of all continuous complex-valued functions on R. Let $\{k_n\}$ be any sequence of positive numbers such that $\sum_{n=1}^{\infty} k_n < \infty$, and let $\|x\|_n = \sup_{t \in [-n, n]} |x(t)|$. Define
$$d(x, y) = \sum_{n=1}^{\infty} k_n \frac{\|x - y\|_n}{1 + \|x - y\|_n}.$$
Show that V is a metric space in which $x_n \to x_0$ if and only if $x_n \to x$ uniformly on every compact subset of R.

Section 6 THE APPROXIMATION THEOREM IN $C_R(M)$; NORMED ALGEBRAS.

The method used in the preceding section to show the separability of $C_R[0, 1]$ does not generalize readily to $C_R(M)$, for the notion of a polynomial function on a compact metric space is meaningless. However, if x is a real-valued continuous function on M, we can consider the class of functions defined for each $t \in M$ by an expression

$$\alpha_0 + \alpha_1[x(t)] + \alpha_2[x(t)]^2 + \cdots + \alpha_n[x(t)]^n,$$

where n is not fixed. Since $C_R(M)$ is closed under multiplication (i.e., the product of a finite number of continuous functions is continuous) as well as under addition and multiplication by scalars, the members of this class form a subset, denoted by $A[x]$, of $C_R(M)$. The Weierstrass approximation theorem states that, if $M = [0, 1]$ and $x_0(t) = t$, then the set $A[x_0]$ is dense in $C_R[0, 1]$, or $\overline{A[x_0]} = C_R[0, 1]$. Now we notice that, in general, $A[x]$ is closed under addition, multiplication, and multiplication by scalars. We are thus led to the following definition.

Definition 4.6.1. An *algebra* A is a vector space with a binary operation of multiplication $(x, y) \to xy$ satisfying the identities

(a) $(xy)z = x(yz)$,
(b) $(x + y)z = xz + yz$, $x(y + z) = xy + xz$,
(c) $\alpha(xy) = (\alpha x)y = x(\alpha y)$.   ($\alpha$ a scalar.)

The algebra A is *commutative* if

(d) $xy = yx$.

A *normed algebra* is a normed vector space which is also an algebra, such that

$$\| xy \| \leq \| x \| \| y \|.$$

A *Banach algebra* is a complete normed algebra.

The norm inequality guarantees that the multiplication is a continuous mapping from $A \times A$ into A, as follows:

$$\| x_n y_n - xy \| = \| x_n y_n - x_n y + x_n y - xy \|$$
$$\leq \| x_n(y_n - y) \| + \| (x_n - x)y \|$$
$$\leq \| x_n \| \| y_n - y \| + \| y \| \| x_n - x \|.$$

Now if $\| x_n - x \| \to 0$ and $\| y_n - y \| \to 0$, then $\{ \| x_n \| \}$ is bounded, say $\| x_n \| \leq M$ for all n, and it follows that $\| x_n y_n - xy \| \to 0$.

A subset B of an algebra A is called a *subalgebra* of A if B is itself an algebra. Now, if $x, y \in C_R(M)$ when M is compact, then

$$| x(t)y(t) | = | x(t) | \cdot | y(t) | \leq \| x \| \| y \|;$$

hence $\|xy\| \leq \|x\| \|y\|$, so that $C_R(M)$ is a commutative Banach algebra. Moreover, for any $x_0 \in C_R(M)$, the set $A[x_0]$ is a subalgebra of $C_R(M)$. Since, in general, the closure $\bar{B}$ of a subalgebra B of a normed algebra A is again a subalgebra of A (see the exercises), it is natural to ask if we can find a function x_0 in $C_R(M)$ such that $\overline{A[x_0]}$ is exactly $C_R(M)$. (This was possible for $M = [0, 1]$.) As there is no natural candidate for x_0, we reverse our point of view and ask if there are circumstances under which $\overline{A[x_0]} \neq C_R(M)$. Certainly, if x_0 is constant, then $\overline{A[x_0]}$ is the set of all constant functions on M, but $C_R(M)$ always contains nonconstant functions. In fact, if $t_2 \neq t_1$, then the function $x(t) = d(t, t_2)$ is in $C_R(M)$ and satisfies $x(t_1) \neq x(t_2)$. In other words, *for any two distinct points, t_1 and t_2, in M there is a function x in $C_R(M)$ such that $x(t_1) \neq x(t_2)$.* Also, if $y(t_1) = y(t_2)$, then $z(t_1) = z(t_2)$ for every function z in $A[y]$, and $\overline{A[y]} \neq C_R(M)$.

Hence, if we are to have $\overline{A[y]} = C_R(M)$, then we must choose y such that $y(t_1) \neq y(t_2)$ for $t_1 \neq t_2$, that is, y must be a one-to-one continuous function from M into R. However, such a function does not, in general, exist, although we cannot prove this here (see, however, the exercises). These considerations lead to the following definition.

Definition 4.6.2. A subalgebra A of $C_R(M)$ is called a *separating* subalgebra if, for any two distinct points t_1 and t_2 of M, there is a function x in A such that $x(t_1) \neq x(t_2)$.

An example of a proper closed separating subalgebra of $C_R[0, 1]$ is the set of all functions in $C_R[0, 1]$ satisfying $x(0) = 0$. This shows that the closure of a separating subalgebra of $C_R(M)$ may still not be the whole of $C_R(M)$, but it also suggests the following theorem.

Theorem 4.6.3. (Stone-Weierstrass). If A is a separating subalgebra of $C_R(M)$ containing the constant functions, then $\bar{A} = C_R(M)$.

We will need the following lemma.

Lemma 4.6.4. If A is a subalgebra of $C_R(M)$ and if $x \in A$, then the function $|x|$ is in $\bar{A}$. Also, if $x, y \in A$ then the functions $x \cup y$ and $x \cap y$, defined by,

$$(x \cup y)(t) = \max [x(t), y(t)], \quad \text{and}$$
$$(x \cap y)(t) = \min [x(t), y(t)]$$

are in $\bar{A}$.

Proof of the Lemma. The second part follows from the first part and the identities

$$x \cup y = \tfrac{1}{2}[x + y + |x - y|]$$
$$x \cap y = \tfrac{1}{2}[x + y - |x - y|],$$

for $\bar{A}$ is itself a subalgebra of $C_R(M)$.

To prove the first part, we recall that there is a sequence of polynomials $\{p_n(u)\}$ in $C_R[0, 1]$ such that $\|\sqrt{u} - p_n(u)\| \to 0$ as $n \to \infty$; this follows from the Weierstrass approximation theorem. Now let $u(t) = \dfrac{x(t)^2}{\|x\|^2}$; then

$$\sqrt{u(t)} = \frac{1}{\|x\|} |x(t)|$$

and the sequence

$$p_n \left[\frac{x(t)^2}{\|x\|^2} \right]$$

in A converges uniformly to $\dfrac{1}{\|x\|} |x(t)|$ in $C_R(M)$. It follows that $|x| \in \bar{A}$.

In the exercises we will indicate a short direct proof that $\sqrt{u}$ can be uniformly approximated by polynomials on $[0, 1]$, so that reliance on the classical Weierstrass theorem can be avoided.

Proof of the theorem. We first assert that if t_1 and t_2 are distinct points of M, and if α_1, and α_2, are any real numbers, then A contains a function z such that $z(t_1) = \alpha_1$ and $z(t_2) = \alpha_2$. To obtain this z, choose a function y in A for which $y(t_1) \neq y(t_2)$, and let

$$z(t) = \alpha_1 + \frac{(\alpha_2 - \alpha_1)[y(t) - y(t_1)]}{y(t_2) - y(t_1)}.$$

Now let x be in $C_R(M)$ but not in A; we show that $x \in \bar{A}$. To do so we first choose a point $t_0 \in M$ and construct a function $y_0 \in \bar{A}$ such that

$$y_0(t_0) = x(t_0) \quad \text{and} \quad y_0(t) \leq x(t) + \epsilon, \qquad \text{for all } t \in M.$$

This is done as follows. For any $t' \in M$ we can, by our foregoing remarks, find a function x' such that

$$x'(t_0) = x(t_0) \quad \text{and} \quad x'(t') \leq x(t') + \frac{\epsilon}{2}.$$

By continuity, there is an open sphere $S(t')$ about t' such that $t \in S(t')$ implies $x'(t) \leq x(t) + \epsilon$. The collection of all such spheres covers M and, by compactness, we obtain spheres $S_1, \ldots, S_n$ covering M. Let $x_1, \ldots, x_n$ be the corresponding members of A, and let

$$y_0 = x_1 \cap x_2 \cap \cdots \cap x_n.$$

Then $y_0 \in \bar{A}$ and $y_0(t_0) = x(t_0)$ since $x_i(t_0) = x(t_0)$ for $i = 1, \ldots, n$. Moreover, $y_0(t) \leq x(t) + \epsilon$ for all $t \in M$ since any $t \in M$ is in some $S(t_j)$ and $y_0(t) \leq x_j(t) \leq x(t) + \epsilon$. This completes the construction of y_0.

Again by continuity, there is a sphere $S(t_0)$ about t_0 such that

$$y_0(t) \geq x(t) - \epsilon, \quad \text{for} \quad t \in S(t_0).$$

Letting t_0 vary, we see that we have proved the following. For any $t' \in M$, there is a function $y' \in \bar{A}$ and a sphere $S(t')$ about t' such that

$$y'(t) \geq x(t) - \epsilon, \quad \text{for} \quad t \in S(t')$$
$$y'(t) \leq x(t) + \epsilon, \quad \text{for all } t \in M.$$

The spheres $S(t')$ cover M, and hence M is covered by finitely many of them, say $S(t'_1), \ldots, S(t'_m)$. Let the corresponding functions be $y'_1, \ldots, y'_m$, and let

$$y = y'_1 \cup \cdots \cup y'_m.$$

We now easily obtain

$$x(t) - \epsilon \leq y(t) \leq x(t) + \epsilon, \quad \text{for all } t \in M$$

with $y \in \bar{A}$. Since, for any $\epsilon > 0$, such a $y \in \bar{A}$ can be found, we see that $\bar{A}$ is dense in $C_R(M)$. But this means $\bar{A} = C_R(M)$, and the theorem is proved.

With the aid of the powerful Stone-Weierstrass theorem, the separability of $C_R(M)$ is not difficult to demonstrate.

Theorem 4.6.5. $C_R(M)$ is separable.

Proof. Since a compact metric space has a countable basis $\{O_n\}$ for open sets, we obtain a countable collection of functions

$$x_0 = 1, x_1, \ldots, x_n, \ldots$$

where, for $n \geq 1$, $x_n(t) = d(t, \tilde{O}_n)$. Then for $t_1 \neq t_2$ there is an x_n with $x_n(t_1) \neq x_n(t_2)$. We now let A be the subalgebra of $C_R(M)$ "generated" by this collection, i.e., A is the set of all functions which are linear combinations of terms of the type

$$x_{i(1)} x_{i(2)} \cdots x_{i(n)}.$$

Alternatively, A is the smallest subalgebra of $C_R(M)$ containing the functions $\{x_n\}$, $n = 0, 1, \ldots$. By the theorem, $\bar{A} = C_R(M)$; to prove the separability of $C_R(M)$ it suffices to exhibit a countable dense subset of A. Such a subset is given by the collection of members of A which employ only *rational* scalars. The detailed proof is left as an exercise.

One further matter can be cleared up with the techniques and results now available. Let $I^{(2)} = [0, 1] \times [0, 1]$ and consider $C_R(I^{(2)})$. We ask if the polynomials in two variables are dense in $C_R(I^{(2)})$. The affirmative answer is a corollary to the following theorem.

Theorem 4.6.6. For any $\epsilon > 0$ and $z \in C_R(M_1 \times M_2)$, there is a function of the form $\sum_{i=1}^{n} x_i y_i$, where $x_i \in C_R(M_1)$ and $y_i \in C_R(M_2)$ such that

$$||z - \sum_{i=1}^{n} x_i y_i || \leq \epsilon.$$

Proof. The set of all finite sums of the type $\sum_{i=1}^{n} x_i y_i$, where n is not fixed, is a subalgebra A of $C_R(M_1 \times M_2)$ containing the constant functions. Further, it is a separating subalgebra, for if $(s_1, t_1) \neq (s_2, t_2)$, then either $s_1 \neq s_2$ or $t_1 \neq t_2$. If $s_1 \neq s_2$, choose $x \in C_R(M_1)$ such that $x(s_1) \neq x(s_2)$; then $x \cdot 1$ separates (s_1, t_1) and (s_2, t_2). If $s_1 = s_2$, $t_1 \neq t_2$, then $1 \cdot y$ will separate these points provided y is chosen so that $y(t_1) \neq y(t_2)$. Theorem 4.6.3 now gives $\bar{A} = C(M_1 \times M_2)$, and the proof is finished.

Corollary. The set of polynomials in two variables is dense in $C_R(I^{(2)})$.

Corollary. If K is a compact subset of $R^{(n)}$ then the set of all polynomials in n variables is dense in $C_R(K)$. (See exercise 8.)

EXERCISES

1. Show that the closure $\bar{B}$ of a subalgebra B of a normed algebra A is again a subalgebra of A.

2. Let B be an incomplete normed algebra, and let $\hat{B}$ be the completion of B when B is considered only as a normed vector space. Show that the multiplication in B can be extended to $\hat{B}$ so that $\hat{B}$ becomes a Banach algebra. (See exercise 5 of section 2.)

3. a. Suppose there is a one-to-one function in $C_R(M)$, where M is compact. Show that M is homeomorphic to a compact subset of R. (See exercise 8, section 2, chapter 3.)

     b. It can be shown that the square $[0, 1] \times [0, 1]$ is not home-
omorphic to any subset of R. Using this fact, show that there
is no function $x \in C_R([0, 1] \times [0, 1])$ such that $C_R([0, 1] \times [0, 1]) = \overline{A[x]}$.

4. Let x and y be real-valued functions on a set S, and define

$$(x \cup y)(t) = \max [x(t), y(t)]$$
$$(x \cap y)(t) = \min [x(t), y(t)].$$

Prove:

$$x \cup y = \tfrac{1}{2} [x + y + |x - y|]$$
$$x \cap y = \tfrac{1}{2} [x + y - |x - y|].$$

5. Define the sequence $\{x_n\}$ in $C_R[0, 1]$ as follows:

$$x_1(t) = 0$$
$$x_{n+1}(t) = x_n(t) + \tfrac{1}{2}[t - x_n(t)^2], \qquad n \geq 1.$$

a. Prove $0 \leq x_n(t) \leq x_{n+1}(t) \leq \sqrt{t}$ for all t.
b. Prove $\| x_n(t) - \sqrt{t} \| \to 0$.

6. Fill in the omitted details in the proof of theorem 4.6.5.

7. Prove the first corollary to theorem 4.6.6.

8. Let K be a compact subset of $R^{(n)}$. Show that the polynomials in n variables are dense in $C_R(K)$.

9. Let P be the set of all real-valued continuous functions on R which
are periodic with period 2π.
a. Show that P is a normed algebra under $\| x \| = \sup_t |x(t)|$.

b. Show that the set of all functions of form

$$\alpha_0 + \sum_{k=1}^{n} (\alpha_k \cos kt + \beta_k \sin kt)$$

is a subalgebra of P. (n is not fixed.)
c. Show that $\bar{A} = P$.

10. Let $C_C(M)$ be the Banach space of all continuous complex-valued
functions on a compact metric space M.
a. Show that $C_C(M)$ is a Banach algebra.
b. Let A be a separating subalgebra of $C_C(M)$ containing the
constant functions, and containing $\bar{x}$ whenever it contains x.
Show that $\bar{A} = C_C(M)$. ($\bar{x}$ is the complex conjugate of x.)
c. Show that $C_C(M)$ is separable.

11. For $x, y, z, C_R(M)$, show
    a. $|x| \leq |y| \Rightarrow \|x\| \leq \|y\|$.
    b. $|(x \cup y) - (x \cup z)| \leq |y - z|$
       $|(x \cap y) - (x \cap z)| \leq |y - z|$.
    c. The operations $\cup$ and $\cap$ are continuous functions from $C_R(M) \times C_R(M)$ onto $C_R(M)$.

12. Show that the cardinality of $C_R(M)$ is equal to that of R (see exercise 20, section 1, chapter 3.)

Section 7   BOUNDED FUNCTIONS AND STEP FUNCTIONS

In section 5, we considered not necessarily continuous functions, and in various other places the hypotheses of continuity and compactness were made only to guarantee the existence of the norm $\|x(t)\| = \sup_{t \in M} |x(t)|$. The following lemma is proved by familiar methods.

Lemma 4.7.1. If M is any metric space and $B_C(M)$ is the set of all bounded functions from M into the complex numbers C, then $B_C(M)$ is a Banach algebra under $\|x(t)\| = \sup_{t \in M} |x(t)|$.

If $M = [0, 1]$, the simplest kind of discontinuous function is a step function, as defined in section 5. The set $S[0, 1]$ of all step functions on $[0, 1]$ is clearly a separating subalgebra of $B_R[0, 1]$ containing the constant functions. The Stone-Weierstrass theorem does not apply here, but we can still determine the closure of $S[0, 1]$ in $B_R[0, 1]$. (From exercise 3 of section 5, we know that $S[0, 1]$ is not closed.) We know that $\overline{S[0, 1]}$ is the class of all functions which can be uniformly approximated by step functions, but the problem is to characterize this class in some other way. It should be clear, at least intuitively, that $\overline{S[0, 1]} \supset C_R[0, 1]$; we could easily write a proof of this, but our final result will be more general. Now, the discontinuities of a real-valued function of a real variable fall into two types, as follows.

(i) The limits $x(t_0^+) = \underset{0 < h \to 0}{\text{limit}} \ x(t_0 + h)$ and $x(t_0^-) = \underset{0 > h \to 0}{\text{limit}} \ x(t_0 + h)$ both exist, but

$$\max [|x(t_0) - x(t_0^+)|, |x(t_0) - x(t_0^-)|] > 0.$$

(ii) One or both of the limits in (i) fails to exist. We will say, in this case, that x has an *oscillatory discontinuity* at the point t_0.

Certainly, functions which are continuous except at a finite number of points of type (i) should be $\overline{S[0, 1]}$ while the function

$$x(t) = \begin{cases} \sin \dfrac{1}{t}, & 0 < t \leq 1 \\ 0, & 0 = t \end{cases}$$

is easily shown not to be in $\overline{S[0, 1]}$ ($t = 0$ is a point where x has a type (ii) discontinuity). We are thus led to conjecture the following theorem.

Theorem 4.7.2. $\overline{S[0, 1]}$ is the set of all functions on $[0, 1]$ which have no discontinuities of type (ii).

Proof. Let $x \in \overline{S[0, 1]}$, and choose $t_0 \in [0, 1]$ We show that $x(t_0^+)$ exists; the existence of the left limit is similarly demonstrated. It suffices to show that, for any $\epsilon > 0$, there is a $\delta(\epsilon)$ such that

$$t_0 < t_1 < t_0 + \delta(\epsilon), \qquad t_0 < t_2 < t_0 + \delta(\epsilon) \quad \text{implies} \quad |x(t_1) - x(t_2)| \leq \epsilon.$$

(This will show that, for any sequence $\{t_k\}$ approaching t_0 from the right, the sequence $\{x(t_k)\}$ is Cauchy and hence has a limit. Moreover, it shows that if $\{t_k'\}$ is a second such sequence, then $\{x(t_k)\}$ and $\{x(t_k')\}$ have the same limit.) Now let $\{x_n\}$ be a sequence of step functions converging uniformly to x. We choose N such that $\|x_N - x\| \leq \epsilon$ and calculate

$$|x(t_1) - x(t_2)| \leq |x(t_1) - x_N(t_1)| + |x_N(t_1) - x_N(t_2)| + |x_N(t_2) - x(t_2)|$$
$$\leq 2\epsilon + |x_N(t_1) - x_N(t_2)|.$$

Since $\lim\limits_{0 < h \to 0} x_N(t_0 + h)$ exists, there is a $\delta(\epsilon)$ such that $|x_N(t_1) - x_N(t_2)| \leq \epsilon$ for $t_0 < t_1 < t_0 + \delta(\epsilon)$ and $t_0 < t_2 < t_0 + \delta(\epsilon)$. Hence $|x(t_1) - x(t_2)| \leq 3\epsilon$ for such t_1 and t_2, and half of the proof is complete. Note that our argument proves that the uniform limit of functions each having no discontinuities of type (ii), can have no discontinuities of type (ii).

Conversely, let x be a function with no discontinuities of type (ii), and choose $\epsilon > 0$. We now construct a step function y such that $\|y - x\| \leq \epsilon$. The compactness of $[0, 1]$ is essential. For any $t_0 \in [0, 1]$ there is a $\delta(t_0, \epsilon)$ such that

$$t_0 < t_1 < t_0 + \delta(t_0, \epsilon), \qquad t_0 < t_2 < t_0 + \delta(t_0, \epsilon)$$
$$\text{implies} \quad |x(t_1) - x(t_2)| < \epsilon,$$

and

$$t_0 - \delta(t_0, \epsilon) < t_1' < t_0, \qquad t_0 - \delta(t_0, \epsilon) < t_2' < t_0$$
$$\text{implies} \quad |x(t_1') - x(t_2')| < \epsilon.$$

The intervals $\{t: \ |t - t_0| < \delta(t_0, \epsilon)\}$, one for each $t_0 \in [0, 1]$, cover $[0, 1]$ and, by compactness, finitely many of them cover $[0, 1]$. The finite point set consisting of the endpoints and midpoints of these intervals lies in some natural order in $[0, 1]$. If it is given by

$$0 = s_1 < s_2 < \cdots < s_n = 1,$$

then we let $J_i = (s_i, s_{i+1})$ and note that for $t, t' \in J_i$, we have $|x(t') - x(t)| \leq \epsilon$. Now choose any $s_i' \in J_i$ for $i = 1, \ldots, n-1$ and define

$$y(t) = \begin{cases} x(s_i') & \text{for} \quad t \in J_i, \quad i = 1, \ldots, n-1 \\ x(s_i) & \text{for} \quad t = s_i, \quad i = 1, \ldots, n. \end{cases}$$

The function y is then in $S[0, 1]$ and, clearly, $\|y - x\| \leq \epsilon$. This completes the proof.

If we use the label $R[0, 1]$ for $\overline{S[0, 1]}$ it is then clear that $R[0, 1]$, as a closed subspace of the Banach space $B_R[0, 1]$ is complete, i.e., $R[0, 1]$ is a Banach space, and indeed it is a Banach algebra. The nonseparability of $S[0, 1]$, and hence of $R[0, 1]$ and $B_R[0, 1]$, will be treated in the exercises.

EXAMPLE. Let

$$x(t) = \begin{cases} 0, & \text{for } t = 0 \\ 0, & \text{for } t \text{ irrational} \\ \dfrac{1}{q}, & \text{for } t = \dfrac{p}{q} > 0. \end{cases}$$

(We assume p and q have no common factor greater than 1.) The function x is continuous at irrational points, discontinuous at rational points, and is a member of $R[0, 1]$. The proofs are left as exercises; this example should caution the reader against too simple a conception of functions in $R[0, 1]$.

Even though the points of discontinuity of a member of $R[0, 1]$ may be dense in $[0, 1]$, the following lemma gives some consolation.

Lemma 4.7.3. If $x \in R[0, 1]$, then x has at most a countable number of discontinuities.

Proof. A point t_0 at which x has a discontinuity of type (i) is characterized by the inequality $\max[|x(t_0) - x(t_0^+)|, |x(t_0) - x(t_0^-)|] > 0$. We assert that, for any n, the set A_n of points at which

$$\max[|x(t) - x(t^+)|, |x(t) - x(t^-)|] \geq \frac{1}{n}$$

is finite. The proof is simple. If there are infinitely many such points, they would have a cluster point t_0. Then for any δ, the interval $(t_0 - \delta, t_0 + \delta)$ contains two points t' and t'' such that $|x(t') - x(t'')| \geq 1/2n$. It follows that either $x(t_0^+)$ or $x(t_0^-)$ fails to exist, but this contradicts the fact that $x \in R[0, 1]$.

Now, the set A of all points where x is discontinuous is clearly given by $A = \bigcup_{n=1}^{\infty} A_n$, a denumerable union of finite sets. Hence A is countable and the proof is finished.

In exercise 9, the reader will meet a generalization of this result.

EXERCISES

1. Prove lemma 4.7.1.

2. Let $B_V(M)$ be the set of all bounded functions from a metric space M into a Banach space V. [x is bounded if $\sup_{t \in M} \| x(t) \| < \infty$].

   Show that $B_V(M)$ is a Banach space. Show that $B_V(M)$ is a Banach algebra if V is a Banach algebra.

3. a. Find an uncountable collection of functions $\{x_\alpha\}$ in $S[0, 1]$ such that $\| x_\alpha \| = 1$ for all α and $\| x_\alpha - x_\beta \| = 1$ for $\alpha \neq \beta$.
   b. Show that $S[0, 1]$ is not separable.

4. Verify the assertions made in the illustrative example in this section.

5. Let $\{\alpha_n\}$ be any sequence of real numbers such that $\lim_{n \to \infty} \alpha_n = 0$, and let $\{t_n\}$ be any sequence of distinct points in $[0, 1]$. Define

$$x(t) = \begin{cases} \alpha_n, & \text{for } t = t_n, n = 1, 2, \ldots, \\ 0, & \text{otherwise.} \end{cases}$$

   Show that x is continuous at all points not in $\{t_n\}$, and discontinuous at points of t_n. Show that $x \in R[0, 1]$.

6. Show that if $x \in R[0, 1]$ then there is an *increasing* sequence of step functions $\{x_n\}$ converging uniformly to x.

7. Let x be a real-valued function on $(-\infty, \infty)$ having no discontinuities of type (ii). Show that x has countably many discontinuities.

8. Show that $|x| \in R[0, 1]$ whenever $x \in R[0, 1]$, and hence that $R[0, 1]$ is closed under the operations of $\cup$ and $\cap$.

9. Let x be any real-valued function on R and let $r_1, r_2, \ldots, r_n, \ldots$ be the rationals in some fixed sequential order. Consider $S = \{t:\ x(t^+)$ and $x(t^-)$ exist, with $x(t^+) > x(t^-)\}$. For $t_0 \in S$, let $r_i(t_0)$ be the first rational in the sequence such that $x(t_0^-) < r_i < x(t_0^+)$. Now define r_j to be the first rational in the sequence satisfying $t \in (r_j, t_0) \Rightarrow x(t) < r_i$, and define r_k similarly. Show that the mapping $t \longrightarrow (i, j, k)$ of S into $I^+ \times I^+ \times I^+$ is one-to-one and, hence, that S is countable. Now prove that the number of discontinuities of x of type (i) is countable.

10. Let the collection $PC[a, b]$ consist of all real-valued functions on $[a, b]$ which are continuous except at finitely many points where they have discontinuities of type (i). Show that $S[a, b] \subset PC[a, b] \subset R[a, b]$ and hence $\overline{PC[a, b]} = R[a, b]$. Show that any function in $PC[a, b]$ is the sum of a continuous function and a step function. Is it true that $PC[a, b] = C[a, b] \oplus S[a, b]$?

Section 8 FUNCTIONS OF BOUNDED VARIATION

The notion of the *variation* of a real-valued function on $[0, 1]$ is designed to measure the amount of oscillation of the values of the function. It is defined as follows. Let $\pi = \{0 = t_0, t_1, \ldots, t_n = 1\}$ be a finite set of points in $[0, 1]$, and let

$$v(\pi, x) = \sum_{k=1}^{n} |x(t_k) - x(t_{k-1})|.$$

The quantity

$$V[x] = \sup_{\pi} v(\pi, x)$$

is called the *total variation* of $x(t)$ on the interval $[0, 1]$. We now define $BV[0, 1]$ to be the set of all real functions x on $[0, 1]$ such that $V[x] < \infty$. Several elementary facts are apparent.

(a) x is constant $\Leftrightarrow V[x] = 0$.
(b) $x \in BV[0, 1] \Rightarrow x$ is bounded.
(c) $V[\alpha x] = |\alpha| V[x]$.
(d) $V[x + y] \leq V[x] + V[y]$.

To prove (d) we write

$$v(\pi, x + y) = \sum_{k=1}^{n} |[x(t_k) + y(t_k)] - [x(t_{k-1}) + y(t_{k-1})]|$$

$$\leq \sum_{k=1}^{n} |x(t_k) - x(t_{k-1})| + \sum_{k=1}^{n} |y(t_k) - y(t_{k-1})|$$

or, $v(\pi, x + y) \leq v(\pi, x) + v(\pi, y) \leq V[x] + V[y]$, whence the desired inequality follows.

(e) $BV[0, 1]$ is a vector space, a subspace of $B_R[0, 1]$. [This follows from (b), (c), and (d).]

(f) $BV[0, 1]$ contains all monotone functions, and for such functions, $V(x) = |x(0) - x(1)|$.

It is intuitively plausible that a function of bounded variation can have no oscillatory discontinuities. The following lemma makes this precise.

Lemma 4.8.1. $BV[0, 1]$ is a proper subspace of $R[0, 1]$.

Proof. It t_0 is an oscillatory discontinuity of x, then there exists an $\epsilon_0 > 0$ such that, for any $\delta > 0$, two points t_1 and s_1 can be found satisfying $|t_0 - t_1| \leq \delta$, $|t_0 - s_1| \leq \delta$, and $|x(t_1) - x(s_1)| \geq \epsilon_0$. By using a finite sequence of δ's, we find a monotone sequence $t_1, s_1, t_2, s_2, \ldots, t_n, s_n$ such that $|x(t_k) - x(s_k)| \geq \epsilon_0$, and if these points are taken, with 0 and 1, for a subdivision π, we get $v(\pi, x) \geq n\epsilon_0$. Since n is arbitrary, x is not in $BV[0, 1]$. Hence $x \in BV[0, 1]$ implies that x has no oscillatory discontinuities. This completes the proof.

The function in the illustrative example of the preceding section, namely,

$$x(t) = \begin{cases} 0, & \text{for } t = 0 \text{ or } t \text{ irrational} \\ \dfrac{1}{q}, & \text{for } t = \dfrac{p}{q} \end{cases}$$

is in $R[0, 1]$ but is not in $BV[0, 1]$, for there exists a sequence $\{\pi_n\}$ such that $v(\pi_n, x) = 1 + 1/2 + 1/3 + \cdots + 1/n$. The details are left as an exercise.

Clearly, $B_R[0, 1] \supset R[0, 1] \supset BV[0, 1] \supset S[0, 1]$, where all of the inclusions are proper. We could use the sup norm on $BV[0, 1]$, in which case it would be an *incomplete* normed vector space dense in $R[0, 1]$. A different norm is usually employed; it is based on the total variation function. Indeed, if we define, for $x \in BV[0, 1]$

$$\|x\| = |x(0)| + V[x]$$

it is trivial, using (a) (b), and (c), to verify that this is a legitimate norm. The $|x(0)|$ term is used to obtain $[\|x\| = 0] \Rightarrow [x = 0]$. The following theorem shows the advantage of using this norm.

Theorem 4.8.2. $BV[0, 1]$ is a Banach space under the norm $\|x\| = |x(0)| + V[x]$.

The completeness remains to be proved; to do this we need the following lemma.

Lemma 4.8.3. If the sequence $\{x_n\}$ is in $BV[0, 1]$, and if it converges *pointwise* to x, then $V[x] \leq \varliminf_{n \to \infty} V[x_n]$.

Proof. Let $\pi = \{0 = t_0, t_1, \ldots, t_n = 1\}$; from

$$|x(t_k) - x(t_{k-1})| \leq |x(t_k) - x_m(t_k)| + |x_m(t_k) - x_m(t_{k-1})| + |x_m(t_{k-1}) - x(t_{k-1})|,$$

we obtain

$$v(\pi, x) \leq v(\pi, x_m) + \sum_{k=1}^{n} [|x(t_k) - x_m(t_k)| + |x_m(t_{k-1}) - x(t_{k-1})|].$$

Since $x_m(t) \to x(t)$ pointwise, we can find, for any $\epsilon > 0$, an $N(\epsilon)$ such that $m \geq N(\epsilon)$ implies

$$v(\pi, x) \leq v(\pi, x_m) + \epsilon \leq V[x_m] + \epsilon.$$

We then have

$$v(\pi, x) \leq \varliminf_{n \to \infty} V[x_n],$$

and hence,

$$V[x] \leq \varliminf_{n \to \infty} V[x_n].$$

as desired.

Proof of Theorem 4.8.2. Let $\{x_n\}$ be a Cauchy sequence in $BV[0, 1]$, i.e., $\{|x_n(0) - x_m(0)| + V[x_m - x_n]\} \to 0$. Then $\{x_n(0)\}$ is Cauchy in R and has a limit which we call $x(0)$. For $t_0 > 0$, we have

$$|x_m(t_0) - x_n(t_0)| \leq |[x_m(t_0) - x_n(t_0)] - [x_m(0) - x_n(0)]| + |x_m(0) - x_n(0)|$$
$$\leq V[x_m - x_n] + |x_m(0) - x_n(0)| = \|x_m - x_n\|,$$

so that $\{x_n(t_0)\}$ is also Cauchy in R and has a limit which we call $x(t_0)$. Hence the sequence $\{x_n\}$ converges not only pointwise but also uniformly to a function x.

Now, since Cauchy sequences in any metric space are bounded, the set $\{|x_n(0)| + V[x_n]\}$ is bounded, and hence $\underset{n \to \infty}{\text{limit}} V[x_n]$ is finite. The lemma then guarantees that x is in $BV[0, 1]$.

We know that $|x_n(0) - x(0)| \to 0$, so that to prove $\|x_n - x\| \to 0$, we must show $V[x_n - x] \to 0$. Now, the function $x_m - x_n$ converges pointwise to the function $x_n - x$ as $m \to \infty$; therefore, according to the lemma, $V[x_n - x] \le \underset{m \to \infty}{\text{limit}} V[x_m - x_n]$. Since $V[x_m - x_n] \to 0$, we can find an $N(\epsilon)$ such that $m, n \ge N(\epsilon)$ implies $V[x_m - x_n] \le \epsilon$; hence $\underset{m \to \infty}{\text{limit}} V[x_m - x_n] \le \epsilon$ for $n \ge N(\epsilon)$. This gives $V[x_n - x] \to 0$ as desired, and the proof is finished.

Corollary. Using $\|x\| = |x(0)| + V|x|$ and $\|x\|_s = \underset{t \in [0, 1]}{\sup} |x(t)|$, we have $\|x\|_s \le \|x\|$, so that $[\|x_n - x\| \to 0] \Rightarrow [\|x_n - x\|_s] \to 0$. That is, convergence in $BV[0, 1]$ is *stronger* than convergence in the sup norm.

It is useful to split $V[x]$, the variation of x, into two parts, the positive and negative variations of x. This is done by defining

$$p(\pi, x) = \sum_{k=1}^{n} \max \{[x(t_k) - x(t_{k-1})], 0\}$$

$$n(\pi, x) = - \sum_{k=1}^{n} \min \{[x(t_k) - x(t_{k-1})], 0\}$$

so that $v(\pi, x) = p(\pi, x) + n(\pi, x)$ and $x(1) - x(0) = p(\pi, x) - n(\pi, x)$ for all π. We now define the positive and negative variations of x by

$$P[x] = \sup_{\pi} p(\pi, x), \quad \text{and}$$

$$N[x] = \sup_{\pi} n(\pi, x).$$

Lemma 4.8.4. For $x \in BV[0, 1]$, $V[x] = P[x] + N[x]$ and $x[1] - x[0] = P[x] - N[x]$.

Proof. From above, we have

$$p(\pi, x) = n(\pi, x) + x(1) - x(0) \le N[x] + x(1) - x(0),$$

so that

$$P[x] \le N[x] + x(1) - x(0) \quad \text{and} \quad P[x] - N[x] \le x(1) - x(0).$$

By a similar argument, we can obtain

$$N[x] \le P[x] - x(1) + x(0) \quad \text{and} \quad P[x] - N[x] \ge x(1) - x(0),$$

so that

$$P[x] - N[x] = x(1) - x(0).$$

From $v(\pi, x) = p(\pi, x) + n(\pi, x)$, we obtain $V[x] \leq P[x] + N[x]$, and

$$p(\pi, x) + n(\pi, x) \leq V[x] \quad \text{or} \quad V[x] \geq 2p(\pi, x) - x(1) + x(0).$$

Hence

$$V[x] \geq 2P[x] - (P[x] - N[x]) = P[x] + N[x],$$

and the proof is complete.

We now define

$$v(t) = V_0^t[x]$$
$$p(t) = P_0^t[x]$$
$$n(t) = N_0^t[x],$$

where the superscript t and subscript 0 indicate that we are restricting x to the interval $[0, t]$. These functions are monotone nondecreasing and take on the value 0 at $t = 0$ (and are hence nonnegative); p and n are called the positive and negative parts of x. Application of the preceding lemma to the interval $[0, t]$ yields the following theorem.

Theorem 4.8.5. A function x is in $BV[0, 1]$ if and only if it can be written as the difference of two bounded nonnegative monotone nondecreasing functions.

In fact, $x(t) = x(0) + p(t) - n(t)$.

To show that $x_1, x_2 \in BV[0, 1] \Rightarrow x_1 x_2 \in BV[0, 1]$ and indeed that $\|x_1 x_2\| \leq \|x_1\| \|x_2\|$ we write

$$x_1(t) = x_1(0) + p_1(t) - n_1(t),$$

$$x_2(t) = x_2(0) + p_2(t) - n_2(t),$$

$$|x_1(0)| + V[x_1] = \|x_1\| = |x_1(0)| + p_1(1) + n_1(1), \quad \text{and}$$

$$|x(0)| + V[x_2] = \|x_2\| = |x_2(0)| + p_2(1) + n_2(1).$$

Moreover,

$$\begin{aligned}
x_1(t)x_2(t) = \ & x_1(0)x_2(0) + x_1(0)p_2(t) - x_1(0)n_2(t) + x_2(0)p_1(t) \\
& + p_1(t)p_2(t) - p_1(t)n_2(t) - x_2(0)n_1(t) - p_2(t)n_1(t) \\
& + n_1(t)n_2(t),
\end{aligned}$$

from which it follows that $x_2x_2 \in BV[0, 1]$, because the right member is the difference of nondecreasing functions. Moreover,

$$
\begin{aligned}
||x_1x_2|| &= |x_1(0)x_2(0)| + V[x_1x_2] \le |x_1(0)x_2(0)| + |x_1(0)|p_2(1) \\
&\quad + |x_1(0)|n_2(1) + |x_2(0)|p_1(1) + p_1(1)p_2(1) \\
&\quad + p_1(1)n_2(1) + |x_2(0)|n_1(1) + p_2(1)n_1(1) \\
&\quad + n_1(1)n_2(1) \\
&= ||x_1|| \, ||x_2||,
\end{aligned}
$$

as desired. (We used (c) and (d) in this calculation.) In summary:

Theorem 4.8.6. $BV[0, 1]$ is a Banach algebra under $||x|| = |x(0)| + V[x]$.

EXERCISES

1. a. Show that if x has a bounded derivative on $[0, 1]$ then x satisfies the Lipschitz condition $|x(t_1) - x(t_2)| \le M|t_1 - t_2|$.
    b. Show that any function satisfying a Lipschitz condition on $[0, 1]$ is in $BV[0, 1]$.

2. a. Show that

    $$
    x(t) = \begin{cases} t \sin \dfrac{1}{t}, & 0 < t \le 1 \\ 0, & 0 = t \end{cases}
    $$

    is in $C[0, 1]$ but not in $BV[0, 1]$.
    b. Show that

    $$
    x(t) = \begin{cases} t^2 \sin \dfrac{1}{t^2}, & 0 < t \le 1 \\ 0, & 0 = t \end{cases}
    $$

    is differentiable at every point of $[0, 1]$ but is not in $BV[0, 1]$. Show that x' is discontinuous at $t = 0$ and is not bounded.
    c. Let

    $$
    x(t) = \begin{cases} t^{3/2} \sin \dfrac{1}{t}, & 0 < t \le 1 \\ 0, & 0 = t. \end{cases}
    $$

    Show that x' exists on $[0, 1]$, $x \in BV[0, 1]$, but x' is not bounded, and hence, is not continuous.

3. Let

$$x(t) = \begin{cases} 0, & \text{for } t \text{ irrational and } t = 0 \\ \dfrac{1}{q^3}, & \text{for } t = \dfrac{p}{q}, \quad t \neq 0. \end{cases}$$

Show that $x \in BV[0, 1]$.

4. Let α_n be any sequence in R such that $\sum\limits_{n=1}^{\infty} |\alpha_n|$ converges, and let $\{t_n\}$ be any sequence of distinct points in $[0, 1]$. Define

$$x(t) = \begin{cases} \sum\limits_{t_n < t} \alpha_n, & 0 < t \leq 1 \\ 0, & t = 0. \end{cases}$$

Prove $x \in BV[0, 1]$, x is continuous at points not in $\{t_n\}$, and x has a jump discontinuity of amount α_n at t_n (i.e., $|x(t_n^+) - x(t_n^-)| = |\alpha_n|$).

5. We proved $R[0, 1] \supset BV[0, 1] \supset S[0, 1]$. Let $S_1[0, 1]$ be the set of step functions satisfying $s(0) = 0$. If the norm $\|x\| = |x(0)| + V[x]$ is used in $S_1[0, 1]$, characterize the closure of $S_1[0, 1]$. (It must be a subspace of $BV[0, 1]$.)

6. Let $\{t_n\}$ be the (countable) set of discontinuities of $x \in BV[0, 1]$. Define

$$s(t) = \begin{cases} x(0^+) - x(0) + \sum\limits_{t_n < t} [x(t_n^+) - x(t_n^-)] + x(t) - x(t^-), \\ \qquad \text{for} \quad 0 < t \leq 1 \\ 0, \qquad \text{for} \quad t = 0. \end{cases}$$

a. Show that $s \in \overline{S_1[0, 1]}$ (see exercise 5).
b. Show that $x - s$ is continuous.
c. Show that $BV[0, 1]$ is the direct sum of $\overline{S_1[0, 1]}$ and $BV[0, 1] \cap C[0, 1]$. Is $BV[0, 1] \cap C[0, 1]$ closed in $BV[0, 1]$? (The norm of $BV[0, 1]$ is being used on $BV[0, 1] \cap C[0, 1]$.)

7. Find a sequence in $BV[0, 1] \cap C[0, 1]$ which converges uniformly to a function not in $BV[0, 1] \cap C[0, 1]$.

8. Let $\{x_n\}$ be a sequence of bounded, monotone nondecreasing functions converging pointwise to x. Show that x is monotone nondecreasing and bounded.

9. Show that $\overline{S_1[0, 1]}$ is not separable (see exercise 5).

10. Let x be a mapping of $[0, 1]$ into $R^{(n)}$ such that $\sup\limits_{\pi} v(\pi, x)$ is finite.

[Here $v(\pi, x) = \sum_{k=1}^{n} \| x(t_k) - x(t_{k-1}) \|$, where the Euclidean norm is used in $R^{(n)}$.]

a. Show that if x is given by the n-tuple of real-valued functions $[x_1, \ldots, x_n]$ then $x_k \in BV[0, 1]$ for $k = 1, \ldots, n$.

b. Show conversely that, if $x = [x_1, \ldots, x_n]$ where $x_k \in BV[0, 1]$ for $k = 1, \ldots, n$, then $\sup_{\pi} v(\pi, x) < \infty$.

c. A continuous function of this type is said to define a *rectifiable* curve C in $R^{(n)}$, with length $L(C) = \sup_{\pi} v(\pi, x)$. Show that x is continuous if and only if x_k is continuous for $k = 1, \ldots, n$.

11. a. Let $y \in BV[0, 1]$, and assume $0 \le \alpha < \beta < \gamma \le 1$. Show that
$$V_\alpha^\gamma[x] = V_\alpha^\beta[x] + V_\beta^\gamma[x].$$

b. Let x define a rectifiable curve in $R^{(n)}$, and let L_α^γ be the length of that part of the curve defined by $\alpha \le t \le \gamma$. Show that
$$L_\alpha^\gamma = L_\alpha^\beta + L_\beta^\gamma, \qquad \text{for} \quad \alpha < \beta < \gamma.$$

12. a. Show that if x' is continuous on $[0, 1]$ then $V[x] = \int_0^1 |x'(t)| \, dt$ (see exercise 1).

b. Formulate and prove a similar result for curves in $R^{(n)}$ (see exercises 10 and 11).

13. Prove that, if $x \in BV[0, 1]$ and if x is continuous at t_0, then v is continuous at t_0, and hence, p and n are also continuous at t_0.

14. Let $x \in BV[0, 1]$ and let $x(t) - x(0) = p(t) - n(t)$ as in the text. Suppose p_1 and n_1 are nondecreasing functions such that, for all t, $x(t) = p_1(t) - n_1(t)$. Show that $V(p) \le V(p_1)$ and $V(n) \le V(n_1)$. In the case that $V(p) = V(p_1)$ and $V(n) = V(n_1)$, can anything further be proved?

15. Let b_v be the set of all sequences $x = \{\alpha_n\}$ of real numbers for which $d(x, 0) = |\alpha_1| + \sum_{k=1}^{\infty} |\alpha_{k+1} - \alpha_k| < \infty$ and define $d(x, y) = d(x - y, 0)$. Show that b_v is a normed vector space under $\|x\| = d(x, 0)$. Is it a Banach algebra?

Section 9   THE $l^{(p)}$ SPACES, $p \ge 1$.

In section 1 (example 4) we defined, for $p > 0$, $l^{(p)}$ to be the set of all sequences $\{\alpha_n\}$ of scalars such that $\sum_{n=1}^{\infty} |\alpha_n|^p$ converges, and we showed that $l^{(p)}$

is a vector space under coordinatewise addition and multiplication by scalars. For $p = 1$, it is trivial to verify that $l^{(1)}$ is a normed vector space under $\|\{\alpha_n\}\| = \sum\limits_{k=1}^{\infty} |\alpha_n|$. For $p > 1$, it is also possible to introduce a norm, although the procedure is not obvious and is based on the inequalities of Hölder and Minkowski.

Lemma 4.9.1. (The Hölder Inequality). Let the positive real numbers p and q satisfy $1/p + 1/q = 1$, and let $(\alpha_1, \ldots, \alpha_n)$ and $(\beta_1, \ldots, \beta_n)$ be any two n-tuples of scalars. Then

$$\sum_{k=1}^{n} |\alpha_k \beta_k| \leq \left[\sum_{k=1}^{n} |\alpha_k|^p \right]^{1/p} \left[\sum_{k=1}^{n} |\beta_k|^q \right]^{1/q}.$$

Proof. We note first that the inequality is homogeneous; therefore, if it is true for $(\alpha_1, \ldots, \alpha_n)$ and $(\beta_1, \ldots, \beta_n)$, then it is also true for $(\lambda\alpha_1, \ldots, \lambda\alpha_n)$ and $(\mu\beta_1, \ldots, \mu\beta_n)$. Hence it suffices to prove that

$$\sum_{k=1}^{n} |\alpha_k \beta_k| \leq 1 \quad \text{whenever} \quad \sum_{k=1}^{n} |\alpha_k|^p = \sum_{k=1}^{n} |\beta_k|^q = 1.$$

(The case where the right member of the inequality vanishes is trivial.)

Consider the curve $\beta = \alpha^{p-1}$ or, equivalently, $\alpha = \beta^{q-1}$. We have

$$A_1 = \int_0^{|\alpha_k|} \alpha^{p-1} \, d\alpha = \frac{|\alpha_k|^p}{p}$$

$$B_1 = \int_0^{|\beta_k|} \beta^{q-1} \, d\beta = \frac{|\beta_k|^q}{q},$$

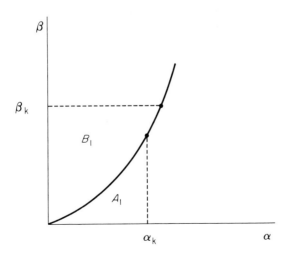

and the inequality $A_1 + B_1 \geq |\alpha_k \beta_k|$ becomes

$$|\alpha_k \beta_k| \leq \frac{1}{p}|\alpha_k|^p + \frac{1}{q}|\beta_k|^q.$$

Summing from $k = 1$ to $k = n$, we obtain

$$\sum_{k=1}^{n} |\alpha_k \beta_k| \leq \frac{1}{p}\left[\sum_{k=1}^{n}|\alpha_k|^p\right] + \frac{1}{q}\left[\sum_{k=1}^{n}|\beta_k|^q\right] = \frac{1}{p} + \frac{1}{q} = 1,$$

as desired, and the proof is complete.

Corollary. If $x = (\alpha_1, \alpha_2, \ldots, \alpha_n, \ldots) \in l^{(p)}$ and $y = (\beta_1, \beta_2, \ldots, \beta_n, \ldots) \in l^{(q)}$, then $(\alpha_1\beta_1, \alpha_2\beta_2, \ldots, \alpha_n\beta_n, \ldots) \in l^{(1)}$.

Corollary. The Hölder inequality becomes an equality if and only if there is a scalar γ such that $|\beta_k|^q = \gamma |\alpha_k|^p$ for $k = 1, 2, \ldots, n$.

The proofs of the corollaries are left as exercises.

Lemma 4.9.2 (The Minkowski Inequality). Let $(\alpha_1, \ldots, \alpha_n)$ and $(\beta_1, \ldots, \beta_n)$ be any two n-tuples of scalars and assume $p \geq 1$. Then

$$\left[\sum_{k=1}^{n}|\alpha_k + \beta_k|^p\right]^{1/p} \leq \left[\sum_{k=1}^{n}|\alpha_k|^p\right]^{1/p} + \left[\sum_{k=1}^{n}|\beta_k|^p\right]^{1/p}.$$

Proof. The case $p = 1$ is left as an exercise. For $p > 1$, we have

$$(|\alpha_k| + |\beta_k|^p) = (|\alpha_k| + |\beta_k|)^{p-1}|\alpha_k| + (|\alpha_k| + |\beta_k|)^{p-1}|\beta_k|$$

and, summing from $k = 1$ to $k = n$, we obtain

$$\sum_{k=1}^{n}(|\alpha_k| + |\beta_k|)^p = \sum_{k=1}^{n}(|\alpha_k| + |\beta_k|)^{p-1}|\alpha_k| + \sum_{k=1}^{n}(|\alpha_k| + |\beta_k|)^{p-1}|\beta_k|.$$

Applying the Hölder inequality to the two sums on the right, and using $q(p - 1) = p$, we have

$$\sum_{k=1}^{n}(|\alpha_k| + |\beta_k|)^{p-1}|\alpha_k| \leq \left[\sum_{k=1}^{n}(|\alpha_k| + |\beta_k|)^p\right]^{1/q}\left[\sum_{k=1}^{n}|\alpha_k|^p\right]^{1/p}$$

and

$$\sum_{k=1}^{n}(|\alpha_k| + |\beta_k|)^{p-1}|\beta_k| \leq \left[\sum_{k=1}^{n}(|\alpha_k| + |\beta_k|)^p\right]^{1/q}\left[\sum_{k=1}^{n}|\beta_k|^p\right]^{1/p}.$$

Hence, by addition,

$$\left[\sum_{k=1}^{n}(|\alpha_k|+|\beta_k|)^p\right] \le \left[\sum_{k=1}^{n}(|\alpha_k|+|\beta_k|)^p\right]^{1/q}\left[\left(\sum_{k=1}^{n}|\alpha_k|^p\right)^{1/p}+\left(\sum_{k=1}^{n}|\beta_k|^p\right)^{1/p}\right],$$

and we easily obtain the Minkowski inequality.

It follows at once from the lemma that, for $p \ge 1$, the space $l^{(p)}$ is a normed vector space under $\|\{\alpha_k\}\| = \left(\sum_{k=1}^{\infty}|\alpha_k|^p\right)^{1/p}$.

Corollary. For $p > 1$, the Minkowski inequality becomes an equality if and only if there is a real nonnegative scalar δ such that $\beta_k = \delta\alpha_k$, $k = 1, \ldots, n$.

Proof of the corollary is left as an exercise.

We come finally to the question of completeness. This is settled in the following theorem.

Theorem 4.9.3. $l^{(p)}$ is a Banach space ($p \ge 1$).

Proof. Let $\{x_n\}$ where $x_n = (\alpha_{n1}, \alpha_{n2}, \ldots, \alpha_{nk}, \ldots)$, be a Cauchy sequence in $l^{(p)}$. Then, for any $\epsilon > 0$, there is an $N(\epsilon)$ such that, for $m, n \ge N(\epsilon)$, we have $\|x_n - x_m\| \le \epsilon$, or

$$\left[\sum_{k=1}^{\infty}|\alpha_{nk} - \alpha_{mk}|^p\right]^{1/p} \le \epsilon.$$

It follows that $|\alpha_{nk} - \alpha_{mk}| \le \epsilon$ for every k whenever $m, n \ge N(\epsilon)$; therefore the sequence $\alpha_{1k}, \alpha_{2k}, \ldots, \alpha_{nk}, \ldots$ is a Cauchy sequence of scalars, and thus has a limit α_k. Let $x = (\alpha_1, \alpha_2, \ldots, \alpha_k, \ldots)$; we now show that $x \in l^{(p)}$ and $\|x_n - x\| \to 0$.

From the preceding we can write, for any M,

$$\|x_m - x_n\|^p = \sum_{k=1}^{M}|\alpha_{nk} - \alpha_{mk}|^p + \sum_{k=M+1}^{\infty}|\alpha_{nk} - \alpha_{mk}|^p \le \epsilon^p$$

whenever $m, n \ge N(\epsilon)$, so that

$$m, n \ge N(\epsilon) \Rightarrow \sum_{k=1}^{M}|\alpha_{nk} - \alpha_{mk}|^p \le \epsilon^p.$$

Since this is a finite sum we can obtain, by letting $n \to \infty$,

$$m \ge N(\epsilon) \Rightarrow \sum_{k=1}^{M}|\alpha_k - \alpha_{mk}|^p \le \epsilon^p,$$

and since this holds for all M, we find

$$(*) \qquad m \ge N(\epsilon) \Rightarrow \sum_{k=1}^{\infty}|\alpha_k - \alpha_{mk}|^p \le \epsilon^p.$$

Applying the Minkowski inequality, we have

$$\left(\sum_{k=1}^{M}|\alpha_k|^p\right)^{1/p}\le\left(\sum_{k=1}^{M}|\alpha_k-\alpha_{mk}|^p\right)^{1/p}+\left(\sum_{k=1}^{M}|\alpha_{mk}|^p\right)^{1/p}\le\epsilon+\|x_m\|$$

so that

$$\left(\sum_{k=1}^{\infty}|\alpha_k|^p\right)^{1/p}\le\epsilon+\|x_m\|$$

and, by the boundedness of Cauchy sequences, we see that $x\in l^{(p)}$.
The relation (*) now yields $\|x-x_m\|\to 0$, and the proof is complete.

EXERCISES

1. Prove the corollaries of lemma 4.9.1.

2. Prove the corollary to lemma 4.9.2.

3. a. Show that $l^{(p)}$, $p>1$, is strongly normed in the sense of exercise 6, section 3.
    b. Discuss $l^{(1)}$ in this regard.

4. In this exercise, $l^{(p)}$, $0<p<1$, is treated.
    a. Prove $|\alpha+\beta|^p\le|\alpha|^p+|\beta|^p$.
    b. Define $d(x,0)=\sum_{n=1}^{\infty}|\alpha_n|^p$ for $x=(\alpha_1,\alpha_2,\ldots,\alpha_n,\ldots)\in l^{(p)}$
    and $d(x,y)=d(x-y,0)$. Show that d is a metric on $l^{(p)}$ but not a norm.
    c. Discuss the completeness question.

5. Let p and q be any real numbers greater than 1. Show that

$$\{\alpha_n\}\in l^{(p)},\qquad\{\beta_n\}\in l^{(q)}\Rightarrow\{\alpha_n\beta_n\}\in l^{(r)}\qquad\text{where}\quad r=\frac{pq}{p+q}.$$

6. Let $l(I)$ be the set of all doubly infinite sequences,

$$(\ldots,\alpha_{-n},\alpha_{-n+1},\ldots,\alpha_{-1},\alpha_0\alpha_1,\ldots,\alpha_n,\ldots),$$

of complex numbers which are absolutely summable in the sense of exercise 6, section 2. Define addition and multiplication by scalars coordinatewise, and let $\|\{\alpha_n\}\|=\sum_n|\alpha_n|$.
    a. Show that $l(I)$ is a Banach space.
    b. For $\{\alpha_n\}$ and $\{\beta_n\}$ in $l(I)$, show that the sequence

$$(\ldots, \alpha_{n+k}\beta_{-k}, \ldots, \alpha_n\beta_0, \alpha_{n-1}\beta_1, \alpha_{n-2}\beta_2, \ldots, \alpha_{n-k}\beta_k, \ldots)$$

is in $l(I)$. Let $\gamma_n = \sum_k \alpha_{n+k}\beta_{-k}$ and show that $\{\gamma_n\}$ is in $l(I)$.

c. Show that $l(I)$ is a Banach algebra under $\{\alpha_n\}\{\beta_n\} = \{\gamma_n\}$.

Section 10 THE SPACES $C_0[0, \infty)$ AND $C_0(-\infty, \infty)$

Definition 4.10.1. The set of all continuous real-valued functions x on $[0, \infty)$ satisfying $\lim_{t \to \infty} x(t) = 0$ will be denoted by $C_0[0, \infty)$. Similarly, $C_0(-\infty, \infty)$ is the set of all real-valued continuous functions on $(-\infty, \infty)$ such that $\lim_{t \to \infty} x(t) = \lim_{t \to -\infty} x(t) = 0$.

Clearly, $C_0[0, \infty)$ and $C_0(-\infty, \infty)$ are Banach algebras under the sup norm. Our goal here is to explore analogs of the Weierstrass approximation theorem for these spaces. In the case of $C_0[0, \infty)$, we have the following theorem.

Theorem 4.10.2. For any fixed $\alpha > 0$, the set of all functions of the form $e^{-\alpha t}p(t)$, where $p(t)$ is a polynomial, is a dense subspace in the Banach algebra $C_0[0, \infty)$.

Note that, since polynomials are not in $C_0[0, \infty)$, the introduction of the exponential factor is a natural modification for the classical Weierstrass theorem, but the use of a fixed α prevents the designated set of functions from being a subalgebra.

We accomplish the proof by making a change of variable to map $[0, \infty)$ onto $(0, 1]$, and by then applying the Weierstrass theorem on $[0, 1]$. To be specific, let $s = e^{-\alpha t}$, or $t = -\dfrac{1}{\alpha} \log s$, and define

$$y(s) = \begin{cases} x\left(-\dfrac{1}{\alpha} \log s\right), & 0 < s \leq 1 \\ 0, & s = 0. \end{cases}$$

Since $\lim_{s \to 0} y(s) = \lim_{t \to \infty} x(t) = 0$, we have $y(s) \in C[0, 1]$. Therefore, for any $\epsilon > 0$, there exists a polynomial $p(s)$ such that

$$|y(s) - p(s)| \leq \epsilon, \qquad 0 \leq s \leq 1.$$

Hence, for $0 \leq t < \infty$,

$$|x(t) - p(e^{-\alpha t})| \le \epsilon,$$

and we have proved that any function $x \in C_0[0, \infty)$ can be uniformly approximated by functions of the type

$$\sum_{k=0}^{n} \beta_k e^{-k\alpha t}.$$

It is now only necessary to show that, for any $\alpha > 0$ and positive integer k, the function $e^{-\alpha k t}$ can be uniformly approximated on $[0, \infty)$ by functions of the type $e^{-\alpha t} p(t)$ where $p(t)$ is a polynomial.

Notice first that it suffices to consider the case $\alpha = 1$, for if

$$|e^{-kt} - e^{-t} p_1(t)| \le \epsilon, \qquad 0 \le t < \infty,$$

then we have

$$|e^{-k\alpha t} - e^{-\alpha t} p_1(\alpha t)| \le \epsilon,$$

and, letting $p_1(\alpha t) = p(t)$, we have

$$|e^{-k\alpha t} - e^{-\alpha t} p(t)| \le \epsilon.$$

Proceeding inductively, we consider the problem of approximating e^{-2t} by functions of the type $e^{-t} p(t)$. Let $P_N(t) = \sum_{k=0}^{N} (-1)^k t^k/k!$ be the Nth partial sum of the Taylor series for e^{-t}; these polynomials do not converge uniformly to e^{-t} on $[0, \infty)$, but it turns out that the sequence $\{e^{-t} P_N(t)\}$ does converge uniformly to e^{-2t} on $[0, \infty)$. This we now prove.

Let $y_N(t) = e^{-2t} - e^{-t} P_N(t)$; then, since $\lim_{t \to \infty} y_N(t) = 0$, $|y_N(t)|$ has an absolute maximum at some $t_0 \in [0, \infty)$, and at this point

$$0 = y_N'(t_0) = -2e^{-2t_0} + 2e^{-t_0} \sum_{k=0}^{N-1} \frac{(-t_0)^k}{k!} + e^{-t_0} \frac{(-t_0)^N}{N!}$$

or

$$0 = -2y_N(t_0) - e^{-t_0} \frac{(-t_0)^N}{N!}.$$

Thus

$$|y_N(t_0)| = \frac{1}{2} e^{-t_0} \frac{t_0^N}{N!},$$

and

$$\|y_N\| \leq \frac{1}{2N!} \sup\{e^{-t}t^N : \ 0 \leq t < \infty\}.$$

For $N > 1$ we have

$$(e^{-t}t^N)' = t^{N-1}e^{-t}(N - t)$$

so that

$$\sup\{e^{-t}t^N : \ 0 \leq t < \infty\} = e^{-N}N^N,$$

and hence

$$\|y_N\| \leq \frac{e^{-N}N^N}{2N!}.$$

At this point we invoke Stirling's formula (see *Differential and Integral Calculus*, Vol. I, R. Courant):

$$\sqrt{2\pi}\, N^{N+1/2}e^{-N} < N! < \sqrt{2\pi}\, N^{N+1/2}e^{-N}\left(1 + \frac{1}{4N}\right).$$

From this we have

$$\frac{1}{N!} < (2\pi)^{-1/2}N^{-N-1/2}e^N,$$

and hence

$$\|y_N\| \leq \tfrac{1}{2}e^{-N}N^N(2\pi)^{-1/2}N^{-N-1/2}e^N = (2\sqrt{2\pi}\,\sqrt{N})^{-1}.$$

Thus $\lim\limits_{N\to\infty} \|y_N\| = 0$, as advertised. This proves that, for any $\alpha > 0$, $\epsilon > 0$, there exists a polynomial $p(t)$ such that

$$|e^{-2\alpha t} - e^{-\alpha t}p(t)| \leq \epsilon$$

for $t \in [0, \infty)$.

Assume now it has been proved that, for any $\alpha > 0$, the function $e^{-n\alpha t}$ can be uniformly approximated by functions of the form $e^{-\alpha t}p(t)$ on $[0, \infty)$, where $p(t)$ is a polynomial.

From the foregoing we have, for any $\epsilon > 0$, a polynomial $p_1(t)$ such that

(a) $\qquad |e^{-2(n+1)(1/2)t} - e^{-(n+1)(1/2)t}p_1(t)| \leq \epsilon, \qquad 0 \leq t < \infty,$

Let M be the supremum of $e^{-(1/2)t}p_1(t)$ on $[0, \infty)$, and let $p_2(t)$ be a polynomial such that

$$|e^{-n(1/2)t} - e^{-(1/2)t}p_2(t)| \leq \frac{\epsilon}{2M}, \qquad 0 \leq t < \infty$$

(using the induction hypothesis).
   Then

$$|e^{-n(1/2)t}e^{-(1/2)t}p_1(t) - e^{-(1/2)t}e^{-(1/2)t}p_1(t)p_2(t)| \leq \frac{\epsilon}{2M} \cdot M,$$

and letting $p(t) = p_1(t)p_2(t)$, we have

(b) $$|e^{-(n+1)(1/2)t}p_1(t) - e^{-t}p(t)| \leq \frac{\epsilon}{2}.$$

Hence, using equations (a) and (b), we obtain

$$|e^{-(n+1)t} - e^{-t}p(t)| \leq \epsilon.$$

This completes the entire proof of theorem 4.10.2.

Turning now to the space $C_0(-\infty, \infty)$, we have this theorem.

Theorem 4.10.3. For any $\alpha > 0$, the set of functions of the form $e^{-\alpha t^2}p(t)$, where $p(t)$ is a polynomial, is a dense subspace of $C_0(-\infty, \infty)$.

To prove the theorem, we note that

$$\Phi: \quad t \longrightarrow (e^{-\alpha t^2}, te^{-\alpha t^2})$$

is a one-to-one bicontinuous mapping of R onto a bounded subset of $R^{(2)}$, and that if the origin is adjoined to the image set, the result, M, is closed. Now if $f(t) \in C_0(-\infty, \infty)$, the function

$$F(x, y) = \begin{cases} f[\Phi^{-1}(x, y)], & (x, y) = (e^{-\alpha t^2}, te^{-\alpha t^2}) \\ 0, & (x, y) = (0, 0) \end{cases}$$

is continuous on M; hence, by the results of section 6 (see exercise 8 of that section), it can be uniformly approximated by polynomials in two variables. Thus, for any $\epsilon > 0$, there exists a polynomial $P(x, y)$ such that

$$|F(x, y) - P(x, y)| \leq \epsilon, \qquad \text{for} \quad (x, y) \in M.$$

But this implies that

$$|f(t) - P(e^{-\alpha t^2}, te^{-\alpha t^2})| \leq \epsilon, \qquad \text{for} \quad t \in R.$$

where $P(x, y)$ is of the form

$$\sum_{m, n=1}^{p, q} \beta_{mn} x^m y^n,$$

and thus

$$P(e^{-\alpha t^2}, te^{-\alpha t^2}) = \sum_{m, n=1}^{p, q} \beta_{mn} e^{-\alpha(m+n)t^2} t^n.$$

It now will suffice to show that functions of the form $e^{-\alpha m t^2} t^n$ can be uniformly approximated by functions of the type $e^{-\alpha t^2} p(t)$ where $p(t)$ is a polynomial.

From our calculations for $C_0[0, \infty)$, we know that, for any $\epsilon > 0$, $\beta > 0$, and positive integer k, there is polynomial $q(s)$ such that

$$|e^{-\beta ks} - e^{\beta s} q(s)| \leq \epsilon$$

whenever $s \geq 0$. Letting $s = t^2$, we have

$$|e^{-\beta k t^2} - e^{-\beta t^2} r(t)| \leq \epsilon$$

for all t, where $r(t) = q(t^2)$. If we multiply by $e^{-\delta t^2} t^n$, we obtain

$$|e^{-(\delta + \beta k)t^2} t^n - e^{-(\beta + \delta)t^2}[t^n r(t)]| \leq \epsilon e^{-\delta t^2} |t|^n.$$

Now choose k, β, and δ so that

$$\beta + \delta = \alpha, \qquad \delta + \beta k = \alpha m;$$

This can be done in various ways, but we let $k = 2m - 1$, from which we find $\beta = \delta = \alpha/2$. This yields

$$|e^{-\alpha m t^2} t^n - e^{-\alpha t^2} p(t)| \leq \epsilon e^{-(\alpha/2)t^2} |t|^n$$

where $p(t) = t^n r(t)$. Since $e^{-(\alpha/2)t^2} |t|^n$ is a bounded function, the proof is complete.

EXERCISES

1. Consider the following spaces under $\|x\| = \sup\{|x(t)| : t \in R\}$:

   $B(-\infty, \infty) = $ all bounded real-valued functions on $(-\infty, \infty)$.
   $BS(-\infty, \infty) = $ all real-valued functions x on $(-\infty, \infty)$ for which there exist points $-\infty < t_1 < \cdots < t_n < \infty$ such that x is constant on each of the $n + 1$ intervals so defined.

$B_0S(-\infty, \infty) =$ those members of $BS(-\infty, \infty)$ which vanish on the intervals $(-\infty, t_1)$ and (t_n, ∞).

$BC(-\infty, \infty) =$ all bounded continuous functions on $(-\infty, \infty)$ $C_0(-\infty, \infty)$.

a. Show that all are normed algebras.

b. Show that $B(-\infty, \infty)$, $BC(-\infty, \infty)$, and $C_0(-\infty, \infty)$ are Banach algebras, and $B(-\infty, \infty) \supset BC(-\infty, \infty) \supset C_0(-\infty, \infty)$.

c. Show that $B(-\infty, \infty) \supset BS(-\infty, \infty) \supset B_0S(-\infty, \infty)$.

d. Show that the completion of $B_0S(-\infty, \infty)$ contains $C_0(-\infty, \infty)$.

e. Characterize (i.e., describe exactly) the completion of $B_0S(-\infty, \infty)$.

f. Characterize the completion of $BS(-\infty, \infty)$. Describe the intersection of this completion with $BC(-\infty, \infty)$.

g. Show that all *bounded* rational functions are in the intersection set obtained in part f.

5 Integration Theory

THE RIEMANN INTEGRAL

We begin by defining a restricted version of the Riemann integral which is applicable to a smaller class of functions than the usual Riemann integral treated in calculus. Recall that $S[a, b]$, the collection of all real-valued step functions on $[a, b]$, is a normed algebra under

$$||x|| = \sup \{|x(t)|: \quad a \le t \le b\},$$

and that it is dense in the Banach algebra $R[a, b]$, the set of all real-valued functions on $[a, b]$ which have no discontinuities of the oscillatory type. Alternatively, $R[a, b]$ consists of all functions which are uniform limits of step functions.

It is obvious how to define integration on $S[a, b]$. If x is a step function with value α_k on (t_{k-1}, t_k), $k = 1, \ldots, n$, then we let

$$I[x] = \sum_{k=1}^{n} \alpha_k(t_k - t_{k-1}),$$

ignoring the values of x at the points $t_0, t_1, \ldots, t_n$.

A little care is needed here to be sure that $I[x]$ is unambiguously defined, for a step function x can have different representations; however, such a function assumes only a finite number of distinct values $\beta_1 < \beta_2 < \cdots \beta_n$ (ignoring the values which are assumed at one point only). Now, each set

$$A_k = \{t: x(t) = \beta_k\}$$

is a finite union of open intervals (plus possibly a finite number of points which we again ignore) of total length λ_k. Then we have

$$I[x] = \sum_{k=1}^{m} \beta_k \lambda_k$$

by the definition above. If we now observe that each α_k in our former computation of $I[x]$ is a β_j, and if we group the terms in that expression, according to increasing magnitude of the α_k's, we find that the two values of $I[x]$ must agree.

Lemma 5.1.1. The function $I: \ x \longrightarrow I[x]$ is a continuous linear mapping of $S[a, b]$ onto R.

Proof. Clearly, $I[\alpha x] = \alpha I[x]$. If x is constant on the intervals (t'_{k-1}, t'_k), $k = 1, \ldots, n$, and y is constant on the intervals (t''_{k-1}, t''_k), $k = 1, \ldots, m$, then both functions are constant on the open intervals defined by all $m + n$ points. The identity $I[x + y] = I[x] + I[y]$ follows easily if this fact is used. The continuity of the linear function I is now a consequence of the following inequality, and theorem 4.3.1.

$$|I[x]| = \left| \sum_{k=1}^{n} \alpha_k(t_k - t_{k-1}) \right| \leq \sum_{k=1}^{n} |\alpha_k| |t_k - t_{k-1}|$$

$$\leq \left(\sup_{k=1}^{n} |\alpha_k| \right) \sum_{k=1}^{n} |t_k - t_{k-1}| = \|x\| (b - a).$$

Since a continuous linear function is automatically uniformly continuous, we can apply theorem 3.3.7 to obtain a continuous linear extension $\hat{I}$ of I mapping $R[a, b]$ onto R. (The theorem yields only the continuous extension $\hat{I}$; the linearity of $\hat{I}$ follows easily from that of I; see the exercises.)

The integral $\hat{I}$ will be called the *restricted* Riemann integral. We now connect it with the usual Riemann integral of calculus, which is defined as follows.

For a bounded function x on $[a, b]$, we associate with every partition

$$\pi: \ a = t_0 < t_1 < \cdots < t_{n-1} < t_n = b$$

the upper and lower sums

$$\bar{S}(\pi) = \sum_{k=1}^{n} M_k(t_k - t_{k-1})$$

$$\underline{S}(\pi) = \sum_{k=1}^{n} m_k(t_k - t_{k-1}),$$

where

$$M_k = \sup \{x(t): \ t_{k-1} \leq t \leq t_k\}$$
$$m_k = \inf \{x(t): \ t_{k-1} \leq t \leq t_k\}.$$

The upper and lower integrals of x are then defined by

$$\overline{\int_a^b} x \, dt = \inf_\pi \overline{S}(\pi)$$

$$\underline{\int_a^b} x \, dt = \sup_\pi \underline{S}(\pi).$$

Clearly,

$$\overline{S}(\pi) \geq \overline{\int_a^b} x \, dt \geq \underline{\int_a^b} x \, dt \geq \underline{S}(\pi)$$

for all π. If the upper and lower integrals are equal, then x is said to be Riemann integrable, and the common value of these integrals is called $\int_a^b x \, dt$, the Riemann integral of x on $[a, b]$.

Note that the upper and lower integrals are defined on the collection of all bounded functions on $[a, b]$, so that the set of Riemann integrable functions is a subset of this collection. We now recall three theorems from advanced calculus.

(I) A bounded function x on $[a, b]$ is Riemann integrable if and only if, for every $\epsilon > 0$, there exists a partition π_ϵ of $[a, b]$ such that

$$\overline{S}(\pi_\epsilon) - \underline{S}(\pi_\epsilon) \leq \epsilon.$$

(II) A bounded function x on $[a, b]$ is Riemann integrable if and only if, for every $\epsilon > 0$, there is a $\delta(\epsilon)$ such that $\overline{S}(\pi) - \underline{S}(\pi) \leq \epsilon$ whenever $\pi = \{a = t_0 < t_1 < \cdots < t_{n-1} < t_n = b\}$ is a partition satisfying

$$\max \{b - t_{n-1}, t_{n-2} - t_{n-3}, \ldots, t_2 - t_1, t_1 - a\} \leq \delta(\epsilon).$$

(III) If $\{x_n\}$ is a sequence of Riemann integrable functions on $[a, b]$ converging uniformly to the function x, then x is Riemann integrable on $[a, b]$ and

$$\int_a^b x \, dt = \lim_{n \to \infty} \int_a^b x_n \, dt.$$

Now, step functions are Riemann integrable in the general and in the restricted sense (to the same value). The continuity of the restricted Riemann integral on $R[a, b]$ and the third result cited above together guarantee that all members of $R[a, b]$ are Riemann integrable in the general sense, and the two integrals agree for such functions. Thus, the Riemann integral is an extension of the restricted Riemann integral. The problem now is to characterize the class of Riemann integrable functions.

The key to the matter lies in the nature of the set D of points of discontinuity of x; if D is not "too large" in some sense, then x is integrable. The precise definition of this notion is as follows.

Definition 5.1.2. A subset D of R is said to have Lebesgue measure zero if, for any $\epsilon > 0$, there is a countable set $\{I_k\}$ of intervals such that

$$\bigcup_{k=1}^{\infty} I_k \supseteq D \quad \text{and} \quad \sum_{k=1}^{\infty} l(I_k) < \epsilon,$$

where $l(I_k)$ is the length of the interval I_k.

EXAMPLE. Any countable set $\{t_n\}$ has Lebesgue measure zero, because the point t_n can be enclosed in an interval of length $\epsilon/2^n$, and $\sum_{n=1}^{\infty} \epsilon/2^n = \epsilon$.

We will need the following lemma.

Lemma 5.1.3. If $E = \bigcup_{n=1}^{\infty} E_n$, where each E_n has Lebesgue measure zero, then E has Lebesgue measure zero.

Proof. Given $\epsilon > 0$, the set E_1 can be enclosed in $\bigcup_{k=1}^{\infty} I_{1k}$, where $\sum_{k=1}^{\infty} l(I_{1k}) \leq \epsilon/2$, and in general, the set E_m can be enclosed in $\bigcup_{k=1}^{\infty} I_{mk}$, where $\sum_{k=1}^{\infty} l(I_{mk}) \leq \epsilon/2^m$. The intervals $\{I_{mk}\}$, $m, k = 1, 2, \ldots$ then satisfy

$$\bigcup_{m,k=1}^{\infty} I_{mk} \supset \bigcup_{m=1}^{\infty} E_m = E, \quad \text{and}$$

$$\sum_{m,k=1}^{\infty} l(I_{mk}) \leq \frac{\epsilon}{2} + \frac{\epsilon}{2^2} + \cdots + \frac{\epsilon}{2^m} + \cdots = \epsilon.$$

This finishes the proof.

In terms of this concept, we can give a very elegant characterization of Riemann integrability.

Theorem 5.1.4. A bounded function x on $[a, b]$ is Riemann integrable if and only if the set D of its discontinuities has measure zero.

Proof. Define the function

$$w(t) = \overline{\lim_{s \to t}} \, x(s) - \underline{\lim_{s \to t}} \, x(s).$$

Clearly, x is continuous at t if and only if $w(t) = 0$. The set $E_n = \{t: \ w(t) \geq 1/n\}$ is closed, and the set D of points where x is discontinuous is $\bigcup_{n=1}^{\infty} E_n$.

Now assume x is integrable, so that for any $\epsilon > 0$ there is a π_ϵ such that $\bar{S}(\pi_\epsilon) - \underline{S}(\pi_\epsilon) \leq \epsilon$. If

$$\pi_\epsilon = \{a = t_0, t_1, \ldots, t_{n-1}, t_n = b\}$$

then the set of intervals

$$I_1 = (t_0, t_1), \ldots, I_n = (t_{n-1}, t_n)$$

can be split into two groups, where the intervals in the first group have non-empty intersection with E_m, and those of the second group do not meet E_m (m is fixed). Then

$$\bar{S}(\pi_\epsilon) - \underline{S}(\pi_\epsilon) = \sum' (M_k - m_k)(t_k - t_{k-1}) + \sum'' (M_k - m_k)(t_k - t_{k-1}) \leq \epsilon,$$

where the prime (double prime) indicates summation over intervals of the first group (second group). On the primed intervals, $M_k - m_k \geq 1/m$, so that

$$\frac{1}{m} \sum'(t_k - t_{k-1}) \leq \sum' (M_k - m_k)(t_k - t_{k-1}) \leq \epsilon$$

and $\sum' (t_k - t_{k-1}) \leq m\epsilon$. Thus E_m can be enclosed in finitely many intervals of length less than or equal to $m\epsilon$, and since ϵ is arbitrary, we see that E_m has measure zero. It now follows from the preceding lemma that D has zero measure.

Conversely, assume that $D = \bigcup_{n=1}^{\infty} E_n$ has measure zero and, hence, that E_n has measure zero for all n. Choose N so that $1/N < \epsilon$. The set $\tilde{E}_N$ is open; therefore $\tilde{E}_N = \bigcup_{k=1}^{\infty} I_k$, where the intervals $\{I_k\}$ are open and disjoint. We now show that $\sum_{k=1}^{\infty} l(I_k) = b - a$.

Suppose, on the contrary, that

$$\sum_{k=1}^{\infty} l(I_k) = \eta < b - a.$$

Since E_N has zero measure, we can find intervals $\{J'_k\}$ such that

$$\bigcup_{k=1}^{\infty} J'_k \supset E_N \quad \text{and} \quad \sum_{k=1}^{\infty} l(J'_k) < \frac{b - a - \eta}{2}.$$

The intervals $\{J'_k\}$ and $\{I_k\}$ together cover $[a, b]$, and hence

$$b - a \leq \sum_{k=1}^{\infty} l(J'_k) + \sum_{k=1}^{\infty} l(I_k) < \frac{b - a - \eta}{2} + \eta = \frac{b - a + \eta}{2} < b - a.$$

This contradiction shows that $\eta = b - a$. We now see that, for some n,

$$\sum_{k=1}^{n} l(I_k) > b - a - \frac{\epsilon}{2}$$

and we can find *closed* intervals $J_1, \ldots, J_n$ such that

$$J_i \subset I_i, \qquad i = 1, \ldots, n \quad \text{and} \quad \sum_{i=1}^{n} l(J_i) > b - a - \epsilon.$$

At each point t_0 of J_1 we have $w(t_0) < 1/N < \epsilon$, and there is $\delta(t_0)$ such that, on the interval $[t_0 - \delta(t_0), t_0 + \delta(t_0)]$, the supremum of x minus the infimum of x is less than ϵ. By the compactness of J_1, finitely many such intervals cover J_1. We can use the endpoints of these intervals to form a partition π_1 of J_1, and similarly, we form partitions $\pi_2, \ldots, \pi_n$ of $J_2, \ldots, J_n$, thus obtaining a partition π of $[a, b]$.

Calculating $\bar{S}(\pi) - \underline{S}(\pi)$, we have

$$
\begin{aligned}
\bar{S}(\pi) - \underline{S}(\pi) &= \sum (M_k - m_k)(t_k - t_{k-1}) \\
&= [\sum_{\pi_1} (M_k - m_k)(t_k - t_{k-1}) + \cdots + \sum_{\pi_n} (M_k - m_k)(t_k - t_{k-1})] \\
&\quad + \sum{}' (M_k - m_k)(t_k - t_{k-1}) \\
&\leq \epsilon \sum_{k=1}^{n} l(J_k) + \epsilon 2M \leq \epsilon(2M + b - a),
\end{aligned}
$$

where $M = \sup \{|x(t)|: \ a \leq t \leq b\}$. It follows that x is Riemann integrable on $[a, b]$. This completes the entire proof.

The set of all Riemann integrable functions on $[a, b]$ is a Banach algebra of functions (see exercise 3).

EXERCISES

1. Let W be a dense subalgebra of a normed algebra V, and let f be a continuous linear function from W into a Banach algebra B.

    a. Show that f has a unique linear continuous extension F mapping V into B. (See theorem 3.3.7.)
    b. Show that if f satisfies the identity $f(xy) = f(x)f(y)$, then F also satisfies this identity.
    c. Apply part a to the mapping I of this section.

2. Prove directly the three theorems cited from advanced calculus.

3. Let V be the set of all Riemann integrable functions.

a. Show that V is a commutative algebra.

b. $V \subset B[a, b]$, the Banach algebra of all bounded functions on $[a, b]$. Show that V is a closed subalgebra of $B[a, b]$ and, thus, that V is itself a Banach algebra.

4. Let

$$x(t) = \begin{cases} \dfrac{1}{q}, & \text{for } 1 \geq t = \dfrac{p}{q} > 0, \text{ where } p \text{ and } q \text{ have no} \\ & \text{common factor} \\ 0, & \text{for } t = 0 \text{ and } t \text{ irrational.} \end{cases}$$

Find a sequence of step functions converging uniformly to x. Find $\displaystyle\int_0^1 x \, dt$.

5. Let V be the set of all Riemann integrable functions on $[a, b]$. Show that $x \in V$ implies that $|x| \in V$, and hence, that $x, y \in V$ imply $x \cup y$ and $x \cap y \in V$. (See lemma 4.6.4 for the definitions of these functions.)

6. Prove that

$$\left| \int_a^b x \, dt \right| \leq \int_a^b |x| \, dt$$

for x Riemann integrable. (HINT: Use $x = x \cup 0 + x \cap 0, |x| = x \cup 0 - x \cap 0$.)

7. a. Let x be a nonnegative Riemann integrable function satisfying $\displaystyle\int_a^b x \, dt = 0$. Give as much information as possible about x.

b. Define $\|x\| = \displaystyle\int_a^b |x| \, dt$; show that $\|x\|$ is a seminorm on V in the sense that it satisfies all of the norm conditions except that $\|x\|$ may be zero for x not identically zero. Follow the procedure of exercise 2, section 1, chapter 3, to obtain a normed vector space V_0 of equivalence classes of functions. Show that V_0 is not a Banach space.

8. Show that, if x' exists and is Riemann integrable on $[a, b]$, then

$$\int_a^b x' \, dt = x(b) - x(a).$$

9. Show that if x is nonnegative and Riemann integrable, then $x^{1/n}$ is also Riemann integrable for any positive integer n.

10.   In this exercise, consider all functions to be defined on $a \le t \le b$.

   a. Let

$$f(t) = \begin{cases} 0, & t \text{ irrational} \\ 1, & t \text{ rational.} \end{cases}$$

   Show that f is not Riemann integrable.

   b. Suppose $|x|$ is Riemann integrable. Is x Riemann integrable?

   c. Exhibit a sequence $\{x_n\}$ of Riemann integrable functions which converges pointwise to a bounded function which is not Riemann integrable, and which satisfies $x_{n+1}(t) \ge x_n(t)$ for all t.

11.   Let x be Riemann integrable on $[a, b]$. Define $F(t) = \int_a^t x(s)\, ds$ for $a < t \le b$, and $F(a) = 0$. Show that F is continuous and has bounded variation.

12.   Exhibit several functions which are Riemann integrable on $[0, 1]$ but are not in $R[0, 1]$.

Section 2   MEASURABLE SETS IN R

In this section we develop Lebesgue measure on the real line. There are various ways of proceeding, but they all begin by generalizing the notion of length from the intervals $[a, b]$, $[a, b)$, $(a, b]$, and (a, b), where we have

$$l(I) = b - a,$$

the *length* of I, to other sets.

One of the first extensions of the notion of length was due to Jordan, whose procedure we now touch upon.

Definition 5.2.1. The characteristic function $C(A; t)$ of a subset A of R is

$$C(A; t) = \begin{cases} 1, & t \in A \\ 0, & t \notin A. \end{cases}$$

Employing the Riemann integral, we introduce

$$\bar{c}(A) = \overline{\int} C(A; t)\, dt$$

and

$$\underline{c}(A) = \underline{\int} C(A; t)\, dt,$$

the *upper* and *lower* content respectively, of A, These are meaningful only for bounded sets in R, due to the nature of the Riemann integral.

A bounded subset A of R is said to have content

$$c(A) = \bar{c}(A) = \underline{c}(A)$$

whenever the characteristic function of A is Riemann integrable.

While this extension of the notion of the length of an interval seems, at first, quite broad, there are some relatively simple sets which do not have content. For example, since the characteristic function of the rational numbers between 0 and 1 is not Riemann integrable, this set does not have content.

Now, from the definition of the Riemann integral, we see that

$$\bar{c}(A) = \inf \left\{ \sum_{k=1}^{n} l(I_k) : \bigcup_{k=1}^{n} I_k \supseteq A \right\}$$

$$\underline{c}(A) = \sup \left\{ \sum_{k=1}^{n} l(I_k) : \bigcup_{k=1}^{n} I_k \subseteq A, \, I_i \cap I_j = \varnothing \right\}.$$

Thus, a set A has content if and only if, for any $\epsilon > 0$, there exist sets B_1 and B_2, each a finite union of intervals, such that

$$B_1 \subseteq A \subseteq B_2 \quad \text{and} \quad c(B_2 - B_1) \leq \epsilon.$$

(It is left as an exercise to show that $B_2 - B_1$ has content.) In other words, a set has content if it can be arbitrarily closely approximated, in a sense, by a finite union of intervals.

The class of Lebesgue measurable sets of finite measure (measure is the generalization of content) is obtained by modification of the foregoing approximation procedure. Throughout the development, the algebraic structure of several classes of sets will play an important role. Accordingly, we begin with a definition.

Definition 5.2.2

(a) A *ring* $\mathscr{S}$ of sets is a collection of subsets of a set X such that, if A and B are in $\mathscr{S}$, then $A - B$ and $A \cup B$ are in $\mathscr{S}$.

(b) An *algebra* of sets $\mathscr{A}$ is a ring of sets in which there is a "unit" set E_0 such that $E_0 \cap A = A$ for all A in $\mathscr{A}$.

(c) A σ-ring (σ-algebra) of sets is a ring (algebra) of sets which is closed under the formation of countable unions.

We note that a ring of sets is closed under the operations $\triangle$ and $\cap$, for

$$A \triangle B = (A - B) \cup (B - A)$$

and

$$A \cap B = (A \cup B) - (A \triangle B).$$

Moreover, the identity

$$\bigcap_{k=1}^{\infty} A_k = A_1 - \bigcup_{k=1}^{\infty} (A_1 - A_k)$$

shows that a σ-ring (σ-algebra) is closed under the formation of countable intersections.

The following lemma will be useful in several places.

Lemma 5.2.3. If $\{A_n\}$ is a sequence of sets in a ring $\mathscr{R}$, then there exists a sequence $\{B_n\}$ in $\mathscr{R}$ such that $B_n \subseteq A_n$ for all n, $B_i \cap B_j = \varnothing$ for $i \neq j$, and

$$\bigcup_{n=1}^{\infty} B_n = \bigcup_{n=1}^{\infty} A_n.$$

Proof. Let $B_1 = A_1$, $B_2 = A_2 - A_1$, $B_3 = A_3 - A_1 \cup A_2$, ..., $B_n = A_n - \bigcup_{k=1}^{n-1} A_k, \ldots$ It should be clear that $\{B_n\}$ fulfills the stipulated conditions.

For the purpose of constructing Lebesgue measure, the basic ring of sets consists of all sets which are finite unions of disjoint, left-closed, right-open intervals. We use the symbol

$$\mathscr{R} = \left\{ \bigcup_{k=1}^{n} I_k: \ I_k = [a_k, b_k), I_j \cap I_k = \varnothing \right\},$$

for this collection.†

For $\bigcup_{k=1}^{n} I_k$ in $\mathscr{R}$, $I_i \cap I_j = \varnothing$, we define

$$m\left(\bigcup_{k=1}^{n} I_k \right) = \sum_{k=1}^{n} l(I_k),$$

the *measure* of the set $\bigcup_{k=1}^{n} I_k$. The reader can verify that the measure of a set in $\mathscr{R}$ is independent of the way in which the set is represented as a finite union of disjoint, left-closed, right-open intervals.

We now prove the following theorem.

---

†The empty set, $[a, a)$, is included in $\mathscr{R}$.

Theorem 5.2.4

(a) $\mathcal{R}$ is a ring of sets.
(b) For any A and B in $\mathcal{R}$,

$$m(A \cup B) + m(A \cap B) = m(A) + m(B).$$

Proof. It is left as an exercise to show that $\mathcal{R}$ is closed under subtraction. Now, if A and C are disjoint sets in $\mathcal{R}$, it is clear that $A \cup C$ is in $\mathcal{R}$. Finally, since $A \cup B = A \cup (B - A)$, we see that $\mathcal{R}$ is closed under union.

To prove (b), we notice that $m(A)$ is simply the Riemann integral of the characteristic function of A. By the additivity of this integral, we have

$$m(A \cup B) = m(A - B) + m(B - A) + m(A \cap B),$$
$$m(A \cup B) + m(A \cap B) = [m(A - B) + m(A \cap B)]$$
$$+ [m(B - A) + m(A \cap B)],$$

and

$$m(A \cup B) + m(A \cap B) = m(A) + m(B).$$

This completes the proof.

Corollary

(a) If $A \in \mathcal{R}$, $B \in \mathcal{R}$, $A \subseteq B$, then $m(A) \leq m(B)$.
(b) If $A_1, A_2, \ldots, A_n \in \mathcal{R}$, $A_i \cap A_j = \varnothing$ $(i \neq j)$, then

$$m\left(\bigcup_{k=1}^{n} A_k\right) = \sum_{k=1}^{n} m(A_k).$$

To describe the procedure for approximating sets by members of $\mathcal{R}$, we must introduce the notion of outer measure.

Definition 5.2.5. For any subset E of R, the *outer measure* $m^*(E)$ is given by

$$m^*(E) = \inf\left\{\sum_{k=1}^{\infty} l(I_k): \bigcup_{k=1}^{\infty} I_k \supseteq E\right\}.$$

The basic properties of m^* are given in the following lemma.

Lemma 5.2.6. The outer measure $m^*(A)$ satisfies the conditions

(a) $m^*(\varnothing) = 0$.
(b) $A \supseteq B \Rightarrow m^*(A) \geq m^*(B)$ (monotonicity).

(c) $m^*(\bigcup\limits_{n=1}^{\infty} A_n) \leq \sum\limits_{n=1}^{\infty} m^*(A_n)$ (countable subadditivity).

Proof. Only (c) requires discussion. For each n, a sequence $\{I_{nk}\}$ of intervals can be found such that $\bigcup\limits_{k=1}^{\infty} I_{nk} \supseteq A_n$ and

$$\sum\limits_{k=1}^{\infty} l(I_{nk}) \leq m^*(A_n) + 2^{-n}\epsilon,$$

provided that $m^*(A_n) < \infty$ for all n. Then

$$\bigcup\limits_{k, n=1}^{\infty} I_{nk} \supseteq \bigcup\limits_{n=1}^{\infty} A_n$$

and

$$m^*\left(\bigcup\limits_{n=1}^{\infty} A_n\right) \leq \sum\limits_{k, n=1}^{\infty} l(I_{nk}) \leq \sum\limits_{n=1}^{\infty} m^*(A_n) + \epsilon.$$

Since this is true for any $\epsilon > 0$, we have

$$m^*\left(\bigcup\limits_{n=1}^{\infty} A_n\right) \leq \sum\limits_{n=1}^{\infty} m^*(A_n)$$

as desired. This completes the proof in the case that $m^*(A_n) < \infty$ for all n. Otherwise the inequality is obvious.

Remark: An equivalent definition of m^* is given by

$$m^*(A) = \inf\left\{\sum\limits_{n=1}^{\infty} m(A_n): \bigcup\limits_{n=1}^{\infty} A_n \supseteq A, A_n \in \mathcal{R}; n = 1, 2, \ldots\right\}.$$

It will be vital, in extending m from $\mathcal{R}$ to a larger class of sets, to know that m and m^* agree on $\mathcal{R}$. We prove this as part of the following.

Lemma 5.2.7

(a) The measure m on $\mathcal{R}$ is countably additive, that is, if $A_n \in \mathcal{R}$, $n = 1, 2, \ldots$, $A_i \cap A_j = \emptyset$, and $\bigcup\limits_{n=1}^{\infty} A_n \in \mathcal{R}$, then

$$m\left(\bigcup\limits_{n=1}^{\infty} A_n\right) = \sum\limits_{n=1}^{\infty} m(A_n).$$

(b) For $A \in \mathcal{R}$, $m(A) = m^*(A)$.

Proof

(a) Let $A = \bigcup\limits_{n=1}^{\infty} A_n$; then $A \supseteq \bigcup\limits_{n=1}^{N} A_n$ and, by the monotonicity and finite additivity of m, we have

$$m(A) \geq \sum_{n=1}^{N} m(A_n).$$

Since this is true for all N, we obtain

$$m(A) \geq \sum_{n=1}^{\infty} m(A_n).$$

To prove the reverse inequality, we choose a representation of A, say $A = \bigcup_{k=1}^{m} J_k$, and notice that the identity

$$\bigcup_{k=1}^{m} J_k = \bigcup_{n=1}^{\infty} A_n, \qquad (A_i \cap A_j = \varnothing)$$

enables us to consider only the case

$$J_0 = \bigcup_{i=1}^{\infty} J_i, \qquad J_i \cap J_j = \varnothing,$$

where $J_i \in \mathscr{R}$, $i = 0, 1, 2, \ldots$, and each J_i is a left-closed, right-open interval. Thus, if $J_i = [a_i, b_i)$, we must show

$$b_0 - a_0 \leq \sum_{k=1}^{\infty} (b_k - a_k).$$

We form the intervals $J_0' = [a_0, b_0 - \epsilon]$ and $J_k' = (a_k - 2^{-k}\delta, b_k)$, $k = 1, 2, \ldots$, and notice that

$$\bigcup_{i=1}^{\infty} J_k' \supseteq J_0'.$$

By the compactness of J_0', there is an N such that

$$\bigcup_{k=1}^{N} J_k' \supseteq J_0'.$$

Thus

$$\sum_{k=1}^{N} [b_k - (a_k - 2^{-k}\delta)] \geq b_0 - \epsilon - a_0,$$

or,

$$\sum_{k=1}^{N} (b_k - a_k) + \delta \geq (b_0 - a_0) - \epsilon,$$

Therefore,

$$\sum_{k=1}^{\infty} (b_k - a_k) + \delta \geq (b_0 - a_0) - \epsilon$$

for all $\delta > 0$, $\epsilon > 0$. It follows that

$$\sum_{k=1}^{\infty} (b_k - a_k) \geq (b_0 - a_0),$$

and the proof of (a) is finished.

(b) By a previous remark,

$$m^*(A) = \inf \left\{ \sum_{n=1}^{\infty} m(A_n): \quad \bigcup_{n=1}^{\infty} A_n \supseteq A, A_n \in \mathcal{R} \right\}$$

and one of the numbers, in the collection whose infimum is indicated, is $m(A)$ (for we can write $A = A \cup \varnothing \cup \varnothing \cup \cdots \cup \varnothing \cup \cdots$). Thus we have

$$m^*(A) \leq m(A).$$

Now let $\{A_n\}$ be a sequence in $\mathcal{R}$ such that $\bigcup_{n=1}^{\infty} A_n \supseteq A$. Then

$$A = A \cap \left(\bigcup_{n=1}^{\infty} A_n \right) = \bigcup_{n=1}^{\infty} (A \cap A_n)$$

and we can find a disjoint sequence $\{B_n\}$ in $\mathcal{R}$ such that $A = \bigcup_{n=1}^{\infty} B_n$ and $B_n \subseteq A \cap A_n$ for all n. By (a) we have

$$m(A) = m\left(\bigcup_{n=1}^{\infty} B_n \right) = \sum_{n=1}^{\infty} m(B_n) \leq \sum_{n=1}^{\infty} m(A_n).$$

Thus

$$m(A) \leq \inf \left\{ \sum_{n=1}^{\infty} m(A_n); \quad \bigcup_{n=1}^{\infty} A_n \supseteq A, A_n \in \mathcal{R} \right\}$$

so that $m(A) \leq m^*(A)$. This completes the proof.

The class of Lebesgue measurable sets of finite measure will be designated $\mathcal{M}_0$. By definition,

$$A \in \mathcal{M}_0 \Leftrightarrow \begin{cases} \text{For any } \epsilon > 0 \text{ there is a set } B \text{ in } \mathcal{R} \text{ such} \\ \text{that } m^*(A \bigtriangleup B) < \epsilon. \end{cases}$$

The measure m is extended from $\mathcal{R}$ to $\mathcal{M}_0$ as follows. For $A \in \mathcal{M}_0$ there is a sequence $\{A_n\}$ in $\mathcal{R}$ such that

$$m^*(A \bigtriangleup A_n) \leq \frac{1}{n}.$$

Consider the sequence $\{m(A_n)\}$. We have

$$|m(A_m) - m(A_n)| \leq m(A_m \triangle A_n) = m^*(A_m \triangle A_n)$$

since $A_m \triangle A_n \in \mathcal{R}$. Applying the relation

$$P \triangle Q \subseteq (P \triangle S) \cup (Q \triangle S),$$

and the subadditivity of m^*, we have

$$m^*(A_m \triangle A_n) \leq m^*(A_m \triangle A) + m^*(A_n \triangle A)$$
$$\leq \frac{1}{m} + \frac{1}{n}.$$

It follows that the sequence $\{m(A_n)\}$ is a Cauchy sequence, and thus it has a limit. We define

$$m(A) = \lim_{n \to \infty} m(A_n).$$

The definition is unambiguous, for if $\{B_n\}$ is a second sequence such that $m^*(B_n \triangle A) \leq 1/n$, we have

$$|m(B_n) - m(A_n)| \leq m(B_n \triangle A_n) = m^*(B_n \triangle A_n)$$
$$\leq m^*[(B_n \triangle A) \cup (A_n \triangle A)]$$
$$\leq m^*(B_n \triangle A) + m^*(A_n \triangle A)$$
$$\leq \frac{2}{n},$$

so that the Cauchy sequences $\{(m(A_n)\}$ and $\{m(B_n)\}$ are equivalent and, hence, have the same limit.

Theorem 5.2.4 extends to $\mathcal{M}_0$. We prove this now.

Theorem 5.2.8

(a) $\mathcal{M}_0$ is a ring of sets.
(b) For $A, B \in \mathcal{M}_0$,

$$m(A \cup B) + m(A \cap B) = m(A) + m(B).$$

Proof

(a) Let A and B be sets in $\mathcal{M}_0$ and let $\{A_n\}$ and $\{B_n\}$ be approximating sequences of sets in $\mathcal{R}$. Using the set identity

$$C \cup D - E \cup F \subseteq (C - E) \cup (D - F)$$

(see the exercises for proof), we have

$$(A_n \cup B_n) - (A \cup B) \subseteq (A_n - A) \cup (B_n - B)$$
$$(A \cup B) - (A_n \cup B_n) \subseteq (A - A_n) \cup (B - B_n).$$

Hence we see that

$$(A_n \cup B_n) \triangle (A \cup B) \subseteq (A \triangle A_n) \cup (B \triangle B_n)$$

and we have

$$m^*[(A_n \cup B_n) \triangle (A \cup B)] \le m^*(A \triangle A_n) + m^*(B \triangle B_n).$$

It follows that $\{A_n \cup B_n\}$ is a sequence in $\mathscr{R}$ approximating $A \cup B$, and $A \cup B \in \mathscr{M}_0$.

Now, using the set identity

$$C \triangle D = \tilde{C} \triangle \tilde{D},$$

we have

$$(A_n - B_n) \triangle (A - B) = (A_n \cap \tilde{B}_n) \triangle (A \cap \tilde{B}) = (\tilde{A}_n \cup B_n) \triangle (\tilde{A} \cup B)$$
$$\subseteq (\tilde{A}_n \triangle \tilde{A}) \cup (B_n \triangle B) = (A_n \triangle A) \cup (B_n \triangle B),$$

and so

$$m^*[(A_n - B_n) \triangle (A - B)] \le m^*(A_n \triangle A) + m^*(B_n \triangle B).$$

It follows that $\{A_n - B_n\}$ is a sequence in $\mathscr{R}$ approximating $A - B$, and $A - B \in \mathscr{M}_0$.

Since $\mathscr{M}_0$ has been shown to be closed under $\cup$ and $-$, it is a ring.

(b) With $\{A_n\}$ and $\{B_n\}$ as in (a), we calculate

$$(A_n \cap B_n) \triangle (A \cap B) = (\tilde{A}_n \cup \tilde{B}_n) \triangle (\tilde{A} \cup \tilde{B})$$
$$\subseteq (\tilde{A}_n \triangle \tilde{A}) \cup (\tilde{B}_n \triangle \tilde{B})$$
$$= (A_n \triangle A) \cup (B_n \triangle B),$$

so that $\{A_n \cap B_n\}$ is a sequence in $\mathscr{R}$ approximating $A \cap B$. Since $\{A_n \cup B_n\}$ was shown to approximate $A \cup B$, we have

$$m(A) + m(B) = \lim_{n \to \infty} [m(A_n) + m(B_n)]$$
$$= \lim_{n \to \infty} [m(A_n \cup B_n) + m(A_n \cap B_n)]$$
$$= m(A \cup B) + m(A \cap B).$$

This completes the proof.

Corollary

(a) If $A \supseteq B$ in $\mathcal{M}_0$, then $m(A) \geq m(B)$ (monotonicity).
(b) If $A_1, A_2, \ldots, A_n$ are disjoint in $\mathcal{M}_0$, then

$$m\left(\bigcup_{k=1}^{n} A_k\right) = \sum_{k=1}^{n} m(A_k) \qquad \text{(finite additivity)}$$

We can now show that m^* and m agree on $\mathcal{M}_0$. For any two sets A and B, we have

$$A \subseteq A \cup B = B \cup (A - B),$$

so that

$$m^*(A) \leq m^*(B) + m^*(A - B) \leq m^*(B) + m^*(A \triangle B).$$

Hence

$$m^*(A) - m^*(B) \leq m^*(A \triangle B),$$

and since A and B are interchangeable, we can write

$$|m^*(A) - m^*(B)| \leq m^*(A \triangle B).$$

If we now let $A \in \mathcal{M}_0$ and $B = A_n \in \mathcal{R}$, where $m^*(A \triangle A_n) \leq 1/n$, we have, since $m = m^*$ on $\mathcal{R}$,

$$|m^*(A) - m(A_n)| \leq \frac{1}{n}.$$

Letting $n \longrightarrow \infty$, we obtain the following.

Lemma 5.2.9. $m(A) = m^*(A)$ for all $A \in \mathcal{M}_0$.

The function m^* has a countable additivity property which we prove in the following lemma.

Lemma 5.2.10. If $\{A_n\}$ is a sequence of disjoint sets in $\mathcal{M}_0$, then

$$m^*\left(\bigcup_{n=1}^{\infty} A_n\right) = \sum_{n=1}^{\infty} m^*(A_n).$$

Proof. Since m^* is monotone and agrees with m on $\mathcal{M}_0$, we have

$$m^*\left(\bigcup_{n=1}^{\infty} A_n\right) \geq m^*\left(\bigcup_{n=1}^{N} A_n\right) = \sum_{n=1}^{N} m^*(A_n)$$

for all N, and hence

$$m^*\left(\bigcup_{n=1}^{\infty} A_n\right) \geq \sum_{n=1}^{\infty} m^*(A_n).$$

The countable subadditivity of m^* gives the reverse inequality, which completes the proof.

The equality in the lemma is valid on a larger class of sets, provided that we allow some sets on which m^* takes the value $+\infty$. We define

$$\mathcal{M} = \left\{\bigcup_{n=1}^{\infty} A_n: \quad A_n \in \mathcal{M}_0, \ n = 1, 2, \ldots\right\}.$$

The ring $\mathcal{M}_0$ is characterized in $\mathcal{M}$ by the following.

Lemma 5.2.11. $A \in \mathcal{M}_0 \Leftrightarrow \{A \in \mathcal{M}$ and $m^*(A) < \infty\}$.

Proof. The implication $\Rightarrow$ is evident. Now assume $A \in \mathcal{M}$ and $m^*(A) < \infty$. By lemma 5.2.3 we can write $A = \bigcup_{n=1}^{\infty} A_n$, where $A_i \cap A_j = \varnothing$ and $A_n \in \mathcal{M}_0$. Let $B_n = \bigcup_{i=1}^{n} A_i$; then

$$m^*(A \triangle B_n) = m^*\left(\bigcup_{i=n+1}^{\infty} A_i\right) = \sum_{i=n+1}^{\infty} m^*(A_i),$$

a quantity which can be made arbitrarily small by choosing n sufficiently large (the series $\sum_{n=1}^{\infty} m^*(A_n)$ converges). Now choose $C_n \in \mathcal{R}$ such that $m^*(B_n \triangle C_n) \leq 1/n$. From the relations

$$A \triangle C_n \subseteq (A \triangle B_n) \cup (B_n \triangle C_n)$$

and

$$m^*(A \triangle C_n) \leq m^*(A \triangle B_n) + m^*(B_n \triangle C_n),$$

we see that A can be approximated by sets in $\mathcal{R}$, and thus $A \in \mathcal{M}_0$. This completes the proof.

Clearly, $\mathcal{M}$ is closed under the formation of countable unions, and from the two preceding lemmas we see that m^* is countably additive on $\mathcal{M}$. We now show that if $A, B \in \mathcal{M}$, then $A - B \in \mathcal{M}$, so that $\mathcal{M}$ is a σ-ring. (See definition 5.2.2.)

Let

$$A = \bigcup_{n=1}^{\infty} A_n \quad \text{and} \quad B = \bigcup_{n=1}^{\infty} B_n$$

be two members of $\mathcal{M}$. We have

$$A_n \cap B = A_n \cap \left(\bigcup_{k=1}^{\infty} B_k \right) = \bigcup_{k=1}^{\infty} (A_n \cap B_k),$$

so that $A_n \cap B \in \mathcal{M}$. Moreover,

$$m^*(A_n \cap B) \leq m^*(A_n) < \infty,$$

so that $A_n \cap B \in \mathcal{M}_0$. It follows that

$$A_n - B = A_n - (A_n \cap B) \in \mathcal{M}_0.$$

Therefore

$$A - B = \bigcup_{n=1}^{\infty} (A_n - B)$$

is in $\mathcal{M}$, as desired. In summary, we state the following theorem.

Theorem 5.2.12. $\mathcal{M}$ is a σ-ring of sets, and m^* is a nonnegative, extended real-valued, countably additive function on $\mathcal{M}$.

Actually, $\mathcal{M}$ is a σ-algebra, for $(-\infty, \infty)$ is in $\mathcal{M}$. Henceforth, we will write $m(A)$ instead of $m^*(A)$ whenever $A \in \mathcal{M}$. The collection $\mathcal{M}$ is the set of Lebesgue measurable subsets of R, and the function $m(A)$ is Lebesgue measure. The function $m^*(A)$ is called Lebesgue outer measure.
The proofs of the following points are left as exercises.

(i) $[m^*(A) = 0] \Rightarrow [A \in \mathcal{M}_0]$.
(ii) $[A \supseteq B, A \in \mathcal{M}, m(A) = 0] \Rightarrow [B \in \mathcal{M}_0, m(B) = 0]$.
(iii) A has Lebesgue measure zero in the sense of definition 5.1.2 if and only if $A \in \mathcal{M}, m(A) = 0$.
(iv) The sets of Lebesgue measure zero form a σ-ring of subsets of R.

We proceed with a closer examination of $\mathcal{M}$. Clearly, since $\mathcal{M}$ is a σ-algebra containing all intervals, it must contain all open sets and all closed sets. Let $\{\mathcal{R}_\alpha\}$ be the collection of all σ-algebras of subsets of R such that each $\mathcal{R}_\alpha$ contains all open sets (and hence also all closed sets). The collection $\mathcal{B} = \bigcap_\alpha \mathcal{R}_\alpha$ is again a σ-algebra (why?) and is the smallest σ-algebra containing the open sets of R. The sets in $\mathcal{B}$ are called Borel sets. It is clear that $\mathcal{M} \supseteq \mathcal{B}$; actually, this inclusion is proper, and not all subsets of R are in $\mathcal{M}$ (i.e., there exist both nonmeasurable sets and measurable sets which are not Borel sets.)† We now state a theorem showing how measurable sets can be approximated by open sets, closed sets, and Borel sets.

†See the appendix.

Theorem 5.2.13

(a) For any measurable set A and any $\epsilon > 0$, there exists an open set O and a closed set F such that

$$O \supseteq A \supseteq F \quad \text{and} \quad m(O - A) < \epsilon, \qquad m(A - F) < \epsilon.$$

(b) For any measurable set A, there exist Borel sets B_1 and B_2 such that

$$B_1 \supseteq A \supseteq B_2 \quad \text{and} \quad m(B_1 - A) = m(A - B_2) = 0.$$

The proof of this theorem is left as an exercise.

Many authors define the *Lebesgue inner measure* function,

$$m_*(A) = \sup \{m(F): \quad F \text{ closed}, F \subseteq A\},$$

where A is an arbitrary subset of R. It is an extended, real-valued, nonnegative function satisfying

(a) $m_*(\varnothing) = 0$,
(b) $[A \supseteq B] \Rightarrow [m_*(A) \geq m_*(B)]$,
(c) $m_*(\bigcup\limits_{n=1}^{\infty} A_n) \geq \sum\limits_{n=1}^{\infty} m_*(A_n)$, if $A_i \cap A_j = \varnothing$, for $i \neq j$.

Properties (a) and (b) are obvious, while the proof of (c) is left as an exercise. Note that Lebesgue outer measure can be written

$$m^*(A) = \inf \{m(O): \quad O \text{ open}, O \supseteq A\}$$

so that m^* and m_* are dual. The proofs of the following assertions are also left as exercises.

(i) $[A \in \mathcal{M}] \Rightarrow [m^*(A) = m_*(A) = m(A)]$
(ii) $[m^*(A) = m_*(A) < \infty] \Rightarrow [A \in \mathcal{M}_0]$.

Finally, to show the role of closed sets in the definition of inner measure and in part (a) of the preceding theorem, we give an example of a measurable set of positive measure which contains no nonempty open set.

EXAMPLE. Let n be a positive integer greater than 2. From the closed interval $[0, 1]$, remove the open interval of length $1/n$ centered at $\frac{1}{2}$. Then remove an open interval of length n^{-2} from the center of each of the two remaining intervals. Next, remove an open interval of length n^{-3} from the center of each of the four remaining intervals, and continue this process indefinitely. Denote the set which remains by C_n; it is measurable since $\tilde{C}_n$ is open. Now

$$m(\tilde{C}_n) = \frac{1}{n} + 2\frac{1}{n^2} + 2^2\frac{1}{n^3} + 2^3\frac{1}{n^4} + \cdots = \frac{1}{n}\sum_{k=0}^{\infty}\left(\frac{2}{n}\right)^k = \frac{1}{n-2}.$$

The set C_3, henceforth denoted by C, is called the *Cantor set* and has measure zero. The set C_n has measure $1 - 1/(n-2)$. It is left as an exercise to show that C_n has empty interior, that is, C_n is a set of positive measure containing no interval. This example shows that the method of approximating sets by intervals employed in the definition of Jordan content cannot be used to construct Lebesgue measure.

In the preceding discussion, the Lebesgue measure on R was developed in detail. The starting point was the ring

$$\mathscr{R} = \left\{\bigcup_{k=1}^{n} I_k : \quad I_k = [a_k, b_k), \quad I_j \cap I_k = \emptyset\right\}$$

on which was defined the measure

$$m\left(\bigcup_{k=1}^{n} I_k\right) = \sum_{k=1}^{n} l(I_k).$$

Lebesgue measure is similarly introduced in $R^{(2)}$, where, instead of intervals, the fundamental sets are basic rectangles of the type

$$J = [a, b) \times [c, d),$$

and, instead of length, the area

$$\mathscr{A}(J) = (b - a)(d - c)$$

is used. The definition of Lebesgue measure begins with the collection

$$\mathscr{R}^{(2)} = \left\{\bigcup_{k=1}^{n} J_k : \quad J_k = [a_k, b_k) \times [c_k, d_k), \; J_i \cap J_j = \emptyset\right\}$$

and the function

$$m\left(\bigcup_{k=1}^{n} J_k\right) = \sum_{k=1}^{n} (b_k - a_k)(d_k - c_k).$$

As before, several things require verification:

(i) $\mathscr{R}^{(2)}$ is a ring of sets.
(ii) m is independent of the particular way in which a set is represented as a union of disjoint basic rectangles.

Once this is done the outer measure function

$$m^*(A) = \inf\left\{\sum_{n=1}^{\infty} m(A_n): \quad \bigcup_{n=1}^{\infty} A_n \supseteq A, A_n \in \mathscr{R}^{(2)}; \quad n = 1, 2, \ldots\right\}$$

is introduced and is shown to satisfy the following:

(iii) m is countably additive on $\mathscr{R}^{(2)}$.
(iv) $m(A) = m^*(A)$ for $A \in \mathscr{R}^{(2)}$.

The proofs of (i)–(iv) are essentially the same as the proofs given in the preceding discussion for $\mathscr{R}$. From this point on, the development may be repreated *verbatim*. The end result is a σ-algebra of sets, $\mathscr{M}^{(2)}$, and a measure m on $\mathscr{M}^{(2)}$. The results on the approximation of measurable sets by open sets, closed sets, and Borel sets are also identical with the one-dimensional case.

The reader should establish these claims, in particular the assertions (i)–(iv), for himself. Once he has done so, he should be ready to state and accept the analogous results on Lebesgue measure in Euclidean N-space.

EXERCISES

1. Prove the following set theoretic identities.

   a. $A \triangle C \subseteq (A \triangle B) \cup (B \triangle C)$.
   b. $[C \cup D - E \cup F] \subseteq (C - E) \cup (D - F)$.
   c. $C \triangle D = \tilde{C} \triangle \tilde{D}$.

2. Let $\{x_n\}$ be the sequence of characteristic functions of a sequence of sets $\{A_n\}$. Find necessary and sufficient conditions that $\{x_n\}$ converge uniformly.

3. Let $\{x_n\}$ and $\{A_n\}$ be as in exercise 2. Then $\{x_n\}$ converges *pointwise* if and only if

   $$x_L(t) = \underline{\lim_{n \to \infty}} x_n(t) = \overline{\lim_{n \to \infty}} x_n(t) = x_U(t).$$

   (See exercise 11, section 3, chapter 2.)

   a. Show that

   $$x_U(t) = \begin{cases} 0, & \text{if } t \text{ is in only finitely many } A_n\text{'s,} \\ 1, & \text{if } t \text{ is in infinitely many } A_n\text{'s.} \end{cases}$$

b. Show that

$$x_L(t) = \begin{cases} 0, & \text{if } t \text{ is in infinitely many } A_n\text{'s,} \\ 1, & \text{if } t \text{ is in only finitely many } A_n\text{'s.} \end{cases}$$

c. Let

$$A_L = \bigcup_{n=1}^{\infty} \bigcap_{m=n}^{\infty} A_m \quad \text{and} \quad A_U = \bigcap_{n=1}^{\infty} \bigcup_{m=n}^{\infty} A_m.$$

Show that $A_L \subseteq A_U$, and that $\{x_n\}$ converges pointwise if and only if $A_L = A_U$. [HINT: Show that

$$A_U = \{t: \quad t \text{ is in infinitely many } A_n\text{'s}\}, \quad \text{and}$$
$$A_L = \{t: \quad t \text{ is in all but finitely many } A_n\text{'s}\}.]$$

4. Show that
$$A \sim B \Leftrightarrow m(A \triangle B) = 0$$
is an equivalence relation in $\mathcal{M}$, and that it has the substitution property with respect to $\cup, \cap -$, and $\triangle$.

5. a. Prove $[m^*(A) = 0] \Rightarrow [A \in \mathcal{M}_0, m(A) = 0]$.
   b. Prove $[A \in \mathcal{M}, A \supseteq B, m(A) = 0] \Rightarrow [B \in \mathcal{M}_0, m(B) = 0]$.

6. Prove theorem 5.2.13.

7. Prove:

   a. $[A \in \mathcal{M}] \Rightarrow [m^*(A) = m_*(A) = m(A)]$.
   b. $[m^*(A) = m_*(A) < \infty] \Rightarrow [A \in \mathcal{M}_0, m(A) = m^*(A) = m_*(A)]$.

8. Prove $m_*\left(\bigcup_{n=1}^{\infty} A_n\right) \geq \sum_{n=1}^{\infty} m_*(A_n)$ for a sequence of disjoint sets $\{A_n\}$.

9. Assume a nonmeasurable set exists.

   a. Show that a bounded nonmeasurable set exists.
   b. Show that $m^*(A)$ can be the same as $m_*(A)$ for a nonmeasurable set A.

10. a. Show that C_n has empty interior. (See the example at the end of this section.)
   b. Show that C_n is a perfect set, i.e., every point of C_n is a cluster point of C_n.
   c. Let the numbers in $[0, 1]$ be represented as ternary decimals, i.e., decimals of the form. $\alpha_1\alpha_2\alpha_3 \cdots$ where each α_i is 0, 1, or 2. Show that $\tilde{C}$, the complement of the Cantor set, consists of exactly those numbers whose ternary decimals contain at least one 1, while C consists of exactly those numbers whose ternary decimals contain no 1's.

11. Prove:

    a. If, for any $\epsilon > 0$, there exists an open set O and a closed set K such that $O \supseteq A \supseteq K$ and $m^*(O - K) \leq \epsilon$, then $A \in \mathcal{M}$.

    b. If there exist Borel sets B_1 and B_2 such that $B_1 \supseteq A \supseteq B_2$ and $m^*(B_1 - B_2) = 0$, then $A \in \mathcal{M}$.

12. Show that the function m defined on $\mathcal{R}^{(2)}$ is countably additive.

Section 3 MEASURABLE FUNCTIONS

In the preceding section we developed the theory of Lebesgue measure on Euclidean spaces. Certain general features of this theory will suffice for the purpose at hand, and accordingly, we introduce the following terminology.

Definition 5.3.1

(a) A *measure space* is a triple $(X, \mathcal{M}, \mu)$, where X is any set, $\mathcal{M}$ is a σ-algebra of subsets of X, and μ is a measure on $\mathcal{M}$. That is, μ is a nonnegative, extended real-valued, countably additive function from $\mathcal{M}$ to R such that $\mu(\varnothing) = 0$.

(b) A measure μ on a σ-algebra $\mathcal{M}$ is called σ-*finite* if every set A in $\mathcal{M}$ is contained in a countable union of sets in $\mathcal{M}$ of finite measure.

(c) A measure μ on a σ-algebra $\mathcal{M}$ is called *complete* if $A \in \mathcal{M}$, $\mu(A) = 0$, $B \subset A$, imply $B \in \mathcal{M}$.

Note that a measure is always *monotone* $[A \subset B \Rightarrow \mu(A) \leq \mu(B)]$, *subtractive* $[A \subset B \Rightarrow \mu(B - A) = \mu(B) - \mu(A)]$, and countably subadditive. Therefore, if μ is complete, then subsets of sets of measure zero are measurable (i.e., are in $\mathcal{M}$) and have measure zero. We also explicitly point out that X is any set, so that it makes no sense in this general setting to talk of open or closed subsets of X. The following lemma, whose proof is left as an exercise, shows that the lack of completeness of a measure can always be remedied.

Lemma 5.3.2. Let $(X, \mathcal{M}, \mu)$ be a measure space, and let $\mathcal{M}_1$ be the collection of all subsets of X of the form $A \cup B$, where $A \in \mathcal{M}$ and B is a subset of a set of measure zero in $\mathcal{M}$. Then $\mathcal{M}_1$ is a σ-algebra of subsets of X, and the function $\mu_1(A \cup B) = \mu(A)$ is a complete measure on $\mathcal{M}_1$.

The measure space $(X, \mathcal{M}_1, \mu_1)$ is called the *completion* of the measure space $(X, \mathcal{M}, \mu)$. We will work with measures that are σ-finite and complete, although many of our results hold without these assumptions.

Lemma 5.3.3

(a) If $\{A_n\}$ is a sequence of measurable sets such that $A_n \subseteq A_{n+1}$, $n = 1$, 2, ..., then

$$\mu\left(\bigcup_{n=1}^{\infty} A_n\right) = \lim_{n\to\infty} \mu(A_n).$$

(b) If $\{A_n\}$ is a sequence of measurable sets such that $A_n \supseteq A_{n+1}$, $n = 1$, 2, ... and $\mu(A_1) < \infty$, then

$$\mu\left(\bigcap_{n=1}^{\infty} A_n\right) = \lim_{n\to\infty} \mu(A_n).$$

Proof.

(a) Letting $A_0 = \varnothing$, we have

$$\mu\left(\bigcup_{n=1}^{\infty} A_n\right) = \mu\left[\bigcup_{n=1}^{\infty} (A_n - A_{n-1})\right] = \sum_{n=1}^{\infty} \mu(A_n - A_{n-1})$$
$$= \lim_{n\to\infty} \sum_{k=1}^{n} [\mu(A_k) - \mu(A_{k-1})] = \lim_{n\to\infty} \mu(A_n).$$

(b) Apply (a) to the increasing sequence of sets $\{A_1 - A_n\}$ to obtain

$$\mu\left[\bigcup_{n=1}^{\infty} (A_1 - A_n)\right] = \lim_{n\to\infty} \mu(A_1 - A_n)$$

or

$$\mu(A_1) - \mu\left(\bigcap_{n=1}^{\infty} A_n\right) = \mu(A_1) - \lim_{n\to\infty} \mu(A_n).$$

Since $\mu(A_1) < \infty$, the desired result follows.

Having dispensed with these measure-theoretic preliminaries, we now consider real-valued functions on X.

A *simple* function is a finite linear combination of characteristic functions of measurable sets and, hence, is of the form

$$x(t) = \sum_{j=1}^{n} \alpha_j \, C(E_j; t).$$

A standard way of representing such a function is obtained if we note that a simple function takes on only finitely many distinct values, $\beta_1, \ldots, \beta_m$. Then, letting $F_k = \{t: \; x(t) = \beta_k\}$, we see that F_k is measurable, $F_i \cap F_j = \varnothing$ for $i \neq j$,

$$X = \bigcup_{k=1}^{m} F_k$$

and

$$x(t) = \sum_{k=1}^{m} \beta_k \, C(F_k; t).$$

The collection of all such simple functions will be denoted by M_1. It is trivial to verify that M_1 *is an algebra of functions closed under the operations of* $\cup$ *and* $\cap$. The natural next class of functions to investigate is the collection of all functions which are pointwise limits of simple functions. This class is of central importance.

Definition 5.3.4. A function x from X into R is called *measurable* if it is the pointwise limit of a sequence of simple functions.

We are *explicitly including* extended real-valued functions as permissible limits. However, simple functions are understood to be finite valued.

We will use the letter M to denote the class of all measurable functions. The reader should have no difficulty in verifying that M *is an algebra of functions which is closed under the operations of* $\cup$ *and* $\cap$; it follows that $|x| \in M$ whenever $x \in M$.

EXAMPLE. For the case $X = R$ with Lebesgue measure on R, all functions in the following categories are measurable.

(a) Continuous functions.
(b) Upper (lower) semicontinuous functions.
(c) Functions having no discontinuities of the oscillatory type.

It is left as an exercise to establish these facts.

The definition of measurability is not always convenient to apply directly. For simple functions, the sets $\{t: \ x(t) > \alpha\}$, $\{t: \ x(t) \geq \alpha\}$, $\{t: \ x(t) < \alpha\}$, and $\{t: \ x(t) \leq \alpha\}$ are measurable, for arbitrary α. The measurablity of such sets is, in fact, characteristic of all measurable functions, as we will demonstrate in theorem 5.3.6. We begin this development as follows.

Lemma 5.3.5. Let x be any function from X into R. Then the following conditions are equivalent:

(i) For any α, $\{t: \ x(t) > \alpha\}$ is measurable.
(ii) For any α, $\{t: \ x(t) \leq \alpha\}$ is measurable.
(iii) For any α, $\{t: \ x(t) \geq \alpha\}$ is measurable.
(iv) For any α, $\{t: \ x(t) < \alpha\}$ is measurable.

Proof. (i) $\Rightarrow$ (ii), for $\{t: \ x(t) \le \alpha\}$ is the complement of $\{t: \ x(t) > \alpha\}$.

(ii) $\Rightarrow$ (iii), for $\{t: \ x(t) \ge \alpha\} = \sim \{t: \ x(t) < \alpha\} = \sim \left[\bigcup_{n=1}^{\infty} \{t: \ x(t) \le \alpha - 1/n\} \right]$, a set which is measurable due to the fact that the measurable sets form a σ-algebra.

(iii) $\Rightarrow$ (iv), for $\{t: \ x(t) < \alpha\} = \sim \{t: \ x(t) \ge \alpha\}$.

(iv) $\Rightarrow$ (i), for $\{t: \ x(t) > \alpha\} = \sim \{t: \ x(t) \le \alpha\} = \sim \left[\bigcap_{n=1}^{\infty} \{t: \ x(t) < \alpha + 1/n\} \right]$, a measurable set. This completes the proof of the lemma.

Now let $x \in M$, and let $\{x_n\}$ be a sequence of simple functions converging pointwise to x. The equality

$$\{t: \ x(t) > \alpha\} = \bigcup_{n=1}^{\infty} \left[\bigcap_{m=n}^{\infty} \{t: \ x_m(t) > \alpha\} \right]$$

together with the fact that $\mathcal{M}$ is a σ-algebra imply that $\{t: \ x(t) > \alpha\}$ is measurable. Hence any measurable function satisfies conditions (i)–(iv). This proves part of the following theorem.

Theorem 5.3.6

(a) A function x is measurable if and only if it satisfies conditions (i)–(iv) of lemma 5.3.5.

(b) A function x is measurable if and only if, for any Borel set A, $x^{-1}(A)$ is measurable.

Proof.

(a) We have left to show that, if x satisfies conditions (i)–(iv) then there exists a sequence $\{x_n\}$ of simple functions converging pointwise to x.

Consider first the case $x(t) \ge 0$ for all t. Define

$$x_n(t) = \begin{cases} n, & \text{if } x(t) \ge n \\ \dfrac{k-1}{2^n}, & \text{if } \dfrac{k-1}{2^n} \le x(t) < \dfrac{k}{2^n} \quad (k = 1, 2, \ldots, n2^n) \end{cases}$$

Conditions (i)–(iv) guarantee that x_n is a simple function, while the increasing fineness of the subdivisions ensures the pointwise convergence of x_n to x.

For an arbitrary measurable function x, we can write $x = x^+ - x^-$, where $x^+ = x \cup O$ and $x^- = (-x) \cup O$, both measurable and nonnegative. By applying the result just proved to x^+ and x^-, we can obtain a sequence of simple functions converging pointwise to x.

(b) If $x^{-1}(A) \in \mathcal{M}$ for any Borel set A, then in particular, $\{t: \ x(t) > \alpha\}$ is measurable for all α, and hence x is measurable by (a). Conversely, the identities

$$x^{-1}\left(\bigcup_{n=1}^{\infty} A_n\right) = \bigcup_{n=1}^{\infty} x^{-1}(A_n), \qquad x^{-1}\left(\bigcap_{n=1}^{\infty} A_n\right) = \bigcap_{n=1}^{\infty} x^{-1}(A_n)$$

and

$$x^{-1}(\tilde{A}) = \sim[x^{-1}(A)]$$

guarantee that the collection $\mathscr{B}$ of all sets B for which $x^{-1}(B) \in \mathscr{M}$ is a σ-algebra. Conditions (i)–(iv), valid for x measurable, then indicate that $\mathscr{B}$ contains all open sets and hence all Borel sets. Therefore, if x is measurable, then $x^{-1}(B)$ is measurable for all Borel sets B. This completes the proof.

Corollary. If x is measurable and bounded below (above), then there is a monotone nondecreasing (nonincreasing) sequence of simple functions converging to x.

Corollary. If x is a bounded measurable function, then there is a monotone sequence of simple functions converging uniformly to x.

The verification of the corollaries is left for the exercises.
For most purposes, sets of measure zero are unimportant. Accordingly, we define

$$x * y \Leftrightarrow \mu\{t: \ x(t) \neq y(t)\} = 0.$$

The relation $*$ is easily shown to be an equivalence relation on the collection of all extended real-valued, measurable functions on X. Moreover, it has the substitution property with respect to the operations of addition, subtraction, multiplication, division (whenever defined), $\cup$, and $\cap$, so that these operations can be applied to the corresponding equivalence classes.

Lemma 5.3.7. If $x \in M$ and $y * x$, then $y \in M$.

Proof. Let $A = \{t: \ x(t) \neq y(t)\}$; then

$$\{t: \ y(t) > \alpha\} = [A \cap \{t: \ y(t) > \alpha\}] \cup [\tilde{A} \cap \{t: \ y(t) > \alpha\}]$$
$$= [A \cap \{t: \ y(t) > \alpha\}] \cup [\tilde{A} \cap \{t: \ x(t) > \alpha\}].$$

The first set, as a subset of a set of measure zero, is measurable, whereas the second set is measurable due to the measurability of x. Therefore, y satisfies condition (i) of lemma 5.3.5, and so $y \in M$.

We now consider pointwise limits of sequences of measurable functions.

Lemma 5.3.8. Let $\{x_n\}$ be a sequence of measurable functions. Then the functions

$$\bar{x}(t) = \sup_{n} x_n(t), \qquad \underline{x}(t) = \inf_{n} x_n(t),$$

$$x_U(t) = \overline{\lim_{n\to\infty}} \, x_n(t), \qquad x_L(t) = \underline{\lim_{n\to\infty}} \, x_n(t),$$

are measurable.

Proof. The measurability of $\bar{x}$ and $\underline{x}$ is a consequence of the set identities

$$\{t:\ \bar{x}(t) > \alpha\} = \bigcup_{n=1}^{\infty} \{t:\ x_n(t) > \alpha\}$$

and

$$\{t:\ \underline{x}(t) < \alpha\} = \bigcup_{n=1}^{\infty} \{t:\ x_n(t) < \alpha\}.$$

The measurability of x_U and x_L follows from the identities

$$x_U(t) = \overline{\lim_{n\to\infty}} \, x_n(t) = \inf_{n} \left[\sup_{m \geq n} x_m(t) \right]$$

and

$$x_L(t) = \underline{\lim_{n\to\infty}} \, x_n(t) = \sup_{n} \left[\inf_{m \geq n} x_m(t) \right],$$

and from the first part of the lemma.

Corollary. If $\lim_{n\to\infty} x_n = x$ pointwise, and if x_n is measurable for $n = 1$, $2, \ldots$, then x is measurable.

The student should note that M is the first useful class of functions encountered which is closed under the operation of taking pointwise limits of sequences.

We will need several types of convergence closely related to pointwise convergence and uniform convergence.

Definition 5.3.9

(a) The sequence $\{x_n\}$ of measurable functions is said to converge to x *almost everywhere* if there exists a set A of measure zero such that, for $t \in \tilde{A}$, $x_n(t) \longrightarrow x(t)$.

(b) The sequence $\{x_n\}$ of almost everywhere finite-valued measurable functions is said to converge *almost uniformly* to x if, for any $\epsilon > 0$, there is a set A_ϵ such that $\mu(A_\epsilon) < \epsilon$ and $\{x_n\}$ converges uniformly to x on $\tilde{A}_\epsilon$.

Note that in (b) the function x must also be almost everywhere finite-valued. The abbreviation "a.e." is standard for "almost everywhere."

Almost uniform convergence is stronger than almost everywhere convergence, as we now show.

Lemma 5.3.10. If $\{x_n\}$ converges almost uniformly to x, then

$$\lim_{n \to \infty} x_n(t) = x(t) \text{ a.e.}$$

Proof. For any m there is a set A_m such that $\mu(A_m) \leq 1/m$ and $\{x_n\}$ converges uniformly, and therefore pointwise, on $\tilde{A}_m$. Hence $x_n \to x$ pointwise on $\bigcup_{m=1}^{\infty} \tilde{A}_m = \sim \left(\bigcap_{m=1}^{\infty} A_m \right)$. Now $\bigcap_{m=1}^{\infty} A_m = \bigcap_{k=1}^{\infty} B_k$, where $B_k = \bigcap_{i=1}^{k} A_i$. But

$$\mu\left(\bigcap_{k=1}^{\infty} B_k \right) = \lim_{k \to \infty} \mu(B_k) = 0$$

since

$$\mu(B_k) \leq \mu(A_k) \leq \frac{1}{k} \quad \text{and} \quad \mu(B_1) = \mu(A_1) < \infty.$$

(See lemma 5.3.3b.) This completes the proof.

For the case where $\mu(X) < \infty$, i.e., every set in $\mathscr{M}$ has finite measure, we give two special results. First we have the following result.

Lemma 5.3.11. If $\mu(X) < \infty$, then every a.e. finite-valued measurable function is the almost uniform limit of a sequence of simple functions.

Proof. Let $E_n = \{t: \ |x(t)| \geq n\}$; then

$$X = E_0 \supseteq E_1 \supseteq \cdots \supseteq E_n \supseteq \cdots \supseteq \bigcap_{n=0}^{\infty} E_n.$$

Since $\mu(E_0) < \infty$, we have

$$0 = \mu\left(\bigcap_{n=0}^{\infty} E_n \right) = \lim_n \mu(E_n).$$

Now x is bounded on $\tilde{E}_n$ and so, by the second corollary to theorem 5.3.6, there is a sequence of simple functions $\{x_{nk}\}$ such that $\lim_{k \to \infty} x_{nk} = x$ *uniformly* on $\tilde{E}_n$. Define x_{nk} to be zero on E_n. Now find an integer $k(n)$ such that

$$|x_{nk(n)}(t) - x(t)| < \frac{1}{n} \quad \text{on} \quad \tilde{E}_n$$

and let $x_n = x_{nk(n)}$. The sequence $\{x_n\}$ then converges almost uniformly to x, and the proof is finished.

Finally, we have the surprising and important result of Egoroff, which implies that, for $\mu(X) < \infty$, almost uniform convergence is equivalent to almost everywhere convergence.

Theorem 5.3.12 (Egoroff). *If $\{x_n\}$ is a sequence of a.e. finite-valued measurable functions converging a.e. to the a.e. finite-valued function x, then in the case that $\mu(X) < \infty$, the sequence converges almost uniformly to x.*

Proof. Let

$$E_k^n = \{t: \ |x_n(t) - x(t)| > 2^{-k}\} \quad \text{and} \quad E_k^{(n)} = \bigcup_{m=n}^{\infty} E_k^m.$$

Clearly,

$$E_k^{(n)} \supseteq E_k^{(n+1)}, \qquad \text{for} \quad n = 1, 2, \ldots,$$

and

$$t_0 \in \bigcap_{n=1}^{\infty} E_k^{(n)}$$

if and only if $|x_n(t_0) - x(t_0)| > 2^{-k}$ for infinitely many values of n. The set

$$E_0 = \bigcup_{k=1}^{\infty} \bigcap_{n=1}^{\infty} E_k^{(n)}$$

is then the set where pointwise convergence fails, and by hypothesis, $\mu(E_0) = 0$. It follows at once that

$$\mu\left[\bigcap_{n=1}^{\infty} E_k^{(n)}\right] = 0, \qquad \text{for all } k,$$

and hence

$$\lim_{n \to \infty} \mu[E_k^{(n)}] = 0, \qquad \text{for all } k$$

(by lemma 5.3.3(b) and $\mu(X) < \infty$).

We can then obtain a strictly increasing sequence of integers $m[k]$ such that

$$\mu[E_k^{(m)}] \leq 2^{-k}, \qquad \text{for} \quad m \geq m[k].$$

Let

$$E_m = \bigcup_{k=m}^{\infty} E_k^{(m[k])};$$

then $E_m \supseteq E_{m+1}$ and

$$\mu(E_m) \leq \sum_{k=m}^{\infty} \mu(E_k^{(m[k])}) \leq \sum_{k=m}^{\infty} 2^{-k} = 2^{-m+1},$$

so that $\lim_{m \to \infty} \mu(E_m) = 0$.

We now show that the convergence is uniform in $\tilde{E}_m$. Now

$$\tilde{E}_m = \sim\left[\bigcup_{k=m}^{\infty} E_k^{(m[k])}\right] = \bigcap_{k=m}^{\infty} \sim[E_k^{(m[k])}]$$

and

$$\sim E_k^{(m[k])} = \{t: \quad |x(t) - x_p(t)| \le 2^{-k} \quad \text{for all } p \ge m[k]\}.$$

Thus, given an $\epsilon > 0$, we choose k such that $2^{-k} < \epsilon$, and we note that any $t_0 \in \tilde{E}_m$ is in $\sim E_k^{(m[k])}$. Hence

$$|x(t_0) - x_p(t_0)| \le 2^{-k} < \epsilon, \quad \text{for } p \ge m(k),$$

which states exactly that $\{x_n\}$ converges uniformly on $\tilde{E}_m$. Since $\mu(E_m) \to 0$ as $m \to \infty$, this shows the almost uniform convergence of the sequence $\{x_n\}$. The proof is now complete.

While the class M of all measurable functions has been investigated in some detail, and various types of convergence have been studied, no metric space or normed vector space has been found. We close this section with a description of a class of measurable functions which is a Banach algebra.

Definition 5.3.13. A measurable function x is *essentially bounded* if there exists a set A of measure zero such that $|x|$ is bounded on $\tilde{A}$.

It is easy to verify that the collection of all essentially bounded functions is a subalgebra of M, closed under the operations $\cup$ and $\cap$. To obtain a norm we let $\{A_\alpha\}$ be the collection of sets of measure zero such that

$$N_\alpha(x) = \sup\{|x(t)|: \quad t \in \tilde{A}_\alpha\} < \infty.$$

This collection is nonempty by hypothesis. We now define

$$\|x\| = \inf_\alpha N_\alpha(x).$$

Lemma 5.3.14. If x is measurable and essentially bounded, then there exists a set A_0 of measure zero such that

$$\|x\| = \sup\{|x(t)|: \quad t \in \tilde{A}_0\}.$$

Moreover, if A_1 is any set of measure zero such that $A_1 \supseteq A_0$, then

$$\|x\| = \sup\{|x(t)|: \quad t \in \tilde{A}_1\}.$$

The proof is left as an exercise. We now verify that $\|x\|$ is a norm function, provided that we identify functions which agree almost everywhere.

(a) $\|x\| \geq 0$ clearly. If $\|x\| = 0$, then $x(t) = 0$ for all $t \in \tilde{A}_0$, and $x(t) = 0$ almost everywhere. Thus we see that we must work with equivalence classes of measurable functions $[x * y \Leftrightarrow x(t) = y(t) \text{ a.e.}]$.

(b) $\|\alpha x\| = |\alpha| \|x\|$, obviously.

(c) Let A_x and A_y be sets of measure zero such that

$$\|x\| = \sup \{|x(t)|: \quad t \in \tilde{A}_x\}$$

and

$$\|y\| = \sup \{|x(t)|: \quad t \in \tilde{A}_y\}.$$

Then we have $\mu(A_x \cup A_y) = 0$ and, for $t \in \sim[A_x \cup A_y]$,

$$|x(t) + y(t)| \leq |x(t)| + |y(t)| \leq \|x\| + \|y\|,$$

so that

$$\|x + y\| \leq \sup \{|x(t) + y(t)|: \quad t \in \sim[A_x \cup A_y]\} \leq \|x\| + \|y\|.$$

(d) With A_x and A_y as above, we have, for $t \in \sim[A_x \cup A_y]$,

$$|x(t)y(t)| = |x(t)||y(t)| \leq \|x\| \|y\|,$$

so that

$$\|xy\| \leq \sup \{|x(t)y(t)|: \quad t \in \sim[A_x \cup A_y]\} \leq \|x\| \|y\|.$$

To prove the completeness, let $\{x_n\}$ be a Cauchy sequence, let

$$A_{mn} = \{t: \quad |x_n(t) - x_m(t)| > \|x_m - x_n\|\},$$

and

$$A_0 = \bigcup_{m,\,n=1}^{\infty} A_{mn}.$$

Then $\mu(A_{mn}) = 0$ and $\mu(A_0) = 0$. On $\tilde{A}_0$ the sequence $\{x_n\}$ is Cauchy in the uniform sense and, hence, has a uniform limit x which is measurable. The function x can be defined arbitrarily on A_0 since we are working with equivalence classes.

We will use the notation $L^{(\infty)}$ to designate the set of equivalence classes of essentially bounded functions. Our results are summarized in the following theorem.

Theorem 5.3.15. $L^{(\infty)}$ is a Banach algebra.

Convergence in the $L^{(\infty)}$ norm is equivalent to almost everywhere uniform convergence, defined in the obvious way.

EXERCISES

1. Let $(X, \mathcal{M}, \mu)$ be a measure space where μ is σ-finite. Show that any set $A \in \mathcal{M}$ can be written $A = \bigcup_{n=1}^{\infty} A_n$ where $A_i \cap A_j = \varnothing$ and $\mu(A_i) < \infty$ for all i and j.

2. Prove that any measure is monotone and subtractive.

3. Prove lemma 5.3.2.

4. Find in R a sequence of Lebesgue measurable sets $A_1 \supset A_2 \supset \cdots \supset A_n \supset \cdots$ such that $\mu\left(\bigcap_{n=1}^{\infty} A_n\right) = \alpha$, an arbitrary preassigned nonnegative number, but $\mu(A_n) = +\infty$ for all n.

5. a. Show that M_1, the collection of all simple functions, is an algebra of functions closed under the operations of $\cup$ and $\cap$.
   b. Do the same for M, the collection of all measurable functions.

6. Verify the assertions made in the example following definition 5.3.4.

7. Suppose, for every α, that the set $\{t: \; x(t) = \alpha\}$ is measurable. Is x measurable? (Assume there exists a nonmeasurable set.)

8. a. Prove the first corollary to theorem 5.3.6.
   b. Prove the second corollary to theorem 5.3.6.

9. Prove lemma 5.3.14.

10. Show that the operations $\cup$ and $\cap$ are continuous functions from $L^{(\infty)} \times L^{(\infty)}$ onto $L^{(\infty)}$.

11. Suppose $|x|$ is measurable. Discuss the question of the measurability of x. (Assume the existence of a nonmeasurable set.)

12. Is the reciprocal of an essentially bounded function also essentially bounded?

13. a. Show that lemma 5.3.11 is false if $\mu(X) = +\infty$.
   b. Show that Egoroff's theorem is false for $\mu(X) = +\infty$.

14. Prove that for $x, y \in M$ the set $\{t: \ x(t) \le y(t)\}$ is measurable.

15. a. If $x(t)$ is a real-valued continuous function on R, show that the inverse image of a Borel set is a Borel set. Is the converse true?
   b. Let $x(t)$ be a continous function on R and $y(t)$ a measurable function on R. What can you say about $x[y(t)]$?

16. a. Let I^+ be the set of all positive integers, and $\mathcal{M}$ the collection of all subsets of I^+. Further, let $\{\alpha_n\}$ be a sequence of nonnegative real numbers such that $\sum_{n=1}^{\infty} \alpha_n < \infty$, and define $\mu(A) = \sum_{k \in A} \alpha_k$ for $A \in \mathcal{M}$. Show that $(I^+, \mathcal{M}, \mu)$ is a measure space. (Define $\mu(\varnothing) = 0$.) Is it complete? σ-finite?
   b. Discuss the problem in the case that $\sum_{k=1}^{\infty} \alpha_k$ diverges.

17. Let $(X, \mathcal{M}, \mu)$ be a measure space, and E an arbitrary fixed set in $\mathcal{M}$. Let $\mathcal{M}_1$ be the collection of all subsets A of E which are in $\mathcal{M}$, and define $\mu_1(A) = \mu(A)$. Show that $(E, \mathcal{M}_1, \mu_1)$ is a measure space.

18. Let $(X, \mathcal{M}, \mu)$ be a σ-finite measure space. Let M_0 be the subspace of M_1 consisting of those simple functions for which $\mu\{t: \ |x(t)| > 0\} < \infty$.

   a. Show that every measurable function is a pointwise limit of functions in M_0.
   b. What happens if the σ-finite assumption is dropped?

19. Let $(X, \mathcal{M}, \mu)$ be a σ-finite measure space.

   a. Characterize the class of all functions which are uniform limits of functions in M_1.
   b. Characterize the class of all functions which are uniform limits of functions in M_0. (See exercise 18.)
   c. Discuss the problem if the σ-finiteness assumption is dropped.

20. Define the term "almost bounded" for measurable functions, and let AB be the collection of all almost bounded functions.

   a. Discuss the closure of AB under various algebraic operations and convergences.
   b. For $\mu(X) < \infty$, what can be said about a function $x \in AB$?

21. Define and discuss a notion of "almost convergent" for sequences in M.

Section 4   CONVERGENCE IN MEASURE

In this section we treat an entirely new kind of convergence which plays an important role in integration theory.

Definition 5.4.1. A sequence of a.e. finite-valued measurable functions $\{x_n\}$ *converges in measure* to an a.e. finite-valued measurable function x if, for any $\epsilon > 0$,

$$\lim_{n \to \infty} \mu\{t: \quad |x_n(t) - x(t)| > \epsilon\} = 0.$$

To see how drastically measure convergence can differ from pointwise convergence, we consider the following example.

EXAMPLE. Partition the interval $[-n, n)$ into $n \cdot 2^{n+1}$ left-closed, right-open subintervals, each of length 2^{-n}. Let x_{kn} be the characteristic function of the kth subinterval. The sequence

$$x_{1,1}, x_{2,1}, x_{3,1}, x_{4,1}, x_{1,2}, \ldots, x_{16,2}; x_{1,3}, \ldots, x_{48,3}; \ldots$$

converges in measure to the zero function although $\{x_{k,n}(t_0)\}$ does not converge for any t_0.

Definition 5.4.2. A sequence of a.e. finite-valued measurable functions $\{x_n\}$ is *fundamental in measure* if, for any $\epsilon > 0$,

$$\lim_{m, n \to \infty} \mu\{t: \quad |x_m(t) - x_n(t)| > \epsilon\} = 0.$$

The definition resembles that of a Cauchy sequence, but we refrain from using this term because M, the set of all measurable functions, is not, in general, a metric space. Some elementary properties of convergence in measure are given in the following lemma.

Lemma 5.4.3

(a) If $\{x_n\}$ converges in measure to both y and z, then $y = z$ almost everywhere.

(b) If $x_n \to x$ almost uniformly, then $x_n \to x$ in measure.

(c) If $\mu(X) < \infty$ and if $x_n \to x_0$ almost everywhere, where each x_n is a.e. finite-valued, then $x_n \to x_0$ in measure.

(d) If $\{x_n\}$ converges in measure, then it is fundamental in measure.

Proof

(a) $\{t: \ |z(t) - y(t)| > \epsilon\} \subseteq \{t: \ |z(t) - x_n(t)| > \epsilon/2\} \cup \{t: \ |x_n(t) - y(t)| > \epsilon/2\}$. Since the measure of the right member of this set inclusion can be made arbitrarily small by taking n sufficiently large, we obtain

$$\mu\{t: \ |y(t) - x(t)| > \epsilon\} = 0, \qquad \text{for any } \epsilon > 0.$$

Finally,

$$\{t: \ |y(t) - z(t)| > 0\} = \bigcup_{k=1}^{\infty} \left\{t: \ |y(t) - z(t)| > \frac{1}{k}\right\},$$

a countable union of sets of measure zero, so that $y = z$ almost everywhere.

(b) Let $x_n \to x$ almost uniformly, and let ϵ and η be arbitrary positive numbers. There is a set F of measure less than η such that $x_n \to x$ uniformly on $\tilde{F}$; hence, there is an integer $N(\epsilon, \eta)$ such that

$$n \geq N(\epsilon, \eta) \Rightarrow |x_n(t) - x(t)| \leq \epsilon, \qquad \text{for} \quad t \in \tilde{F}.$$

Therefore

$$\mu\{t: \ |x_n(t) - x(t)| > \epsilon\} < \eta$$

whenever $n \geq N(\epsilon, \eta)$, and $x_n \to x$ in measure.

(c) This proof follows at once from (b) and from the theorem of Egoroff.

(d) The set inclusion

$$\{t: \ |x_n(t) - x_m(t)| > \epsilon\} \leq \left\{t: \ |x_n(t) - x(t)| > \frac{\epsilon}{2}\right\}$$
$$\cup \left\{t: \ |x(t) - x_m(t)| > \frac{\epsilon}{2}\right\}$$

yields (d), since the measure of the right member tends to zero as $m, n \to \infty$.

The converse of (d) is our immediate goal; we will need the following lemma.

Lemma 5.4.4. If $\{x_n\}$ is a sequence of a.e. finite-valued measurable functions which is fundamental in measure, then some subsequence is almost uniformly convergent.

Proof. Since $\{x_n\}$ is fundamental in measure, we can, for any k, find an $N(k)$ such that $m, n \geq N(k)$ implies

$$\mu\{t: \ |x_n(t) - x_m(t)| \geq 2^{-k}\} < 2^{-k},$$

and without loss we can assume $N(1) < N(2) < \cdots < N(k) \cdots$. Let

$$y_k = x_{N(k)},$$
$$A_k = \{t: \ |y_k(t) - y_{k+1}(t)| \geq 2^{-k}\},$$

and

$$B_k = \bigcup_{j=k}^{\infty} A_j.$$

Then

$$\mu(B_k) \leq \sum_{j=k}^{\infty} \mu(A_j) < \sum_{j=k}^{\infty} 2^{-j} = 2^{-k+1},$$

so that $\mu(B_k) \to 0$ as $k \to \infty$. Hence we need only show that $\{y_k\}$ converges uniformly on $\tilde{B}_k$. Now, for $p \leq q$,

$$|y_p(t) - y_q(t)| \leq |y_p(t) - y_{p+1}(t)|$$
$$+ |y_{p+1}(t) - y_{p+2}(t)| + \cdots + |y_{q-1}(t) - y_q(t)|,$$

and if $k \leq p$, $t \in \tilde{B}_k$, then

$$t \in \bigcap_{j=k}^{\infty} \tilde{A}_j \quad \text{and} \quad |y_p(t) - y_q(t)| \leq \sum_{i=p}^{\infty} 2^{-i} = 2^{-p+1}.$$

Therefore, $\{y_k\}$ satisfies a uniform Cauchy condition on $\tilde{B}_k$ and, hence, converges uniformly on $\tilde{B}_k$. This completes the proof.

It is now easy to obtain the converse of lemma 5.4.3(d).

Theorem 5.4.5. If $\{x_n\}$ is a sequence of a.e. finite-valued measurable functions which is fundamental in measure, then there exists a measurable function x, almost everywhere uniquely determined, such that $\{x_n\}$ converges in measure to x.

Proof. The uniqueness is a consequence of lemma 5.4.3(a). By the preceding lemma, a subsequence $\{y_k\}$ of $\{x_n\}$ converges almost uniformly to a measurable function x. Now,

$$\{t: \ |x_n(t) - x(t)| > \epsilon\} \subseteq \left\{t: \ |x_n(t) - y_k(t)| > \frac{\epsilon}{2}\right\}$$
$$\cup \left\{t: \ |y_k(t) - x(t)| > \frac{\epsilon}{2}\right\}.$$

Since $\{y_k\}$ converges in measure to x by lemma 5.4.3(b), $\mu\{t: \ |y_k(t) - x(t)| > \epsilon/2\}$ can be made arbitrarily small by taking k sufficiently large. Moreover, $\mu\{t: \ |x_n(t) - y_k(t)| > \epsilon/2\}$ can similarly be made small, since $\{x_n\}$ is fundamental in measure. Hence

$$\underset{n \to \infty}{\text{limit}} \ \mu\{t: \ |x_n(t) - x(t)| > \epsilon\} = 0,$$

and $\{x_n\}$ converges in measure to x. This completes the proof.

The parallel with metric space convergence can be drawn further. If we define

$$\{x_n\} \sim \{y_n\} \Leftrightarrow [\text{for any } \epsilon > 0, \ \underset{n \to \infty}{\text{limit}} \ \mu\{t: \ |x_n(t) - y_n(t)| > \epsilon\} = 0],$$

then $\sim$ is an equivalence relation on the set of fundamental in measure sequences, and two such sequences have the same limit if and only if they are equivalent. The proofs are left as exercises. We now turn our attention to some algebraic properties of convergence in measure.

Theorem 5.4.6. Let $\{x_n\}$ and $\{y_n\}$ be sequences converging in measure to x and y, respectively. Then

(a) $\{x_n + y_n\}$ converges in measure to $x + y$.
(b) $\{\alpha x_n\}$ converges in measure to αx.
(c) For $\mu(X) < \infty$, $\{x_n y_n\}$ converges in measure to xy.
(d) $\{|x_n|\}$ converges in measure to $|x|$.

Proof
(a) is an immediate consequence of the set inclusion

$$\{t: \ |[x_n(t) + y_n(t)] - [x(t) + y(t)]| > \epsilon\} \subseteq \left\{t: \ |x_n(t) - x(t)| > \frac{\epsilon}{2}\right\}$$
$$\cup \left\{t: \ |y_n(t) - y(t)| > \frac{\epsilon}{2}\right\}.$$

(b) is left as an exercise.
(c) can be treated by writing

$$x_n(t)y_n(t) - x(t)y(t) = [x_n(t) - x(t)][y_n(t) - y(t)]$$
$$+ x(t)[y_n(t) - y(t)] + y(t)[x_n(t) - x(t)],$$

and dealing with the individual terms. Let $u_n = x_n - x$ and $v_n = y_n - y$; then $\{u_n\}$ and $\{v_n\}$ converge in measure to the zero function, and

$$\{t: \ |u_n(t)v_n(t)| > \epsilon\} \subseteq \{t: \ |u_n(t)| > \sqrt{\epsilon}\} \cup \{t: \ |v_n(t)| > \sqrt{\epsilon}\}.$$

From this it follows that $\{u_n v_n\}$ converges in measure to the zero function. The other two terms are handled alike; hence we treat only the last one. We have

$$\{t:\ |y(t)u_n(t)| > \epsilon\} \subseteq \{t:\ |y(t)| > N\} \cup \left\{t:\ |u_n(t)| > \frac{\epsilon}{N}\right\}.$$

If $E_N = \{t:\ |y(t)| > N\}$, then $E_N \supseteq E_{N+1}$ and $\bigcap_{N=1}^{\infty} E_N = \varnothing$. Because $\mu(X) < \infty$, we have $\lim_{N \to \infty} \mu(E_N) = 0$. Now, we first choose N so that $\mu\{t:\ |y(t)| > N\} \leq \eta/2$ and then choose $N(\eta)$ so that $n \geq N(\eta)$ implies $\mu\{t:\ |u_n(t)| > \epsilon/N\} \leq \eta/2$. It follows that

$$\mu\{t:\ |y(t)u_n(t)| > \epsilon\} \leq \frac{\eta}{2} + \frac{\eta}{2} = \eta$$

for $n \geq N(\eta)$, and the proof of (c) is complete.
(d) is left for the exercises.

For $\mu(X) < \infty$, parts (a) and (c) of the preceding theorem indicate that $x_n \to x$, $y_n \to y$, $\alpha_n \to \alpha$ implies $(x_n + y_n) \to (x + y)$ and $\alpha_n x_n \to \alpha x$. It is reasonable, therefore, to seek a metric on the class of a.e. finite-valued measurable functions under which metric convergence is equivalent to convergence in measure. For $\mu(X) = +\infty$, the second of the foregoing convergence properties does not always hold, and in that case, a metric of the desired type certainly cannot be found. These matters are taken up in the exercises.

EXERCISES

1. Let $\{x_n\}$ and $\{y_n\}$ be sequences converging in measure to x and y, respectively.
   a. Show that $|x_n|$ converges in measure to $|x|$.
   b. Show that $\{x_n \cup y_n\}$ and $\{x_n \cap y_n\}$ converge in measure to $x \cup y$ and $x \cap y$, respectively.

2. a. Show that the relation
   $$\{x_n\} \sim \{y_n\} \Leftrightarrow [\text{for any } \epsilon > 0, \lim_{n \to \infty} \mu\{t:\ |x_n(t) - y_n(t)| > \epsilon\} = 0]$$
   is an equivalence relation on the collection of all fundamental in measure sequences.
   b. Show that two such sequences are equivalent if and only if they have the same limit.

3. Prove part (b) of theorem 5.4.6.

4. a. Show that, for $\mu(X) = +\infty$, it can happen that $\alpha_n \to \alpha$ and $\{x_n\}$ converges in measure to x without $\{\alpha_n x_n\}$ converging in measure to αx.

b. Show that if $x \in L^{(\infty)}$ and if $\{y_n\}$ converges in measure to y, then $\{xy_n\}$ converges in measure to xy.

c. Obtain a weaker condition on x to ensure that $\{xy_n\}$ converges in measure to xy when $\{y_n\}$ converges in measure to y. (See exercise 20 of section 3.)

5. Show that part (c) of lemma 5.4.3 is false for $\mu(X) = +\infty$.

6. a. For $\mu(X) < \infty$, let $\{x_n\}$ be a sequence converging in measure to x_0. Assume $\mu\{t\colon x_n(t) = 0\} = 0$ for $n = 0, 1, 2, \ldots$. Show that $\{x_n^{-1}\}$ converges in measure to x_0^{-1}.

b. Can part a be generalized to the case $\mu(X) = +\infty$?

7. Let $\{x_n\}$ be fundamental in measure, and let $\{y_k\}$ and $\{z_j\}$ be subsequences converging almost everywhere to y and z, respectively. Show that $y = z$ almost everywhere.

8. For $\mu(X) < \infty$, let M' be the vector space of all a.e. finite-valued measurable functions, where functions are identified if they differ only on a set of measure zero. Define $F_x(\epsilon) = \mu\{t\colon |x(t)| > \epsilon\}$ and $d(x, y) = \inf\{\epsilon\colon F_{x-y}(\epsilon) \le \epsilon\}$. Prove the following.

a. $F_x(\epsilon)$ maps $[0, \infty)$ into $[0, \mu(X)]$.

b. $\eta < \epsilon \Rightarrow F_x(\eta) \ge F_x(\epsilon)$ so that $F_x(\epsilon)$ is monotone nonincreasing.

c. $F_x(\epsilon)$ is right continuous but not necessarily left continous.

d. $\mu\{t\colon |x(t)| > d(x, 0) + \epsilon\} \le d(x, 0) + \epsilon$.

e. $d(x, y)$ is a translation invariant metric on M'.

f. Convergence under this metric is equivalent to convergence in measure; hence, M' is complete under the metric.

g. For $X = [0, 1]$ and $x(t) = t^2$, show $d(2x, 0) \ne 2d(x, 0)$, so that the metric is not a norm.

Section 5 THE INTEGRAL

Let us recall some notation. If $(X, \mathcal{M}, \mu)$ is a measure space, then

$$M = \text{the measurable functions on } X, \quad \text{and}$$
$$M_1 = \text{the simple functions on } X.$$

We now define

$$M_0 = \{s \in M_1: \quad \mu\{t: \quad s(t) \neq 0\} < \infty\}.$$

For $s \in M_0$, the integral is defined in the obvious way:

$$\int s \, d\mu = \sum_{j=1}^{n} \alpha_j \mu(E_j),$$

where $\alpha_1, \ldots, \alpha_n$ are the nonzero values assumed by s, and $E_j = \{t: \quad s(t) = \alpha_j\}$, $j = 1, \ldots, n$. The following lemma is equally obvious.

Lemma 5.5.1. For $s, s_1, s_2 \in M_0$, we have

$$\int (\alpha_1 s_1 + \alpha_2 s_2) \, d\mu = \alpha_1 \int s_1 \, d\mu + \alpha_2 \int s_2 \, d\mu$$

$$s \geq 0 \Rightarrow \int s \, d\mu \geq 0$$

$$\left| \int s \, d\mu \right| \leq \int |s| \, d\mu.$$

The integral on M_0 is extended to a wide class of measurable functions by the following procedure.

Definition 5.5.2

(a) For $x \in M$, $x \geq 0$, let

$$\int x \, d\mu = \sup \left\{ \int s \, d\mu: \quad 0 \leq s \leq x, \ s \in M_0 \right\}.$$

(b) For $x \in M$, write $x = x^+ - x^-$; then, if either

$$\int x^+ \, d\mu < \infty \quad \text{or} \quad \int x^- \, d\mu < \infty,$$

let

$$\int x \, d\mu = \int x^+ \, d\mu - \int x^- \, d\mu.$$

Note that in the case $\int x^+ \, d\mu = \int x^- \, d\mu = \infty$, the symbol $\int x \, d\mu$ is not defined. If exactly one of the numbers $\int x^+ \, d\mu$, $\int x^- \, d\mu$ is finite, then $\int x \, d\mu$ is either $+\infty$ or $-\infty$.

The symbol $L^{(1)}(\mu)$ is used to denote the class of functions for which both $\int x^+ \, d\mu$ and $\int x^- \, d\mu$ are finite.

Several properties of the integral are immediate consequences of the definition; we list them below and leave the proofs as exercises.

1. $\int \alpha x \, d\mu = \alpha \int x \, d\mu.$

2. $0 \le x \le y \Rightarrow 0 \le \int x \, d\mu \le \int y \, d\mu.$

We now define the set function

$$\int_A x \, d\mu = \int x \, C(A) \, d\mu$$

where $C(A)$ is the characteristic function of the (measurable) set A. The following theorem and corollary about this function are of central importance.

Theorem 5.5.3. For any $x \in M$, $x \ge 0$, the function

$$\nu(A) = \int_A x \, d\mu$$

is countably additive; that is, if $\{A_n\}$ is a sequence of disjoint sets in $\mathcal{M}$, then

$$\nu\left(\bigcup_{n=1}^{\infty} A_n\right) = \sum_{n=1}^{\infty} \nu(A_n).$$

Proof. If x is the characteristic function of a measurable set E, then the theorem is a restatement of the countable additivity of μ. If $x \in M_0$, $x \ge 0$, i.e.,

$$x(t) = \sum_{j=1}^{m} \alpha_j C(E_j; t), \qquad \alpha_j \ge 0 \, (j = 1, \ldots, m),$$

then

$$\nu(A_n) = \int_{A_n} x \, d\mu = \sum_{j=1}^{m} \alpha_j \mu(E_j \cap A_n)$$

and

$$\sum_{n=1}^{\infty} \nu(A_n) = \sum_{n=1}^{\infty} \left(\int_{A_n} x \, d\mu \right) = \sum_{j=1}^{m} \alpha_j \sum_{n=1}^{\infty} \mu(E_j \cap A_n)$$

$$= \sum_{j=1}^{m} \alpha_j \mu\left(E_j \cap \left[\bigcup_{n=1}^{\infty} A_n\right]\right) = \int_{\bigcup_{n=1}^{\infty} A_n} x \, d\mu$$

$$= \nu\left(\bigcup_{n=1}^{\infty} A_n\right),$$

as desired.

In the general case we have, for any simple function $s \in M_0$ satisfying $0 \le s \le x$,

$$\int_{\bigcup_{n=1}^{\infty} A_n} s \, d\mu = \sum_{n=1}^{\infty} \int_{A_n} s \, d\mu \leq \sum_{n=1}^{\infty} \int_{A_n} x \, d\mu$$

so that

$$\int_{\bigcup_{n=1}^{\infty} A_n} x \, d\mu \leq \sum_{n=1}^{\infty} \int_{A_n} x \, d\mu,$$

or

$$\nu\left(\bigcup_{n=1}^{\infty} A_n\right) \leq \sum_{n=1}^{\infty} \nu(A_n).$$

If, for any n, $\nu(A_n) = \infty$, then the reverse inequality obviously holds. We now assume $\nu(A_n) < \infty$ for all n. Fixing our attention on A_1 and A_2, we find simple functions s_1 and s_2 in M_0 such that s_i vanishes on $\tilde{A}_i$, $0 \leq s_i \leq x$, and

$$\int_{A_1} x \, d\mu \leq \int_{A_1} s_1 \, d\mu + \frac{\epsilon}{2},$$

$$\int_{A_2} x \, d\mu \leq \int_{A_2} s_2 \, d\mu + \frac{\epsilon}{2}.$$

Then

$$\int_{A_1} x \, d\mu + \int_{A_2} x \, d\mu \leq \int_{A_1} s_1 \, d\mu + \int_{A_2} s_2 \, d\mu + \epsilon$$

and, letting $s = s_1 + s_2$, we see that $0 \leq s \leq x$ and

$$\int_{A_1} x \, d\mu + \int_{A_2} x \, d\mu \leq \int_{A_1 \cup A_2} s \, d\mu + \epsilon$$

$$\leq \int_{A_1 \cup A_2} x \, d\mu + \epsilon.$$

Since this holds for every $\epsilon > 0$, we have

$$\nu(A_1) + \nu(A_2) \leq \nu(A_1 \cup A_2)$$

and, by induction,

$$\sum_{n=1}^{N} \nu(A_n) \leq \nu\left(\bigcup_{n=1}^{N} A_n\right)$$

for all N. But $\nu\left(\bigcup_{n=1}^{N} A_n\right) \leq \nu\left(\bigcup_{n=1}^{\infty} A_n\right)$; therefore

$$\sum_{n=1}^{N} \nu(A_n) \leq \nu\left(\bigcup_{n=1}^{\infty} A_n\right)$$

for all N; this implies

$$\sum_{n=1}^{\infty} \nu(A_n) \le \nu\left(\bigcup_{n=1}^{\infty} A_n\right)$$

as desired, and the proof is complete.

Corollary. For $x \in L^{(1)}(\mu)$, the function $\nu(A) = \int_A x \, d\mu$ is countably additive.

The corollary is easily proved if we write $x = x^+ - x^-$ and apply the theorem.

The additivity of ν has several immediate and useful consequences.

(a) If $\mu(B) = 0$, then

$$\int_A x \, d\mu = \int_{A \cup B} x \, d\mu$$

for

$$\int_{A \cup B} x \, d\mu = \int_A x \, d\mu + \int_{B-A} x \, d\mu,$$

and the second term on the right is zero.

(b) Letting

$$A_+ = \{t: \quad x(t) \ge 0\}$$
$$A_- = \{t: \quad x(t) < 0\},$$

we have

$$\int |x| \, d\mu = \int_{A_+} |x| \, d\mu + \int_{A_-} |x| \, d\mu$$
$$= \int x^+ \, d\mu + \int x^- \, d\mu$$

so that

$$|x| \in L^{(1)} \Leftrightarrow x \in L^{(1)}.$$

Moreover,

$$\left| \int x \, d\mu \right| = \left| \int x^+ \, d\mu - \int x^- \, d\mu \right|$$
$$\le \left| \int x^+ \, d\mu \right| + \left| \int x^- \, d\mu \right|,$$

and thus

$$\left| \int x \, d\mu \right| \le \int |x| \, d\mu$$

for any $x \in L^{(1)}$.

(c) If $x \in M$, $y \in L^{(1)}$ and $|x| \le y$, then $x \in L^{(1)}$. (The inequality implies that $|x| \in L^{(1)}$, and so, by (b), $x \in L^{(1)}$.)

(d) From the inequality $|x \pm y| \le |x| + |y|$, we see that

$$x, y \in L^{(1)} \Rightarrow (x \pm y) \in L^{(1)}.$$

(e) Referring to definition 5.3.1, we see that the set function

$$\nu(A) = \int_A x \, d\mu$$

is a measure on $\mathcal{M}$ whenever $x \ge 0$, $x \in M$. In particular, $\nu(A)$ satisfies the conclusion of lemma 5.3.3.

The additivity of the integral has not yet been demonstrated. The following result, known as the *Lebesgue monotone convergence theorem*, will enable us to prove the desired additivity.

Theorem 5.5.4. Let $\{x_n\}$ be a sequence of nonnegative measurable functions such that $x_n(t) \le x_{n+1}(t)$. If $x(t) = \lim_{n \to \infty} x_n(t)$, then

$$\lim_{n \to \infty} \int x_n \, d\mu = \int x \, d\mu.$$

Proof. If $\lim_{n \to \infty} \int x_n \, d\mu = \infty$, then clearly, $\int x \, d\mu = \infty$. Hence we assume

$$\lim_{n \to \infty} \int x_n \, d\mu = \alpha < \infty.$$

Let β be any number such that $0 < \beta < 1$, and define

$$E_n = \{t: \ x_n(t) \ge \beta x(t)\}.$$

We have

$$\beta \int_{E_n} x \, d\mu = \int_{E_n} \beta x \, d\mu \le \int_{E_n} x_n \, d\mu \le \int x_n \, d\mu \le \alpha$$

for all n. Applying lemma 5.3.3 to the measure

$$\nu(A) = \int x \, d\mu,$$

we obtain, by letting $n \to \infty$,

$$\beta \int x \, d\mu \le \alpha,$$

since $E_n \subseteq E_{n+1}$ for $n = 1, 2, \ldots$. Thus,

$$\int x \, d\mu \le \frac{1}{\beta} \operatorname*{limit}_{n \to \infty} \int x_n \, d\mu$$

and, letting $\beta \uparrow 1$, we have

$$\int x \, d\mu \le \operatorname*{limit}_{n \to \infty} \int x_n \, d\mu.$$

Since the reverse inequality is obvious, the proof is complete.

We are now ready to prove the additivity.

Theorem 5.5.5. If $x, y \in L^{(1)}(\mu)$, then

$$\int (x + y) \, d\mu = \int x \, d\mu + \int y \, d\mu.$$

Proof. Consider first the case $x, y \ge 0$. By theorem 5.3.6, there exist sequences of simple functions $\{s_n\}$ and $\{s'_n\}$ such that $s_n \uparrow x$ and $s'_n \uparrow y$, and by the Lebesgue monotone convergence theorem,

$$\int s_n \, d\mu \to \int x \, d\mu \quad \text{and} \quad \int s'_n \, d\mu \to \int y \, d\mu.$$

Thus, we have

$$\int (s_n + s'_n) \, d\mu = \int s_n \, d\mu + \int s'_n \, d\mu \to \int x \, d\mu + \int y \, d\mu,$$

since the integral is additive on simple functions. But $(s_n + s'_n) \uparrow (x + y)$, and hence, we also have

$$\int (s_n + s'_n) \, d\mu \to \int (x + y) \, d\mu.$$

It follows that

$$\int (x + y) \, d\mu = \int x \, d\mu + \int y \, d\mu$$

for $x, y \ge 0$. Clearly, this result also holds when $x, y \le 0$.

Now consider the case $x \geq 0$, $y \leq 0$, Let

$$A = \{t: \quad x(t) + y(t) \geq 0\}$$
$$B = \{t: \quad x(t) + y(t) < 0\};$$

we then have

$$\int_A x \, d\mu = \int_A [(x + y) + (-y)] \, d\mu = \int_A (x + y) \, d\mu + \int_A (-y) \, d\mu$$

and

$$\int_B y \, d\mu = \int_B [(x + y) + (-x)] \, d\mu = \int_B (x + y) \, d\mu + \int_B (-x) \, d\mu.$$

Rearranging these equalities we obtain

$$\int_A (x + y) \, d\mu = \int_A x \, d\mu + \int_A y \, d\mu$$

and

$$\int_B (x + y) \, d\mu = \int_B x \, d\mu + \int_B y \, d\mu.$$

Finally, by addition we obtain

$$\int (x + y) \, d\mu = \int x \, d\mu + \int y \, d\mu.$$

Thus the theorem is proved for functions which do not change sign. In the general case, X can be partitioned into four sets on each of which neither x nor y change sign; and addition of the four resulting equalities yields the final desired result. This completes the proof.

Corollary. If $x \leq y$, $x, y \in L^{(1)}$, then

$$\int x \, d\mu \leq \int y \, d\mu.$$

The proof of the corollary is left as an exercise.

We come now to a generalization of the Lebesgue monotone convergence theorem which applies to arbitrary sequences of nonnegative measurable functions. It is known as the *theorem of Fatou*.

Theorem 5.5.6. Let $\{x_n\}$ be any sequence of nonnegative measurable functions, and let

$$x(t) = \underset{n\to\infty}{\underline{\text{limit}}}\, x_n(t).$$

Then

$$\int x\, d\mu \le \underset{n\to\infty}{\underline{\text{limit}}} \int x_n\, d\mu.$$

Proof. By definition,

$$x(t) = \underset{n\to\infty}{\underline{\text{limit}}}\, x_n(t) = \sup_{n}\, [\inf_{m\ge n} x_m(t)];$$

hence, the sequence $\{y_n\} = \{\inf\limits_{m\ge n} x_m\}$ is nondecreasing with limit x. Moreover,

$$\int y_n\, d\mu \le \inf_{m\ge n} \int x_m\, d\mu \le \underset{n\to\infty}{\underline{\text{limit}}} \int x_n\, d\mu$$

and, letting $n \to \infty$, we have, by the monotone convergence theorem,

$$\int x\, d\mu \le \underset{n\to\infty}{\underline{\text{limit}}} \int x_n\, d\mu,$$

as desired. This completes the proof.

The theorem of Fatou enables us to prove the *Lebesgue-dominated convergence theorem*, which gives a very general condition sufficient for interchanging the order of taking a limit and integrating.

Theorem 5.5.7. If $\{x_n\}$ is a sequence of functions in $L^{(1)}$ converging to x, and if $|x_n| \le y$ for all n and some $y \in L^{(1)}$, then

$$\lim_{n\to\infty} \int x_n\, d\mu = \int x\, d\mu.$$

Proof. Since $-y \le x_n \le y$ for all n, we see that $|x| \le y$ and so $x \in L^{(1)}$. The sequences $\{y - x_n\}$ and $\{x_n + y\}$ are nonnegative, and by applying Fatou's theorem, we find

$$\int \underset{n\to\infty}{\underline{\text{limit}}}\, (y - x_n)\, d\mu \le \underset{n\to\infty}{\underline{\text{limit}}} \int (y - x_n)\, d\mu$$

and

$$\int \underset{n\to\infty}{\underline{\text{limit}}}\, (x_n + y)\, d\mu \le \underset{n\to\infty}{\underline{\text{limit}}} \int (x_n + y)\, d\mu.$$

These inequalities are equivalent to

$$\int y \, d\mu - \int x \, d\mu \le \int y \, d\mu - \overline{\lim_{n \to \infty}} \int x_n \, d\mu$$

and

$$\int x \, d\mu + \int y \, d\mu \le \underline{\lim_{n \to \infty}} \int x_n \, d\mu + \int y \, d\mu,$$

which can be combined to yield

$$\overline{\lim_{n \to \infty}} \int x_n \, d\mu \le \int x \, d\mu \le \underline{\lim_{n \to \infty}} \int x_n \, d\mu.$$

Hence we have

$$\int x \, d\mu = \lim_{n \to \infty} \int x_n \, d\mu$$

as desired. This completes the proof.

Clearly, the theorem is true if we assume only that $x_n \longrightarrow x$ a.e.
This completes the development of the definition and major properties of the Lebesgue integral. We close this section with two observations of a (now) more elementary nature, leaving the verifications to the reader.

(a) If x is bounded, measurable, and satisfies

$$\mu\{t: \ |x(t)| > 0\} < \infty,$$

then $x \in L^{(1)}(\mu)$.
(b) The set function

$$\nu(A) = \int_A x \, d\mu,$$

for $x \in L^{(1)}(\mu)$, is *absolutely continuous*, i.e., for any $\epsilon > 0$, there is a $\delta(\epsilon)$ such that $|\nu(E)| \le \epsilon$ whenever $\mu(E) \le \epsilon$.

[REMARK: It suffices to consider (b) for $x \ge 0$, and to use the fact that (b) is true for simple functions.]

EXERCISES

1. Establish lemma 5.5.1.

2. Establish the following:

a. $\int \alpha x \, d\mu = \alpha \int x \, d\mu$

b. $0 \le x \le y \Rightarrow 0 \le \int x \, d\mu \le \int y \, d\mu$

(Use only the definition of the integral.)

3. Establish the corollary to theorem 5.5.5.

4. Prove that if $x \in M$, x is bounded, and
$$\mu\{t: \quad |x(t)| > 0\} < \infty,$$
then $x \in L^{(1)}(\mu)$.

5. Prove that for $x \in L^{(1)}$ the set function
$$\nu(A) = \int_A x \, d\mu$$
is absolutely continuous. (See the definition at the end of this section.)

6. Prove the following:

a. $\left[x \in L^{(1)}(\mu), \ x \ge 0, \ \int x \, d\mu = 0 \right] \Rightarrow [x(t) = 0 \text{ a.e.}]$

b. $[x \in L^{(1)}(\mu), \ \mu(A) = 0] \Rightarrow \left[\int_A x \, d\mu = 0 \right]$

c. $\left[x \in L^{(1)}(\mu), \ \int_A x \, d\mu = 0 \quad \text{for all } A \in \mathcal{M} \right] \Rightarrow [x(t) = 0 \text{ a.e.}]$

7. Let $x \in L^{(1)}(\mu)$, $y \in L^{\infty}(\mu)$, with $\alpha \le y(t) \le \beta$ a.e. Show that there exists $\gamma \in [\alpha, \beta]$ such that
$$\int y \, |x| \, d\mu = \gamma \int |x| \, d\mu.$$

8. Consider Lebesgue measure on R. Prove:

a. $L^{(1)}(\mu)$ is not closed under multiplication.

b. If $x \in L^{(1)}(\mu)$ and $\int_I x \, d\mu = 0$ for all intervals I, then $x(t) = 0$ a.e.

9. A complex-valued function $x(t) = x_1(t) + i x_2(t)$ on a measure space $(X, \mathcal{M}, \mu)$ is defined to be measurable if $x_1(t)$ and $x_2(t)$ are measurable. It is defined to be integrable if $x_1(t)$ and $x_2(t)$ are integrable, in which case
$$\int x \, d\mu = \int x_1 \, d\mu + i \int x_2 \, d\mu.$$

a. Show that $x(t)$ is measurable if and only if $x^{-1}(B) \in \mathcal{M}$ for every Borel set B in the complex plane. (The Borel sets are the members of the smallest σ-algebra containing the open sets.)

b. Show that $|x| = (x_1^2 + x_2^2)^{1/2}$ is integrable if and only if x is integrable.

c. Show that

$$\int [\alpha x + \beta y]\, d\mu = \alpha \int x\, d\mu + \beta \int y\, d\mu$$

and

$$\left| \int x\, d\mu \right| \le \int |x|\, d\mu.$$

Section 6 THE SPACE $L^{(1)}$

As we have seen, sets of measure zero are negligible for purposes of integration. To make this quite precise, we define on M, the set of measurable functions on the measure space $(X, \mathcal{M}, \mu)$, the relation $\sim$, as follows:

$$x \sim y \Leftrightarrow \mu\{t: \;\; |x(t) - y(t)| > 0\} = 0$$

It is easy to prove that $\sim$ is an equivalence relation and that it has the substitution property with respect to all algebraic operations. Rather than introduce a new name for the resulting set of equivalence classes, we will continue to work with members of M, with the understanding that equivalent members of M are to be identified.

With this agreement, we can prove the following lemma about $L^{(1)}(\mu)$.

Lemma 5.6.1. The vector space $L^{(1)}(\mu)$ is normed by

$$\|x\| = \int |x|\, d\mu.$$

Proof. Clearly, $\|x\| \ge 0$, and by the results of section 5,

$$\left[\|x\| = \int |x|\, d\mu = 0 \right] \quad \text{implies} \quad [|x|(t) = 0 \quad \text{for almost all } t];$$

thus, $x(t)$ is identified with the zero function. The relations

$$\|\alpha x\| = |\alpha|\,\|x\| \quad \text{and} \quad \|x + y\| \le \|x\| + \|y\|$$

are immediate, and the lemma is proved.

A sequence $\{x_n\}$ in $L^{(1)}(\mu)$ converges to a function $x \in L^{(1)}(\mu)$ if and only if

$$\|x_n - x\| = \int |x_n - x| \, d\mu \to 0 \quad \text{as} \quad n \to \infty.$$

This kind of convergence is called *convergence in the mean*. The reader should refer to the example at the beginning of section 4 to see that a sequence in $L^{(1)}(\mu)$ may converge in the mean without converging at any point whatsoever. However, the following lemma will make up for this defect.

Lemma 5.6.2. Let $\{x_n\}$ be a sequence in $L^{(1)}(\mu)$.

(a) If $\{x_n\}$ converges in the mean to x, then $\{x_n\}$ converges in measure to x.

(b) If $\int |x_n - x_m| \, d\mu \to 0$ as $m, n \to \infty$, that is, $\{x_n\}$ is a *mean Cauchy sequence*, then $\{x_n\}$ is fundamental in measure.

Proof. (a) Let

$$E_n = \{t: \ |x_n(t) - x(t)| > \epsilon\};$$

then

$$\int |x_n - x| \, d\mu \geq \int_{E_n} |x_n - x| \, d\mu \geq \epsilon \mu(E_n),$$

Since the left member tends to zero as $n \to \infty$, we have $\mu(E_n) \to 0$ as $n \to \infty$ also. The proof of (b) is identical.

With this lemma we are now prepared to prove the completeness of $L^{(1)}(\mu)$.

Theorem 5.6.3. $L^{(1)}(\mu)$ is a Banach space.

Proof. In view of lemma 5.6.1, we need only show that every mean Cauchy sequence $\{x_n\}$ in $L^{(1)}$ converges. By lemma 5.6.2, $\{x_n\}$ is fundamental in measure, and accordingly, by theorem 5.4.5, $\{x_n\}$ converges in measure to a function $x \in M$; moreover, x is uniquely determined a.e. Also, we know that some subsequence $\{x_{n(k)}\}$ of $\{x_n\}$ converges almost uniformly, and therefore a.e., to x.

Now choose $N(\epsilon)$ such that

$$\int |x_m - x_n| \, d\mu < \epsilon$$

for $m, n > N(\epsilon)$. For $n(k)$ and $n(j)$ larger than $N(\epsilon)$, we apply Fatou's theorem to obtain

$$\int \underset{k\to\infty}{\underline{\text{limit}}} \, |x_{n(j)} - x_{n(k)}| \, d\mu \le \underset{k\to\infty}{\underline{\text{limit}}} \int |x_{n(j)} - x_{n(k)}| \, d\mu,$$

and thus

$$\int |x_{n(j)} - x| \, d\mu \le \epsilon.$$

Now,

$$\int |x| \, d\mu \le \int |x - x_{n(j)}| \, d\mu + \int |x_{n(j)}| \, d\mu,$$

a finite number; hence $x \in L^{(1)} \, (\mu)$, and the subsequence $\{x_{n(j)}\}$ converges in the mean to x. Finally, from the inequality

$$\int |x_n - x| \, d\mu \le \int |x_n - x_{n(j)}| \, d\mu + \int |x_{n(j)} - x| \, d\mu,$$

we see that $\{x_n\}$ converges in the mean to x. This completes the proof.

We close this section with a review of the four principal types of convergence for sequences of a.e. finite-valued measurable functions:

(a) Almost everywhere convergence.
(b) Almost uniform convergence.
(c) Convergence in measure.
(d) Convergence in the mean.

In general, they are related as shown in the following diagram. (The arrows indicate implication.)

$$
\begin{array}{ccc}
\text{mean} \quad \cdot & & \cdot \quad \text{a.u.} \\
\downarrow & & \downarrow \\
\text{measure} \quad \cdot & & \cdot \quad \text{a.e.}
\end{array}
$$

In case $\mu(X) < \infty$, the theorem of Egoroff adds two implications:

$$
\begin{array}{ccc}
\text{mean} \quad \cdot & & \cdot \quad \text{a.u.} \\
\downarrow & & \uparrow \\
\text{measure} \quad \cdot \longleftarrow & & \cdot \quad \text{a.e.}
\end{array}
$$

Finally, for sequence $\{x_n\}$ such that $|x_n| \le y \in L^{(1)}(\mu)$ for all n, we have three additional implications:

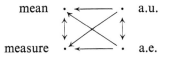

Here, the Lebesgue-dominated convergence theorem is responsible for two of these.

The reader should establish the validity of these diagrams as well as the existence of counter-examples for omitted implications.

EXERCISES

1. a. Show that if $x \in L^{(1)}(\mu)$, $y \in L^{(\infty)}(\mu)$ then $xy \in L^{(1)}$.
   b. Show that, if $\{x_n\}$ is a sequence in $L^{(1)}(\mu)$ converging to x in the mean, then $x_n y$ converges in the mean to xy for $y \in L^{(\infty)}(\mu)$.

2. Let $x \in L^{(1)}(\mu)$. Define

$$x_n(t) = \begin{cases} n, & \text{if } x(t) \geq n \\ x(t), & \text{if } |x(t)| \leq n \\ -n, & \text{if } x(t) \leq -n. \end{cases}$$

   Prove that $\| x_n - x \| \to 0$.

3. Let $\mu(X)$ be finite in the measure space $(X, \mathcal{M}, \mu)$.

   a. Show that, if $\{x_n\}$ is a uniformly convergent sequence of functions in $L^{(1)}(\mu)$, then $\{x_n\}$ converges in the mean.
   b. Give an example to show that an almost uniformly convergent sequence in $L^{(1)}(\mu)$ need not converge in the mean.

4. Let $\{x_n\}$ be a sequence of measurable functions such that $|x_n| \leq y \in L^{(1)}(\mu)$.
   a. Prove that if $\{x_n\}$ converges a.e., then it converges in the mean.
   b. Prove that if $\{x_n\}$ converges almost uniformly, then it converges in the mean.
   c. Prove that if $\{x_n\}$ converges in measure, then it converges in the mean.
   d. Prove that if $\{x_n\}$ converges a.e., then it converges almost uniformly.

5. Exhibit a sequence of functions converging in measure but not converging in the mean.

6. Let $\{x_n\}$ be a sequence in $L^{(1)}(\mu)$ converging almost uniformly to x. Is x in $L^{(1)}(\mu)$?

Section 7 THE $L^{(p)}$ SPACES

For a real nonnegative number p we define $L^{(p)}$ to be the set of all measurable functions x for which $|x|^p$ is in $L^{(1)}$ $(X, \mathcal{M}, \mu)$. The space $L^{(\infty)}$ was discussed in section 3.

Clearly, if $x \in L^{(p)}$, then αx is in $L^{(p)}$ for any scalar α. Moreover, the inequality

$$|x(t) + y(t)|^p \leq [|x(t)| + |y(t)|]^p \leq \{2 \max [|x(t)|, |y(t)|]\}^p$$
$$= 2^p \max [|x(t)|^p, |y(t)|^p] \leq 2^p[|x(t)|^p + |y(t)|^p]$$

shows that $L^{(p)}$ *is a vector space*, for $p > 0$. To obtain further new results, we now restrict our attention to the case $p > 1$, where we will have available the Hölder and Minkowski inequalities.

Lemma 5.7.1

(a) (Hölder) Let $x \in L^{(p)}$ and $y \in L^{(q)}$, where $1/p + 1/q = 1$; then $xy \in L^{(1)}$ and

$$\int |xy| \, d\mu \leq \left\{\int |x|^p \, d\mu\right\}^{1/p} \left\{\int |y|^q \, d\mu\right\}^{1/q}.$$

(b) (Minkowski) Let x and y be in $L^{(p)}$. Then

$$\left\{\int |x + y|^p \, d\mu\right\}^{1/p} \leq \left\{\int |x|^p \, d\mu\right\}^{1/p} + \left\{\int |y|^p \, d\mu\right\}^{1/p}.$$

Proof. It was shown earlier that, for any two nonnegative real numbers α and β,

$$\alpha\beta \leq \frac{1}{p}\alpha^p + \frac{1}{q}\beta^q;$$

hence we have

$$|x(t)y(t)| \leq \frac{1}{p}|x(t)|^p + \frac{1}{q}|y(t)|^q,$$

from which it follows that $xy \in L^{(1)}$. If, in this inequality, we replace $x(t)$ and $y(t)$ by

$$\frac{x(t)}{\left\{\int |x|^p d\mu\right\}^{1/p}} \quad \text{and} \quad \frac{y(t)}{\left\{\int |y|^q d\mu\right\}^{1/q}},$$

respectively, we obtain

$$\frac{|x(t)y(t)|}{\left(\int |x|^p d\mu\right)^{1/p} \left(\int |y|^q d\mu\right)^{1/q}} \leq \frac{1}{p}\frac{|x(t)|^p}{\int |x|^p d\mu} + \frac{1}{q}\frac{|y(t)|^q}{\int |y|^q d\mu}.$$

Finally, we integrate this inequality and find

$$\frac{\int |xy| d\mu}{\left(\int |x|^p d\mu\right)^{1/p} \left(\int |y|^q d\mu\right)^{1/q}} \leq \frac{1}{p} + \frac{1}{q} = 1,$$

which yields (a).

To obtain (b), we calculate

$$\int |x+y|^p d\mu = \int |x+y||x+y|^{p-1} d\mu$$

$$\leq \int |x||x+y|^{p-1} d\mu + \int |y||x+y|^{p-1} d\mu,$$

noting that both integrals exist by virtue of (a) and the fact that $|x+y|^{p-1} \in L^{(q)}$. Applying (a) we find

$$\int |x+y|^p d\mu \leq \left(\int |x|^p d\mu\right)^{1/p} \left(\int |x+y|^p d\mu\right)^{1/q}$$

$$+ \left(\int |y|^p d\mu\right)^{1/p} \left(\int |x+y|^p d\mu\right)^{1/q}.$$

If $(\int |x+y|^p d\mu)^{1/q} = 0$, then (b) is obvious; otherwise we divide by this quantity to obtain

$$\left(\int |x+y|^p d\mu\right)^{1-(1/q)} \leq \left(\int |x|^p d\mu\right)^{1/p} + \left(\int |y|^q d\mu\right)^{1/q},$$

which is (b). This completes the proof.

Thus we see that for $p > 1$ the space $L^{(p)}$ is a normed vector space under

$$\|x\|_p = \left(\int |x|^p d\mu\right)^{1/p}.$$

(As usual, we identify functions which agree almost everywhere.)

Theorem 5.7.2. For $p \geq 1$ the space $L^{(p)}$ is a Banach space.

Proof. The cases $L^{(1)}$ and $L^{(\infty)}$ have already been treated. It remains to be shown that, for $1 < p < \infty$, the space $L^{(p)}$ is complete.

Let $\{x_n\}$ be a Cauchy sequence in $L^{(p)}$, that is,

$$\int |x_m - x_n|^p \, d\mu \longrightarrow 0 \quad \text{as} \quad m, n \longrightarrow \infty.$$

Let

$$E_{mn}(\epsilon) = \{t: \ |x_m(t) - x_n(t)| > \epsilon\}.$$

Then

$$\int |x_m - x_n|^p \, d\mu \geq \int_{E_{mn}(\epsilon)} |x_m - x_n|^p \, d\mu \geq \epsilon^p \mu \left[E_{mn}(\epsilon) \right],$$

from which it follows that $\{x_m\}$ is fundamental in measure. Hence there exists a measurable function x_0 such that $x_n \longrightarrow x_0$ in measure, and a subsequence $\{x_{n(k)}\}$ of $\{x_n\}$ converging almost uniformly, and thus almost everywhere, to x_0. For a fixed k we then have

$$\lim_{j \to \infty} |x_{n(j)}(t) - x_{n(k)}(t)|^p = |x_0(t) - x_{n(k)}(t)|^p$$

almost everywhere. An application of Fatou's theorem now yields

$$|x_0 - x_{n(k)}|^p \in L^{(1)}$$

and

$$\int |x_0 - x_{n(k)}|^p \, d\mu \leq \varliminf_{j \to \infty} \int |x_{n(j)} - x_{n(k)}|^p \, d\mu;$$

also,

$$x_0 = (x_0 - x_{n(k)}) + x_{n(k)},$$

the sum of two functions in $L^{(p)}$, is in $L^{(p)}$. It now follows that

$$\lim_{k \to \infty} \|x_0 - x_{n(k)}\|_p^p = \lim_{k \to \infty} \int |x_0 - x_{n(k)}|^p \, d\mu = 0.$$

To see that $\|x_n - x_0\|_p \longrightarrow 0$, we need only write

$$\|x_n - x_0\|_p \leq \|x_n - x_{n(k)}\|_p + \|x_{n(k)} - x_0\|_p$$

and employ the result just proved together with the fact that $||x_n - x_m||_p \rightarrow$ 0. This finishes the proof.

The reader should notice that the proof is essentially identical with the one already given for $p = 1$.

It is left as an exercise to show that M_0 is dense in $L^{(p)}$.

EXERCISES

1. Show that M_0 is dense in $L^{(p)}$.

2. Show, for $x \in M$, $x \in L^{(p)} \Leftrightarrow |x| \in L^{(p)}$.

3. Consider the case $\mu(X) < \infty$. Show:
    a. $L^{(\infty)} \subseteq L^{(p)} \subseteq L^{(q)} \subseteq L^{(1)}$ whenever $q \leq p$.
    b. $x \in L^{(p)} \subseteq L^{(q)}$, $(q \leq p)$ implies $A_q(x) \leq A_p(x)$, where
    $$A_p(x) = \mu(X)^{-1/p} ||x||_p.$$

4. Let μ be the Lebesgue measure on $[a, b]$. Show that $\lim_{p \to \infty} A_p(x) = ||x||_\infty$ for $x \in L^{(\infty)}$, and hence that $||x||_p \rightarrow ||x||_\infty$ as $p \rightarrow \infty$.

5. Let $X = [0, 1]$ and μ be Lebesgue measure. Define $L^{(w)} = \bigcap_{\infty > p \geq 1} L^{(p)}$.
    a. Show that $L^{(\infty)} \subset L^{(w)} \subset L^{(p)} \subset L^{(q)} \subset L^{(1)}$, where all inclusions are *proper* $(q < p)$. [Try $x(t) = |\log t|$ to show $L^\infty \subset L^{(w)}$].
    b. Calculate $||\log t||_p$ for p a positive integer.
    c. Show that $L^{(w)}$ is closed under multiplication.
    d. For $q < p$, show that the identity mapping of $L^{(p)}$ into $L^{(q)}$ is continuous.

6. Let y be a fixed member of $L^{(q)}$ where $1/p + 1/q = 1$. Show that the mapping
    $$T: \quad x \longrightarrow \int xy \, d\mu$$
    is a continuous linear function from $L^{(p)}$ to R.

7. Do the inclusions of exercise 3a continue to hold if $\mu(X) = \infty$? Explain.

8. (Refer to exercise 9, section 5). Let $L_c^{(p)}$ be the collection of all complex-valued measurable functions on X such that

$$\|x\|_p = \left[\int |x|^p \, d\mu \right]^{1/p} < \infty.$$

Show that $L_c^{(p)}$ is a Banach space. ($1 < p < \infty$).

9. Let $(X, \mathcal{M}, \mu)$ be any measure space. Show that $\mathcal{M}$ is a metric space under $d(A, B) = \mu(A \triangle B)$. [Identify A and B if $\mu(A \triangle B) = 0$.] Show that $L^{(p)}(X, \mathcal{M}, \mu)$, $1 \le p < \infty$, is separable if and only if $\mathcal{M}$ is separable.

Section 8 SOME SPECIAL RESULTS ON R

We begin with some considerations which lead to a comparison of the Riemann and Lebesgue integrals. If x is a bounded real-valued function on a measure space $(X, \mathcal{M}, \mu)$ where $\mu(X) < \infty$, then the upper and lower Lebesgue integrals can be defined as follows:

$$L \overline{\int} x \, d\mu = \inf \left\{ \int y \, d\mu : \ y \ge x, y \in M_1 \right\}$$

$$L \underline{\int} x \, d\mu = \sup \left\{ \int y \, d\mu : \ y \le x, y \in M_1 \right\}.$$

If we now specialize our measure space to Lebesgue measure on a *finite* interval $[a, b]$, then these integrals differ from the upper and lower Riemann integrals in that the sup and inf are taken over a wider class of functions, for only step functions were previously allowed. It is obvious, then, that

$$R \overline{\int} x \, dt \ge L \overline{\int} x \, dt \ge L \underline{\int} x \, dt \ge R \underline{\int} x \, dt$$

for any bounded function x on $[a, b]$.

The following theorem establishes the connection between Lebesgue and Riemann integration on $[a, b]$.

Theorem 5.8.1. For x bounded on $[a, b]$, we have:

(a) $\left[L \overline{\int} x \, dt = L \underline{\int} x \, dt \right] \Leftrightarrow [x \text{ is measurable}].$

(b) If

$$L \overline{\int} x \, dt = L \underline{\int} x \, dt,$$

then

$$x \in L^{(1)}[a, b]$$

and

$$L \int x \, dt = L \overline{\int} x \, dt = L \underline{\int} x \, dt.$$

(c) If $R \int x \, dt$ exists, then x is measurable, integrable, and

$$R \int x \, dt = L \int x \, dt.$$

Proof. Parts (b) and (c) are easy consequences of part (a) and our general results on integration. They are left as exercises.

Assume now that x is measurable on $[a, b]$ as well as bounded. Then by our general results there exist sequences $\{y_n\}$ and $\{z_n\}$ of simple functions converging *uniformly* to x and, for all n and t,

$$y_n(t) \leq y_{n+1}(t), \qquad z_n(t) \geq z_{n+1}(t).$$

It easily follows that

$$L \overline{\int} x \, dt = L \underline{\int} x \, dt.$$

Conversely, if

$$L \overline{\int} x \, dt = L \underline{\int} x \, dt,$$

then sequences of simple functions $\{y_n\}$ and $\{z_n\}$ exist such that, for all n and t,

$$y_n(t) \leq x(t) \leq z_n(t) \quad \text{and} \quad \int [z_n - y_n] \, dt < \frac{1}{n}.$$

Let

$$y(t) = \sup_n y_n(t) \quad \text{and} \quad z(t) = \inf_n z_n(t);$$

then y and z are measurable, and

$$y_k(t) \leq y(t) \leq x(t) \leq z(t) \leq z_k(t).$$

Let

$$E = \{t: \ x(t) > y(t)\} \quad \text{and} \quad E_n = \left\{t: \ [z(t) - y(t)] > \frac{1}{n}\right\}.$$

Then

$$E \subseteq \bigcup_n E_n \quad \text{and} \quad E_n \subseteq \left\{t: \ [z_k(t) - y_k(t)] > \frac{1}{n}\right\} = F_{nk}, \qquad \text{for all } k.$$

Now

$$\frac{1}{k} > \int [z_k - y_k] \, dt > \int_{F_{nk}} [z_k - y_k] \, dt > \frac{1}{n} \mu(F_{nk}),$$

and hence

$$\mu(E_n) \le \mu(F_{nk}) \le \frac{n}{k}, \qquad \text{for all } k.$$

It follows that $\mu(E_n) = 0$ for all n, and therefore $\mu(E) = 0$. Thus $x(t) = y(t)$ almost everywhere, and x is measurable, since it agrees with a measurable function except on a set of measure zero. This completes the proof.

We turn now to the relations between continuity and measurability for real-valued functions on R.

Lemma 5.8.2. If x is continuous on a closed set F of R, then there exists a continuous extension $\hat{x}$ of x to R. Moreover, if $|x| \le M$ on F, then $|\hat{x}| \le M$ on R.

Proof. There exist disjoint open intervals $\{I_n\}$ such that $\tilde{F} = \bigcup_{n=1}^{\infty} I_n$. Define x on each $I_n = (a_n, b_n)$ linearly by

$$x(t) = x(a_n) + (t - a_n) \, [x(b_n) - x(a_n)] \, (b_n - a_n)^{-1}.$$

Lemma 5.8.3. For any $\epsilon > 0$ and any simple function y, there exists a continuous function x and a closed set F such that $\mu(\tilde{F}) \le \epsilon$ and x agrees with y on F.

Proof. Let $y = \sum_{k=1}^{n} \alpha_k z_k$, where z_k is the characteristic function of a measurable set E_k and the numbers $\alpha_1, \ldots, \alpha_n$ are the distinct values assumed by y. Let F_k, $k = 1, \ldots, n$, be closed sets such that $E_k \supseteq F_k$ and $\mu(E_k - F_k) \le \epsilon/n$. Then the set $F = \bigcup_{k=1}^{n} F_k$ is closed and $\mu(\tilde{F}) = \sum_{k=1}^{n} \mu(E_k - $

$F_k) \leq \epsilon$. Construct the continuous function x by letting it agree with y on F (it is then continuous on F) and extending it to R by the preceding lemma.

Thus we can approximate simple functions by continuous functions in a sense. This will now be extended.

Theorem 5.8.4 Lusin. Let x be an a.e. finite-valued function in M. Then, for any $\epsilon > 0$, there exists a closed set F such that x is continuous on F and $\mu(\tilde{F}) \leq \epsilon$.

Proof. Consider first the case where x is defined only on a set E of finite measure. Let $\{x_n\}$ be a sequence of simple functions converging pointwise to x. By the preceding lemma, there exist closed sets $\{F_n\}$ such that $\mu(E - F_n) \leq \epsilon/2^{n+1}$ and x_n is continuous on F_n. Let $F_0 = \bigcap_{n=1}^{\infty} F_n$; then all x_n are continuous on F_0 and

$$\mu(E - F_0) = \mu\left(E - \bigcap_{n=1}^{\infty} F_n\right) = \mu\left(\bigcup_{n=1}^{\infty}[E - F_n]\right) \leq \sum_{n=1}^{\infty} \frac{\epsilon}{2^{n+1}} = \frac{\epsilon}{2}.$$

By Egoroff's theorem, which applies since $\mu(E) < \infty$, a set F_1 can be found such that $F_0 \supseteq F_1$, $\mu(F_0 - F_1) \leq \epsilon/2$ and $x_n \rightarrow x$ *uniformly* on F_1. Then x, as the uniform limit of continuous functions on F, is also continuous on F; moreover

$$\mu(E - F) = \mu(E - F_0) + \mu(F_0 - F_1) \leq \frac{\epsilon}{2} + \frac{\epsilon}{2} = \epsilon.$$

We can extend this result from E to R by writing $R = \bigcup_{n=1}^{\infty} E_n$ where the E_n's are disjoint sets of finite measure. Each E_n contains a set G_n, closed in R, such that $\mu(E_n - G_n) \leq \epsilon/2^n$ and x is continuous on each G_n. If we specify

$$E_{2n} = [n, n+1), \quad E_{2n+1} = [-n, -n+1),$$

then the set $G_0 = \bigcup_n G_n$ is closed and x is continuous on G_0; moreover

$$\mu(\tilde{G}_0) = \sum_{n=1}^{\infty} \mu(E_n - G_n) \leq \sum_{n=1}^{\infty} \frac{\epsilon}{2^n} = \epsilon.$$

This completes the proof of Lusin's theorem.

We now restrict our attention to $L^{(1)}(-\infty, \infty)$ and discuss various dense sets in that space. Of course, the space M_0 of integrable simple functions is dense. Now let $x = \sum_{k=1}^{n} \alpha_k x_k$ be in M_0, where the functions $x_1, \ldots,$ x_n are characteristic functions of measurable sets $E_1, \ldots, E_n$ of finite measure, and $\alpha_1, \ldots, \alpha_n$ are the distinct nonzero values assumed by x. By the results of section 2, we can find sets $J_1, \ldots, J_n$, each consisting of a finite disjoint union of bounded intervals, such that $\mu(E_k \triangle J_k) \le \epsilon$, $k = 1, \ldots, n$. If y_k is the characteristic function of J_k, and $y = \sum_{k=1}^{n} \alpha_k y_k$, a *step function*, then

$$\int |x - y|\, dt = \int \left| \sum_{n=1}^{n} \alpha_k[x_k - y_k] \right| dt \le \sum_{k=1}^{n} |\alpha_k|\, \mu(E_k \triangle J_k) \le \epsilon \sum_{k=1}^{n} |\alpha_k|.$$

Thus any integrable simple function can be approximated arbitrarily closely in $L^{(1)}(-\infty, \infty)$ by integrable step functions, and consequently, the integrable step functions are dense in $L^{(1)}(-\infty, \infty)$.

We call a step function *rational* if it assumes rational values on intervals whose endpoints are rational. Since any integrable step function can be arbitrarily closely approximated in $L^{(1)}(-\infty, \infty)$ by integrable rational step functions, and since these form a denumerable set, we see that $L^{(1)}(-\infty, \infty)$ is separable. Finally, if z is the characteristic function of $[\alpha, \beta)$, then the continuous function

$$x(t) = \begin{cases} 0, & t \le \alpha \quad \text{or} \quad t \ge \beta \\ 1, & \alpha + \epsilon \le t \le \beta - \epsilon \\ \dfrac{1}{\epsilon}(t - \alpha), & \alpha \le t \le \alpha + \epsilon \\ -\dfrac{1}{\epsilon}(t - \beta), & \beta - \epsilon \le t \le \beta \end{cases}$$

satisfies $\int |x - z|\, dt = \epsilon$. From this we see that the continuous functions are dense in $L^{(1)}(-\infty, \infty)$, and indeed, the continuous functions with compact support are dense. We summarize these results in the next theorem.

Theorem 5.8.5. The following subspaces of $L^{(1)}(-\infty, \infty)$ are dense:

(a) The continuous functions with compact support.
(b) The integrable simple functions.
(c) The integrable step functions.

Moreover, the rational step functions are dense, so that $L^{(1)}(-\infty, \infty)$ is separable.

If we employ the results of sections 5 and 10 of chapter 4, we can obtain

Theorem 5.8.6

(a) Polynomials are dense in $L^{(1)}[a, b]$.
(b) For any $\alpha > 0$, the set of functions of type $e^{-\alpha t} p(t)$, where $p(t)$ is a polynomial, are dense in $L^{(1)}[0, \infty)$.

(c) For any $\alpha > 0$, the set of functions of type $e^{-\alpha t^2}p(t)$, where $p(t)$ is a polynomial, are dense in $L^{(1)}(-\infty, \infty)$.

Proof. Part (a) is obvious. To prove (b), let x be any function in $L^{(1)}[0, \infty)$ and let y be a continuous function with compact support such that $\|x - y\| \le \epsilon$. By the results of chapter 4, section 10, there exists a polynomial $p(t)$ such that

$$|e^{(\alpha/2)t} y(t) - e^{-(\alpha/2)t} p(t)| \le \epsilon, \qquad \text{for all } t \in [0, \infty).$$

Hence

$$|y(t) - e^{-\alpha t}p(t)| \le \epsilon e^{-(\alpha/2)t}$$

and

$$\int_0^\infty |y(t) - e^{-\alpha t}p(t)| \, dt \le \frac{2}{\alpha} \epsilon.$$

Thus

$$\|x(t) - e^{-\alpha t}p(t)\| \le \epsilon \left(1 + \frac{2}{\alpha}\right),$$

and the proof of (b) is complete.

The proof of (c) is similar. If $x \in L^{(1)}(-\infty, \infty)$, we can find a y, continuous with compact support, such that $\|x - y\| \le \epsilon$. Using again the results of section 10, chapter 4, we can find a polynomial $p(t)$ such that

$$|e^{(\alpha/2)t^2} y(t) - e^{-(\alpha/2)t^2} p(t)| \le \epsilon, \qquad \text{for all } t.$$

Hence

$$|y(t) - e^{-\alpha t^2}p(t)| \le e^{-(\alpha/2)t^2}\epsilon,$$

$$\|y(t) - e^{-\alpha t^2}p(t)\| \le \epsilon \sqrt{\frac{2\pi}{\alpha}},$$

and

$$\|x(t) - e^{-\alpha t^2}p(t)\| < \epsilon \left(1 + \sqrt{\frac{2\pi}{\alpha}}\right).$$

This completes the proof.

The conclusions of the two theorems are, in fact, valid in $L^{(p)}, 1 \le p < \infty$. Details are left for the exercises.

EXERCISES

1. a. Prove part (b) of theorem 5.8.1.
    b. Prove part (c) of theorem 5.8.1.

2. Prove that the "rational" step functions are countable.

3. a. Extend theorem 5.8.5 to $L^{(p)}$, $1 < p < \infty$.
    b. Extend theorem 5.8.6. to $L^{(p)}$, $1 < p < \infty$.

4. Is $L^{(\infty)}$ separable?

Section 9 DIFFERENTIATION

In this section we restrict our attention to Lebesgue measure on R and ask in exactly what sense integration and differentiation are inverse operations. An example will indicate some of the difficulties involved.

The Cantor set C (see Section 2) is obtained by removal of the following sequence of intervals from $[0, 1]$;

$$I\left(\frac{1}{2}\right) = \left(\frac{1}{3}, \frac{2}{3}\right)$$

$$I\left(\frac{1}{4}\right) = \left(\frac{1}{3^2}, \frac{2}{3^2}\right)$$

$$I\left(\frac{3}{4}\right) = \left(\frac{2}{3} + \frac{1}{3^2}, \frac{2}{3} + \frac{2}{3^2}\right)$$

$$I\left(\frac{1}{8}\right) = \left(\frac{1}{3^3}, \frac{2}{3^3}\right)$$

$$I\left(\frac{3}{8}\right) = \left(\frac{2}{3^2} + \frac{1}{3^3}, \frac{2}{3^2} + \frac{2}{3^3}\right)$$

$$I\left(\frac{5}{8}\right) = \left(\frac{2}{3} + \frac{1}{3^3}, \frac{2}{3} + \frac{2}{3^3}\right) \quad I\left(\frac{7}{8}\right) = \left(\frac{2}{3} + \frac{2}{3^2} + \frac{1}{3^3}, \frac{2}{3} + \frac{2}{3^2} + \frac{2}{3^3}\right)$$

etc.

The intervals are indexed so that $r > s$ if and only if $I(r)$ lies to the right of $I(s)$. The *Cantor function* $c(t)$ is now defined as follows:

$$c(t) = \frac{m}{2^n} \quad \text{if} \quad t \in I\left(\frac{m}{2^n}\right)$$

(This defines $c(t)$ on the complement of the Cantor set.)

$$c(0) = 0$$
$$c(t) = \sup\{c(s): \ s \in \tilde{C}, s < t\}, \qquad 0 \neq t \in C.$$

Clearly,

$$0 = c(0) \leq c(t) \leq c(1) = 1.$$

We now show that $c(t)$ *is nondecreasing.* Assume $s < t$. If s and t are both in $\tilde{C}$, then either they are in the same interval $I(r)$, in which case $c(s) = c(t) = r$, or $s \in I(r_1)$, $t \in I(r_2)$ with $r_1 < r_2$, so that

$$c(s) = r_1 < r_2 = c(t).$$

If $s \in \tilde{C}$ and $t \in C$, then

$$c(s) \leq \sup \{c(u): \quad u \in \tilde{C}, u < t\} = c(t).$$

If $s \in C$, $t \in \tilde{C}$, then $r < s$, $r \in \tilde{C}$ imply $c(r) < c(t)$; hence

$$c(s) = \sup \{c(r): \quad r \in \tilde{C}, r < s\} \leq c(t).$$

Finally, if $s \in C$, $t \in C$, then there is an $r \in \tilde{C}$ such that $s < r < t$, and the foregoing results yield $c(s) \leq c(r) \leq c(t)$.

Lastly, we show that $c(t)$ *is continuous.* It is clear that $c(t)$ is continuous on $\tilde{C}$. Now let $t_0 \in C$; then

$$c(t_0) = \sup \{c(s): \quad s \in \tilde{C}, s < t_0\}.$$

Therefore, for any $\epsilon > 0$, there is an s such that $s < t_0$ and $c(t_0) - c(s) < \epsilon$. The monotonicity of $c(t)$ now yields the left continuity of $c(t)$ at t_0. To show that $c(t)$ is right continuous at t_0, we note that there is a number of the form $m/2^m$ such that

$$c(t_0) < \frac{m}{2^n} < c(t_0) + \epsilon, \qquad \text{with } m \text{ odd.}$$

But for some $s > t_0$ we know that $c(s) = m/2^n$, and we need only apply the monotonicity again to obtain the right continuity of $c(t)$ at t_0.

Now, $c'(t)$ exists and has the value zero almost everywhere, so that $\int_a^t c'(s)\, ds$ is the zero function. In other words, $c(t)$ is a function which has a derivative almost everywhere, but is *not* recoverable from its derivative by integration. *One of our problems will then be to identify the class of almost everywhere differentiable functions, $x(t)$, which do satisfy the equation*

$$x(t) = x(a) + \int_a^t x'(s)\, ds.$$

Another problem is: Given $x \in L^{(1)}[a, b]$, *is the function* $\int_a^t x(s) \, ds$ *almost everywhere differentiable, and if so, do we have*

$$\frac{d}{dt} \int_a^t x(s) \, ds = x(t)$$

almost everywhere? We attack this question first. (It is affirmatively answered in theorem 5.9.7.)

Lemma 5.9.1. For $x \in L^{(1)}[a, b]$, the function $F(t) = \int_a^t x(s) \, ds$ is continuous and of bounded variation.

Proof. The continuity of $F(t)$ is an immediate consequence of the absolute continuity of the indefinite integral as a set function. Now, for any partition $\pi = \{a = t_0 < t_1 < \ldots < t_n = b\}$, we have

$$\sum_{k=1}^n |F(t_k) - F(t_{k-1})| = \sum_{k=1}^n \left| \int_{t_{k-1}}^{t_k} x(s) \, ds \right| \leq \sum_{k=1}^n \int_{t_{k-1}}^{t_k} |x(s)| \, ds = \int_a^b |x(s)| \, ds$$

so that $F(t)$ is of bounded variation.

One of the principal theorems of this section asserts that a function of bounded variation is almost everywhere differentiable, but considerable preliminary work must be done before we can prove it.

Definition 5.9.2. Let $x(t)$ be any real-valued function on $[a, b]$. The four *derivates* of $x(t)$ at a point t_0 are defined as follows (the values $+\infty$ and $-\infty$ are allowed):

$$D^+[x; t_0] = \overline{\lim_{h \to 0^+}} \frac{1}{h}[x(t_0 + h) - x(t_0)]$$

$$D_+[x; t_0] = \lim_{h \to 0^+} \frac{1}{h}[x(t_0 + h) - x(t_0)]$$

$$D^-[x; t_0] = \overline{\lim_{h \to 0^+}} \frac{1}{h}[x(t_0) - x(t_0 - h)]$$

$$D_-[x; t_0] = \lim_{h \to 0^+} \frac{1}{h}[x(t_0) - x(t_0 - h)].$$

If these four numbers are identical and finite, then their common value is denoted by $x'(t_0)$ and is called the *derivative* of $x(t)$ at t_0.

We first need the notion of a Vitali covering and a fundamental lemma about such coverings.

Definition 5.9.3. A collection of closed intervals $\{I_\alpha\}$ is a Vitali covering of a set E if, for any $\epsilon > 0$ and $t \in E$, there is an I_α in the covering containing t and having length less than ϵ.

Lemma 5.9.4. If $\mu^*(E) < \infty$ and $\{I_\alpha\}$ is a Vitali covering of E, then for any $\epsilon > 0$ there exist finitely many disjoint intervals, $I_1, \ldots, I_n$ from $\{I_\alpha\}$ such that $\mu^*\left(E - \bigcup_{k=1}^{n} I_k\right) < \epsilon$.

Proof. Let O be an open set of finite measure containing E. We then discard all members of $\{I_\alpha\}$ which are not contained in O; the remaining I_α's still form a Vitali covering of E, and it suffices to use only them. By induction we select a sequence $\{I_n\}$ of intervals from $\{I_\alpha\}$ as follows. Choose I_1 at random, and assume $I_1, \ldots, I_m$ have been chosen. Let λ_m be the supremum of the lengths of the I_α's which do not intersect $\bigcup_{k=1}^{m} I_k$; then $\lambda_m \leq \mu(O)$. Now choose I_{m+1} so that $l(I_{m+1}) > \frac{1}{2}\lambda_m$ and $I_{m+1} \cap \left(\bigcup_{k=1}^{n} I_k\right) = \varnothing$. This completes the selection of the sequence $\{I_n\}$.

Since

$$\bigcup_{n=1}^{\infty} I_n \subseteq O \quad \text{and} \quad \sum_{n=1}^{\infty} l(I_n) = \mu\left(\bigcup_{n=1}^{\infty} I_n\right) \leq \mu(O) < \infty,$$

the series $\sum_{n=1}^{\infty} l(I_n)$ converges, and for any $\epsilon > 0$, there is an $N(\epsilon)$ such that

$$\sum_{n=N(\epsilon)+1}^{\infty} l(I_n) < \epsilon.$$

Let

$$F = E - \bigcup_{n=1}^{N(\epsilon)} I_n;$$

we will show $\mu^*(F) < 5\epsilon$.

Now, if $t_0 \in F$ then there exists an interval I_0 in the Vitali covering containing t_0 such that

$$I_0 \cap \left(\bigcup_{n=1}^{N(\epsilon)} I_n\right) = \varnothing.$$

Thus I_0 must have nonempty intersection with some member of the sequence $\{I_n\}$, for if $I_0 \cap I_k = \varnothing$ when $k \leq m$, then $l(I_0) \leq \lambda_m < 2l(I_{m+1})$, an inequality which cannot hold for all m, since $l(I_m) \to 0$. Let p be the smallest integer for which $I_0 \cap I_p \neq \varnothing$; then $p > N(\epsilon)$ and $l(I_0) \leq \lambda_{p-1} \leq 2l(I_p)$. Denoting the midpoint of I_p by t_p, we have

$$|t_0 - t_p| \leq l(I_0) + \tfrac{1}{2}l(I_p) \leq \tfrac{5}{2}l(I_p).$$

If J_p is an interval with midpoint t_p and length $5l(I_p)$, then $t_0 \in J_p$; in other words, every point of F is contained in some J_p, $p > N(\epsilon)$, or

$$F \subseteq \bigcup_{p=N(\epsilon)+1}^{\infty} J_p.$$

Hence

$$\mu^*(F) \le \mu^*\left(\bigcup_{p=N(\epsilon)+1}^{\infty} J_p\right) \le 5 \sum_{p=N(\epsilon)+1}^{\infty} l(I_p) < 5\epsilon.$$

This completes the proof of the lemma.

Theorem 5.9.5. If $x(t)$ is a finite-valued nondecreasing function on $[a, b]$, then it has a derivative $x'(t)$ almost everywhere. Moreover, $x'(t)$ is integrable and satisfies the inequality

$$\int_a^b x'(t)\, dt \le x(b) - x(a).$$

Corollary. A function of bounded variation is almost everywhere differentiable.

Proof of the Theorem. We will show that the set

$$E = \{t: \quad D_-[x; t] < D^+[x; t]\}$$

has measure zero. An entirely similar proof, which we leave as an exercise, can be given to show that the set

$$G = \{t: \quad D_+[x; t] < D^-[x; t]\}$$

has measure zero. Since $D^+[x; t) \ge D_+[x; t]$ and $D^-[x; t] \ge D_-[x; t]$ everywhere, it will then follow that

$$D^+[x; t] \ge D_+[x; t] \ge D^-[x; t] \ge D_-[x; t] \ge D^+[x; t]$$

for almost all t, i.e., the four derivates are equal almost everywhere.

Let $E_{rs} = \{t: \quad D_-[x; t] < r < s < D^+[x; t]\}$ for rational r and s. Then $E = \bigcup_{r,s} E_{rs}$, a countable union, and to show $\mu(E) = 0$, it suffices to show $\mu^*(E_{rs}) = 0$. For a fixed $\epsilon > 0$ let O be an open set containing E_{rs} such that $\mu(O) \le \mu^*(E_{rs}) + \epsilon$.

We proceed to obtain a Vitali covering of E_{rs}. Now, $t \in E_{rs}$ means

$$D_-[x; t] = \varliminf_{h \to 0^+} \frac{1}{h}[x(t) - x(t - h)] < r < s$$

$$< \varlimsup_{h \to 0^+} \frac{1}{h}[x(t + h) - x(t)] = D^+[x; t];$$

hence there are arbitrarily small positive values of h for which

(a) $\frac{1}{h}[x(t+h) - x(t)] > s$, and

(b) $\frac{1}{h}[x(t) - x(t-h)] < r$.

Using (b), we see that the intervals of the form $[t - h, t]$ form a Vitali covering of E_{rs}, and there is no loss in requiring these intervals to be contained in O. The lemma now yields disjoint intervals $I_j = [t_j - h_j, t_j]$, $j = 1, \ldots, n$ such that

$$\bigcup_{j=1}^{n} I_j \subset O \quad \text{and} \quad \mu^*\left(E_{rs} - \bigcup_{j=1}^{n} I_j\right) < \epsilon.$$

We now examine the set

$$F = E_{rs} \cap \left[\bigcup_{j=1}^{n} I_j\right]$$

and apply (a) to obtain a Vitali covering of F by disjoint intervals

$$J_k = [t'_k, t'_k + h'_k], \qquad k = 1, \ldots, m$$

such that

$$\bigcup_{k=1}^{m} J_k \subset \bigcup_{j=1}^{n} I_j \quad \text{and} \quad \mu^*\left(F - \bigcup_{k=1}^{m} J_k\right) < \epsilon.$$

We now have

$$\mu^*(E_{rs}) = \mu^*\left[\left(E_{rs} - \bigcup_{k=1}^{n} I_k\right) \cup \left(E_{rs} \cap \left[\bigcup_{k=1}^{n} I_k\right]\right)\right] \leq \epsilon + \mu^*(F)$$

$$\leq \epsilon + \mu^*\left[\left(F - \bigcup_{k=1}^{m} J_k\right) \cup \left(F \cap \left[\bigcup_{k=1}^{m} J_k\right]\right)\right] \leq \mu\left(\bigcup_{k=1}^{m} J_k\right) + 2\epsilon.$$

Furthermore,

$$\sum_{k=1}^{m} [x(t'_j + h'_j) - x(t'_j)] \leq \sum_{k=1}^{n} [x(t_k) - x(t_k - h_k)] \leq r \sum_{k=1}^{n} h_k \leq r\mu(O)$$

$$\leq r[\mu^*(E_{rs}) + \epsilon]$$

and

$$\sum_{k=1}^{m} [x(t'_j + h'_j) - x(t'_j)] \geq s \sum_{j=1}^{m} h'_j = s\mu\left(\bigcup_{k=1}^{m} J_k\right) \geq s[\mu^*(E_{rs}) - 2\epsilon].$$

Hence,

$$s[\mu^*(E_{rs}) - 2\epsilon] \leq r[\mu^*(E_{rs}) + \epsilon], \qquad \text{for all} \quad \epsilon > 0,$$

and thus

$$s\mu^*(E_{rs}) \leq r\mu^*(E_{rs}).$$

But $r < s$, and we must conclude $\mu^*(E_{rs}) = 0$.

We now know that the four derivates of a nondecreasing function are equal almost everywhere, and thus the derivative $x'(t)$ is defined almost everywhere. To see that $x'(t)$ is measurable, consider the sequence

$$y_n(t) = n\left[x\left(t + \frac{1}{n}\right) - x(t)\right]$$

of nonnegative measurable functions. Since $\{y_n(t)\}$ converges pointwise to $x'(t)$ almost everywhere, we see that $x'(t)$ is measurable. (Define $x(t) = x(b)$ for $t > b$ so that the $y_n(t)$'s are well defined for all n and t.) Now

$$\underset{n \to \infty}{\underline{\lim}} \int_a^b y_n(t)\, dt = \underset{n \to \infty}{\underline{\lim}} n\left[\int_a^b x\left(t + \frac{1}{n}\right) dt - \int_a^b x(t)\, dt\right]$$

$$= \underset{n \to \infty}{\underline{\lim}}\left[n\int_b^{b+1/n} x(t)\, dt - n\int_a^{a+1/n} x(t)\, dt\right]$$

$$= x(b) - \underset{n \to \infty}{\overline{\lim}}\, n\int_a^{a+1/n} x(t)\, dt \leq x(b) - x(a),$$

and Fatou's theorem applies to the sequence $\{y_n(t)\}$. Thus

$$x'(t) = \underset{n \to \infty}{\lim}\, y_n(t) = \underset{n \to \infty}{\underline{\lim}}\, y_n(t)$$

is integrable, and

$$\int_a^b x'(t)\, dt \leq x(b) - x(a).$$

This completes the proof of the theorem.

The corollary is an immediate consequence of the theorem and the fact that a function of bounded variation can be written as the difference of two nondecreasing functions.

These results guarantee that, for every x in $L^{(1)}[a, b]$, the function $F(t) = \int_a^t x(s)\, ds$ is almost everywhere differentiable, although at this point we cannot yet assert that differentiation of $F(t)$ will yield $x(t)$. Hence we seek additional information on functions of the type $\int_a^t x(s)\,ds$. We derive such information in the following lemma by using the absolute continuity of the set function $v(E) = \int_E |x(t)|\,dt$.

Lemma 5.9.6. If the function $F(t) = \int_a^t x(s)\,ds$ is identically zero on $[a, b]$, then $x(t)$ is zero almost everywhere.

Proof. From

$$\int_{t_1}^{t_2} x(s)\,ds = \int_a^{t_2} x(s)\,ds - \int_a^{t_1} x(s)\,ds,$$

we see that $\int_I x(t)\,dt$ is zero for all intervals I; therefore, by countable additivity, $\int_A x(t)\,dt = 0$ for all open sets A. Let $E^+ = \{t:\ x(t) \geq 0\}$, and let A be an open set such that $A \supset E^+$ and $\mu(A - E^+) < \epsilon$. Then

$$\int_a^b x^+(t)\,dt = \left| \int_a^b x^+(t)\,dt \right| = \left| \int_{E^+} x(t)\,dt \right| = \left| \int_A x(t)\,dt - \int_{A - E^+} x(t)\,dt \right|$$

$$\leq \left| \int_A x(t)\,dt \right| + \left| \int_{A - E^+} x(t)\,dt \right| = \left| \int_{A - E^+} x(t)\,dt \right|.$$

The absolute continuity of $v(B) = \int_B x(t)\,dt$ now shows that $\int_a^b x^+(t)\,dt = 0$, and hence $x^+(t) = 0$ almost everywhere. Since $x^-(t)$ can be treated similarly, we obtain $x(t) = 0$ almost everywhere, and the proof is complete.

We will now see that differentiation is the inverse of integration in the sense of the following theorem.

Theorem 5.9.7. If $x \in L^{(1)}[a, b]$ and if $F(t) - F(a) = \int_a^t x(s)\,ds$, then $F'(t) = x(t)$ almost everywhere.

Proof. We consider first the case where x is bounded, say $|x| \leq M$ on $[a, b]$. Then we have

$$\left| \frac{1}{h}[F(t + h) - F(t)] \right| = \left| \frac{1}{h} \int_t^{t+h} x(s)\,ds \right| \leq M,$$

so that the difference quotient is bounded, besides tending to the derivative $F'(t)$ almost everywhere. We define the sequence

$$x_n(t) = n\left[F\left(t + \frac{1}{n}\right) - F(t) \right]$$

and then apply the Lebesgue-dominated convergence theorem to obtain the fact that $\| x_n - F' \| \to 0$ in $L^{(1)}[a, b]$. But this implies

$$\int_a^t x_n(s) \, ds \longrightarrow \int_a^t F'(s) \, ds$$

for all $t \in [a, b]$. We now calculate

$$\int_a^t x_n(s) \, ds = n \left[\int_a^t F\left(s + \frac{1}{n} \right) ds - \int_a^t F(s) \, ds \right]$$

$$= n \left[\int_t^{t+1/n} F(s) \, ds - \int_a^{a+1/n} F(s) \, ds \right],$$

noting that $F(s)$ is continuous. Hence, by the mean-value theorem of integral calculus,

$$\int_a^t F'(s) \, ds = \lim_{n \to \infty} \left[F\left(t + \theta \frac{1}{n} \right) - F\left(a + \theta' \frac{1}{n} \right) \right]$$

$$= F(t) - F(a) = \int_a^t x(s) \, ds, \qquad (\theta, \theta_1 \in [0, 1]),$$

and we have

$$\int_0^t [F'(s) - x(s)] \, ds = 0$$

for all $t \in [a, b]$. The preceding lemma now yields $F'(t) = x(t)$ almost everywhere, and the proof is finished in the bounded case.

In the general case, we can write $x(t) = x^+(t) - x^-(t)$, so that it is sufficient to prove the theorem for a nonnegative function $z \in L^{(1)}[a, b]$. Let $z_n(t) = z(t) \cap n$; then the function $z(t) - z_n(t)$ is nonnegative,

$$\int_a^t [z(s) - z_n(s)] \, ds$$

is nondecreasing, and therefore its derivative is nonnegative. Thus we obtain

$$\frac{d}{dt} \left[\int_a^t z(s) \, ds \right] \geq \frac{d}{dt} \left[\int_a^t z_n(s) \, ds \right] = z_n(t),$$

where the validity of our theorem for the bounded case has been used. Since the inequality holds for all n, we find

$$\frac{d}{dt} \left[\int_a^t z(s) \, ds \right] \geq z(t).$$

Let

$$G(t) - G(a) = \int_a^t z(s) \, ds;$$

thus $G'(t) \geq z(t)$ and

$$\int_a^b G'(t) \, dt \geq \int_a^b z(t) \, dt.$$

But $G(t) - G(a)$ is nondecreasing [$z(s)$ is nonnegative]; therefore

$$\int_a^b G'(t) \, dt \leq G(b) - G(a) = \int_a^b z(t) \, dt.$$

Thus

$$\int_a^b [G'(t) - z(t)] \, dt = 0$$

and, since $G'(t) - z(t)$ is nonnegative, we conclude $G'(t) = z(t)$ almost every-where. This completes the entire proof.

The foregoing theorem has the following implication: If there exists a function $x \in L^{(1)}[a, b]$ such that

$$F(t) - F(a) = \int_a^t x(s) \, ds,$$

then $F(t)$ has a derivative almost everywhere,

$$F(t) - F(a) = \int_a^t F'(s) \, ds,$$

and $F'(t) = x(t)$ a.e.

Definition 5.9.8. $F(t)$ is absolutely continuous on $[a, b]$ if it satisfies the equation

$$F(t) - F(a) = \int_a^t F'(s) \, ds, \qquad \text{for all} \quad t \in [a, b].$$

The collection of all absolutely continuous functions on $[a, b]$ is denoted by $A[a, b]$.

Our goal now is to give a characterization of $A[a, b]$. The absolute continuity of the set function $v(E) = \int_E x(t) \, dt$ yields immediately the fact that any function F in $A[a, b]$ must satisfy the condition:

A_1:   For any $\epsilon > 0$ there is a $\delta(\epsilon)$ such that

$$\left| \sum_{k=1}^{n} [F(t_k) - F(t_k')] \right| \leq \epsilon$$

whenever the intervals

$$I_k = (t_k', t_k), \qquad k = 1, \ldots, n$$

are disjoint and satisfy

$$\sum_{k=1}^{n} |t_k - t_k'| \leq \delta(\epsilon).$$

It is interesting and useful to show that A_1 is equivalent to the condition:

A: For any $\epsilon > 0$, there is a $\delta(\epsilon)$ such that

$$\sum_{k=1}^{n} |F(t_k) - F(t_k')| \leq \epsilon$$

whenever the intervals

$$I_\kappa = (t_k', t_k), \qquad k = 1, \ldots, n$$

are disjoint and satisfy

$$\sum_{k=1}^{n} |t_k - t_k'| \leq \delta(\epsilon).$$

To prove $A \Rightarrow A_1$, we have only to use the triangle inequality

$$\left| \sum_{k=1}^{n} [F(t_k) - F(t_k')] \right| \leq \sum_{k=1}^{n} |F(t_k) - F(t_k')|.$$

To show that $A_1 \Rightarrow A$, we let $I_1, \ldots, I_n$ be n disjoint open intervals such that

$$\sum_{k=1}^{n} |t_k - t_k'| \leq \delta\left(\frac{\epsilon}{2}\right),$$

and we assume that they are labeled so that

$$F(t_k) - F(t_k') \geq 0, \qquad \text{for} \quad k = 1, \ldots, i$$

and

$$F(t_k) - F(t'_k) < 0, \qquad \text{for} \quad k = i+1, \ldots, n.$$

Since

$$\sum_{k=1}^{i} |t_k - t'_k| \leq \delta\left(\frac{\epsilon}{2}\right)$$

and

$$\sum_{k=i+1}^{n} |t_k - t'_k| \leq \delta\left(\frac{\epsilon}{2}\right),$$

we have from A_1

$$\left| \sum_{k=1}^{i} [F(t_k) - F(t'_k)] \right| = \sum_{k=1}^{i} |F(t_k) - F(t'_k)| \leq \frac{\epsilon}{2}$$

and

$$\left| \sum_{k=i+1}^{n} [F(t_k) - F(t'_k)] \right| = \sum_{k=i+1}^{n} |F(t_k) - F(t'_k)| \leq \frac{\epsilon}{2}.$$

Thus

$$\sum_{k=1}^{n} |F(t_k) - F(t'_k)| \leq \epsilon$$

and the proof that A is equivalent to A_1 is finished.

The proof of the following lemma is left as an exercise.

Lemma 5.9.9. The class of all functions satisfying condition A is an algebra of functions closed under the operations $\cup$ and $\cap$.

It is clear that a function satisfying condition A must be continuous (take $n = 1$). Moreover, since its total variation in an interval of length $\delta(\epsilon)$ cannot exceed ϵ, its total variation on $[a, b]$ cannot exceed $\epsilon \frac{b - a}{\delta(\epsilon)}$. We summarize these facts in a lemma.

Lemma 5.9.10. If $x(t)$ satisfies condition A on $[a, b]$, then $x \in BV[a, b] \cap C[a, b]$, and therefore $x'(t)$ exists almost everywhere.

The Cantor function is an example of a continuous function of bounded variation which does not satisfy condition A. To see this, choose any $\epsilon < 1$.

Since the Cantor set C has measure zero, there is an open set O containing C with $\mu(O)$ arbitrarily small. Let $O = \bigcup_{n=1}^{\infty} I_n$, where $I_n = (t_n', t_n)$ and $I_i \cap I_j = \varnothing$. Then

$$\sum_{n=1}^{\infty} |c(t_n) - c(t_n')| = 1,$$

and for some large N, we have

$$\sum_{n=1}^{N} |c(t_n) - c(t_n')| \geq \epsilon.$$

But

$$\sum_{n=1}^{n} |t_n - t_n'| \leq \mu(O),$$

which can be arbitrarily small. Thus $c(t)$ does not satisfy condition A.

Loosely speaking, $c(t)$ does its increasing on a set of measure zero and is constant on the complement of this set. We will now see the role of condition A.

Lemma 5.9.11. If $x(t)$ satisfies condition A and if $x'(t) = 0$ almost everywhere, then $x(t)$ is constant.

Proof. Let η and ϵ be arbitrary positive numbers, and let $\delta(\epsilon)$ be determined by condition A. Choose $\hat{t} \in (a, b)$ arbitrarily, and let $E = \{t \in (a, \hat{t}): x'(t) = 0\}$. For every $t \in E$ there exist arbitrarily small positive numbers h such that

$$\frac{1}{h}|x(t + h) - x(t)| < \eta \quad \text{and} \quad [t, t + h] \subset [a, b].$$

The collection of closed intervals $[t, t + h]$, so determined, constitutes a Vitali covering of E. Thus there exist finitely many disjoint intervals

$$I_k = [t_k', t_k], \qquad k = 1, \ldots, n$$

such that

$$\mu\left(E - \bigcup_{k=1}^{n} I_k\right) \leq \delta(\epsilon);$$

hence

$$\mu\left([a, \hat{t}] - \bigcup_{k=1}^{n} I_k\right) \leq \delta(\epsilon).$$

Assume that the intervals I_k, $k = 1, \ldots, n$ are labeled so that

$$a = t_0 \leq t_1' < t_1 \leq t_2' < t_2 \leq \cdots \leq t_n' < t_n \leq t_{n+1}' = \hat{t}.$$

We calculate

$$|x(\hat{t}) - x(a)| \leq \sum_{k=1}^{n+1} |x(t_k') - x(t_{k-1})| + \sum_{k=1}^{n} |x(t_k) - x(t_k')|.$$

Now,

$$\sum_{k=1}^{n+1} |t_k' - t_{k-1}| = \mu\left([a, \hat{t}] - \bigcup_{k=1}^{n} I_k\right) \leq \delta(\epsilon),$$

so that

$$\sum_{k=1}^{n+1} |x(t_k') - x(t_{k-1})| \leq \epsilon$$

by the absolute continuity of $x(t)$. Moreover,

$$\sum_{k=1}^{n} |x(t_k) - x(t_k')| \leq \sum_{k=1}^{n} \eta |t_k - t_k'| \leq \eta(b - a)$$

by the definition of the Vitali covering, and hence

$$|x(\hat{t}) - x(a)| \leq \epsilon + \eta(b - a)$$

for all ϵ and η. It follows that $x(\hat{t}) = x(a)$ for all $\hat{t} \in (a, b)$, and the proof is complete.

We can now prove the characterization theorem.

Theorem 5.9.12. $F(t)$ is absolutely continuous, that is,

$$F(t) - F(a) = \int_a^t F'(t) \, dt,$$

if and only if $F(t)$ satisfies condition A.

Proof. We need only show the "if" part of the theorem, since the converse was derived earlier.

Since $F(t)$ is of bounded variation it has an integrable derivative and $\int_a^t F'(s) \, ds$ satisfies condition A. Hence (see lemma 5.9.9) the function

$$G(t) = F(t) - \int_a^t F'(s) \, ds$$

satisfies condition A. But $G'(t) = 0$ almost everywhere, and the preceding lemma applies. We conclude that $G(t)$ is constant, and that

$$F(t) - F(a) = \int_a^t F'(s)\, ds.$$

The proof is now complete.

For emphasis, we restate our basic conclusions on the relation of integration and differentiation.

(a) For any $x \in L^{(1)}[a, b]$ we have

$$\frac{d}{dt}\int_a^t x(s)\, ds = x(t)$$

almost everywhere.

(b) $\int_a^t F'(s)\, ds = F(t) - F(a)$ if and only if $F(t)$ satisfies condition A.

For many applications, it is not sufficient to know that a function is differentiable, but it is also necessary to know that it can be recovered from its derivative by integration. This indicates the importance of the class $A[a, b]$.

Condition A is not always easy to apply directly; the following lemma gives a simple criterion sufficient for x to be in $A[a, b]$.

Lemma 5.9.13. If $x'(t)$ exists and satisfies $|x'(t)| \le M$ for all t in $[a, b]$ then $x \in A[a, b]$.

Proof

$$\sum_{k=1}^n |x(t_k) - x(t_k')| = \sum_{k=1}^n |t_k - t_k'|\,|x'(t_k'')| \le M \sum_{k=1}^n |t_k - t_k'|$$

for some $t_k'' \in (t_k', t_k)$, by the mean-value theorem. It follows that $x(t) \in A[a, b]$ as desired.

Two very general results, which we state without proof, are as follows.

1. If $x(t)$ is continuous, $x'(t)$ exists at all but a countable number of points in $[a, b]$, and $x' \in L^{(1)}[a, b]$, then $x \in A[a, b]$.
2. $x \in A[a, b] \Leftrightarrow \{x \in C[a, b] \cap BV[a, b]$ and $\mu[x(E)] = 0$ whenever $\mu(E) = 0.\}$

In section 5 the theorems of Lebesgue gave conditions under which sequences of functions can be termwise integrated. We close this section with a special result due to Fubini on the termwise differentiation of series.

Theorem 5.9.14. Let the finite-valued functions $x_n(t), n = 1, 2, \ldots,$ be nondecreasing on $[a, b]$ such that the series $\sum_{n=1}^{\infty} x_n(t)$ converges everywhere to a finite-valued function $x_0(t)$ on $[a, b]$. Then $\sum_{n=1}^{\infty} x'_n(t) = x'_0(t)$ almost everywhere.

Proof. Without loss of generality we can assume $x_n(0) = 0$ for all n, so that $x_n(t) \geq 0$ for $t \in [a, b]$. Clearly, $x_0(t)$ is also nondecreasing; hence, on the complement of some set E of measure zero, all $x_n(t)$ $(n = 0, 1, \ldots)$ are differentiable and $x'_n(t) \geq 0$. It follows that the partial sums

$$s'_n(t) = \sum_{k=1}^{n} x'_k(t)$$

satisfy

$$s'_n(t) \leq s'_{n+1}(t)$$

and also

$$s'_n(t) \leq x'_0(t),$$

for

$$\frac{1}{h}[x_0(t + h) - x_0(t)] = \frac{1}{h} \sum_{k=1}^{\infty} [x_k(t + h) - x_k(t)]$$

$$\geq \frac{1}{h} \sum_{k=1}^{n} [x_k(t + h) - x_k(t)] = \frac{1}{h}[s_n(t + h) - s_n(t)].$$

Hence the sequence of partial sums $\{s'_n(t)\}$ converges on $\tilde{E}$.

It suffices now to show that on $\tilde{E}$ some subsequence converges to $x'_0(t)$. We choose a sequence of integers $\{n(j)\}$ such that $x_0(b) - s_{n(j)}(b) < 2^{-j}$. Then

$$\sum_{j=1}^{\infty} [x_0(t) - s_{n(j)}(t)] \leq \sum_{j=1}^{\infty} [x_0(b) - s_{n(j)}(b)] \leq \sum_{j=1}^{\infty} 2^{-j}$$

and the series

$$\sum_{j=1}^{\infty} [x_0(t) - s_{n(j)}(t)]$$

converges. Its terms are nonnegative and nondecreasing; therefore by the results just obtained, the termwise differentiated series

$$\sum_{j=1}^{\infty} [x'_0(t) - s'_{n(j)}(t)]$$

converges a.e. But this implies

$$[x_0'(t) - s_{n(j)}'(t)] \to 0 \quad \text{a.e. as} \quad j \to \infty,$$

and the proof is complete.

EXERCISES

1. Prove in detail that, if $x(t)$ is nondecreasing on $[a, b]$, then

$$\mu\{t: \quad D_+[x; t] < D^-[x; t]\} = 0.$$

2. Prove lemma 5.9.9.

3. Consider the function

$$x(t) = \begin{cases} t^2 \sin \dfrac{1}{t^2}, & 0 < t \le 1 \\ 0, & t = 0. \end{cases}$$

Show that $x'(t)$ exists for all $t \in [0, 1]$ but is unbounded. Show $x'(t) \notin L^{(1)}[a, b]$, but $x'(t)$ is integrable in the improper Riemann sense.

4. Consider

$$x(t) = \begin{cases} t^{3/2} \sin \dfrac{1}{t}, & 0 < t \le 1 \\ 0, & t = 0. \end{cases}$$

Show that $x'(t)$ exists for all $t \in [0, 1]$ but is unbounded. Show $x' \in L^{(1)}[a, b]$. Show directly that $x \in A[a, b]$.

5. Extend the Cantor function to the entire real line by defining $c(t + 1) = c(t) + 1$, and let $C(t) = \sum_{n=1}^{\infty} c(nt)/2^n$.

a. Show that $C(t)$ is continuous on $[0, 1]$ and that it has zero derivative almost everywhere.

b. For $0 \le t_1 < t_2 \le 1$ show that there exists an n such that $c(nt_1) < c(nt_2)$. Use this to show that $t_1 < t_2 \Rightarrow C(t_1) < C(t_2)$.

c. Is $C(t)$ absolutely continuous?

6. Let the real-valued function $x(t)$ on R satisfy the *Lipschitz condition L*: $|x(t_1) - x(t_2)| \le M|t_1 - t_2|$.

a. Show that $x(t)$ is absolutely continuous.

b. Show that there are absolutely continuous functions which do not satisfy a Lipschitz condition (see exercise 4).

7.   Let a condition A' be formed from condition A by omitting the requirement that the intervals be disjoint. Prove or disprove the following statement: $(A' \Leftrightarrow L)$ for a function $x(t)$ on $[a, b]$ (see exercise 6).

8.   Assume that the four derivates of $x(t)$ are bounded on $[a, b]$. Show that $x(t)$ satisfies a Lipschitz condition (see exercise 6).

9.   Let E be the set of points at which $x(t)$ possesses finite but distinct left and right derivatives. Show that E is countable (see exercise 9, section 7, chapter 4).

10.   Prove the following intermediate-value theorem for derivatives. Let $x(t)$ have a finite derivative $x'(t)$ everywhere on $[a, b]$ and suppose, for $a \le t_1 < t_2 \le b$, that $x'(t_1) < C < x'(t_2)$. Show that there exists a $t_3 \in (t_1, t_2)$ such that $x'(t_3) = C$. [HINT: Define the functions

$$x_L(t) = \begin{cases} \dfrac{x(t) - x(t_1)}{t - t_1}, & t_1 < t \le t_2 \\ x'(t_1), & t = t_1 \end{cases}$$

and

$$x_R(t) = \begin{cases} \dfrac{x(t_2) - x(t)}{t - t_2}, & t_1 \le t < t_2 \\ x'(t_2), & t = t_2 \end{cases}$$

on $[t_1, t_2]$. Consider their ranges and apply the ordinary intermediate-value theorem as well as the mean-value theorem.] What is the implication of this result for the nature of the discontinuities of $x'(t)$?

11.   The *density* of a measurable set E at a point t is defined as

$$\underset{h \to 0}{\text{limit}}\ \frac{1}{2h}\mu(E \cap [t - h, t + h]).$$

Show that the density of E is 1 almost everywhere on E and 0 almost everywhere on $\tilde{E}$. [HINT: Let $t_0 \in E$ and consider

$$F(t) = \int_{t_0}^{t} y(s)\,ds$$

where $y(s)$ is the characteristic function of E.]

12.   Prove the following integration-by-parts theorem: If

$$x \in L^{(1)}[a, b] \quad \text{and} \quad y \in A[a, b],$$

then

$$\int_a^b x(t)y(t)\, dt = F(b)y(b) - F(a)y(a) - \int_a^b F(t)y'(t)\, dt,$$

where $F'(t) = x(t)$.

13.   Let $x \in L^{(1)}[a, b]$. Show how the set function $v(E) = \int_E x(t)\, dt$ is determined by the point function $F(t) = \int_a^t x(s)\, ds$, and conversely.

Section 10   THE LEBESGUE-STIELTJES INTEGRAL

Lebesgue measure on the real line is a special case of a more general type of measure on R. The development of this type of measure is similar to that of Lebesgue measure, the principal difference coming at the very beginning.

We start with a finite-valued, nondecreasing function x on R, and assume that for every $t_0 \in R$.

$$\lim_{t \to t_0^-} x(t) = x(t_0),$$

that is, x is left-continuous. We then assign measure

$$\mu([a, b)] = x(b) - x(a)$$

to left-closed, right-open intervals, zero measure to the empty set, and measure

$$\mu\left(\bigcup_{k=1}^{n} I_k\right) = \sum_{k=1}^{n} [x(b_k) - x(a_k)]$$

to members of the ring $\mathscr{R}$ of finite disjoint unions of left-closed, right-open intervals $I_k = [a_k, b_k)$.

It is not difficult to show that μ is a nonnegative, finitely additive, monotone function on $\mathscr{R}$. We omit a formal proof, but make the following remarks.

(a) The nonnegativity of μ is a consequence of the monotonicity of x.

(b) It must be verified that $\mu(E)$, for $E \in \mathscr{R}$, is independent of the way in which E is written as a finite union of disjoint intervals in $\mathscr{R}$.

(c) The left-continuity of the function x is necessary in obtaining a proof of the following lemma.

Lemma 5.10.1. The function μ on $\mathscr{R}$ is countably additive.

Proof. Let $\{A_n\}$ be a sequence of disjoint sets in $\mathscr{R}$ such that

$$\bigcup_{n=1}^{\infty} A_n = A_0 \in \mathscr{R}.$$

Since $A_0 \supset \bigcup_{n=1}^{N} A_n$ we have, by the finite additivity and monotonicity of μ,

$$\mu(A_0) \geq \mu\left(\bigcup_{n=1}^{N} A_n\right) = \sum_{n=1}^{N} \mu(A_n).$$

Thus

$$\mu(A_0) \geq \sum_{n=1}^{\infty} \mu(A_n).$$

The reverse inequality is somewhat harder to prove. To simplify things, we notice that it suffices to show that, if $\{I_n\}, n = 0, 1, 2, \ldots$, is a sequence of left-closed, right-open intervals with

$$I_0 = \bigcup_{n=1}^{\infty} I_n, \qquad I_i \cap I_j = \varnothing, \qquad (i > j > 0),$$

then

$$\mu(I_0) \leq \sum_{n=1}^{\infty} \mu(I_n).$$

[See the proof of lemma 5.2.7(a)] Let $I_n = [a_n, b_n)$ and define

$$I_0' = [a_0, b_0 - \epsilon], \qquad I_n' = (a_n - \delta_n, b_n), \qquad n > 0.$$

By the compactness of I_0', there is an N such that

$$I_0' \subseteq \bigcup_{n=1}^{N} I_n',$$

and by the monotonicity of x, we have

$$[x(b_0 - \epsilon) - x(a_0)] \leq \sum_{n=1}^{N} [x(b_n) - x(a_n - \delta_n)].$$

Using the left-continuity of x we can find for any $\eta > 0$ an $\epsilon(\eta)$ such that

$$\mu(I_0) - [x(b_0 - \epsilon) - x(a_0)] \leq \eta$$

whenever $\epsilon < \epsilon(\eta)$. For such an ϵ, we have

$$[\mu(I_0) - \eta] \leq [x(b_0 - \epsilon) - x(a_0)] \leq \sum_{n=1}^{N} [x(b_n) - x(a_n - \delta_n)].$$

Again applying the left-continuity of x we have, for any $\delta > 0$, a δ'_n such that, for $\delta_n \leq \delta'_n$,

$$[x(b_n) - x(a_n - \delta_n)] - \mu(I_n) < 2^{-n}\delta,$$

$n = 1, 2, \ldots, N$. Hence

$$[\mu(I_0) - \eta] \leq \sum_{n=1}^{N} [\mu(I_n) + 2^{-n}\delta] \leq \sum_{n=1}^{\infty} \mu(I_n) + \delta,$$

and it follows that

$$\mu(I_0) \leq \sum_{n=1}^{\infty} \mu(I_n)$$

as desired This completes the proof.

In summary, we have established that the function μ is a measure on the ring $\mathscr{R}$.

From this point on, the construction of the Lebesgue-Stieltjes measure proceeds exactly as in the case of Lebesgue measure. We will not present all of the details, but will only list the steps.

1. Define the function

$$\mu^*(A) = \inf \left\{ \sum_{n=1}^{\infty} \mu(A_n): \bigcup_{n=1}^{\infty} A_n \supseteq A, A_n \in \mathscr{R} \right\}$$

and show that it is nonnegative, monotone, countably subadditive, and that $\mu^*(\varnothing) = 0$.
2. Show that $\mu(A) = \mu^*(A)$ for $A \in \mathscr{R}$.
3. Define the class $\mathscr{M}_0$ of sets of finite measure by stipulating that $A \in \mathscr{M}_0$ if and only if, for any $\epsilon > 0$, there is a set $B \in \mathscr{R}$ such that $\mu^*(A \triangle B) \leq \epsilon$.
4. Define μ on $\mathscr{M}_0$ and show that $\mu(A) = \mu^*(A)$ for $A \in \mathscr{M}_0$.
5. Define

$$\mathscr{M} = \left\{ \bigcup_{n=1}^{\infty} A_n: A_n \in \mathscr{M}_0, \quad n = 1, 2, \ldots \right\},$$

and show that it is a σ-algebra.

6. Show that μ^* is a measure on $\mathcal{M}$.

The function μ^* (henceforth we drop the asterisk) on $\mathcal{M}$ is the Lebesgue-Stieltjes measure defined by x. The σ-algebra $\mathcal{M}$ contains the Borel sets, and μ is finite on the bounded sets in $\mathcal{M}$. We state without proof the fact that every measure on the Borel sets which is finite on bounded sets is a Lebesgue-Stieltjes measure, i.e., it is derived from a nondecreasing function x in the manner outlined in this section.

The material on integration developed in sections 3 through 7 applies without modification if a Lebesgue-Stieltjes measure is used, for those sections apply to functions on an arbitrary measure space.

EXERCISES

1. Show that a Lebesgue-Stieltjes measure μ is an *absolutely continuous* set function with respect to ordinary Lebesgue measure if and only if the nondecreasing function associated with it is continuous. (See section 5, chapter 5.)

2. A nondecreasing function x on R is continuous except for a countable number of jump discontinuities. Let x_L and x_R be the left-continuous and right-continuous functions obtained from x by redefining x in the obvious ways at its discontinuities. Let μ_L be the measure as constructed in this section.

    Define $\mathcal{R}'$ as the ring of all finite disjoint unions of left-open, right-closed intervals, and use x_R to construct a measure μ_R. Compare μ_L and μ_R.

Section 11   PRODUCT MEASURES AND THE FUBINI
THEOREM

Let $(L, \mathcal{L}, \lambda)$ and $(M, \mathcal{M}, \mu)$ be two measure spaces where both measures are σ-finite. We wish to define a measure on a σ-algebra $\mathcal{N}$ of subsets of $N = L \times M$ where we require that $\mathcal{N}$ contain all sets of the form $A \times B$ for $A \in \mathcal{L}$, $B \in \mathcal{M}$. Accordingly, we introduce the following notation:

A *measurable rectangle* is a subset of $N = L \times M$ of the form $A \times B$ where $A \in \mathcal{L}, B \in \mathcal{M}$.

$\mathcal{N}_0$ = the smallest algebra of subsets of N which contains all measurable rectangles.

$\mathcal{N}$ = the smallest σ-algebra of subsets of N which contain all measurable rectangles.

The algebra $\mathcal{N}_0$ is characterized as follows.

Lemma 5.11.1. $\mathcal{N}_0$ is the collection of all finite disjoint unions of measurable rectangles.

Proof. $\mathcal{N}_0$ must contain the collection of all finite disjoint unions of measurable rectangles, so we need only prove that this collection is an algebra of sets, and for this it suffices to show that $\mathcal{N}_0$ is closed under the operations of complementation and intersection.

Since $\sim(A \times B) = (\tilde{A} \times Y) \cup (A \times \tilde{B})$, we see that the complement of a measurable rectangle is in $\mathcal{N}_0$. Hence from

$$\sim\left[\bigcup_{i=1}^{n} (A_i \times B_i)\right] = \bigcap_{i=1}^{n} [\sim(A_i \times B_i)]$$

it follows that we have now only to show that $\mathcal{N}_0$ is closed under intersection.

Let $(A_1 \times B_1), \ldots, (A_m \times B_m)$ be disjoint measurable rectangles and similarly let $(C_1 \times D_1), \ldots, (C_n \times D_n)$ be disjoint measurable rectangles, so that $J = \bigcup_{i=1}^{m} (A_i \times B_i)$ and $K = \bigcup_{j=1}^{n} (C_j \times D_j)$ are typical sets in $\mathcal{N}_0$. Using $(A \times B) \cap (C \times D) = (A \cap C) \times (B \cap D)$ and distributivity, we calculate that

$$J \cap K = \left[\bigcup_{i=1}^{m} (A_i \times B_i)\right] \cap \left[\bigcup_{j=1}^{n} (C_j \times D_j)\right] = \bigcup_{i,j=1}^{m,n} [(A_i \cap C_j) \times (B_i \cap D_j)],$$

which is a union of sets in $\mathcal{N}_0$. To see that these sets are disjoint we calculate

$$[(A_i \cap C_j) \times (B_i \cap D_j)] \cap [(A_k \cap C_l) \times (B_k \cap D_l)]$$
$$= (A_i \cap A_k \cap C_j \cap C_l) \times (B_i \cap B_k \cap D_j \cap D_l).$$

If $A_i \cap A_k \cap C_j \cap C_l \neq \varnothing$ then $B_i \cap B_k \cap D_j \cap D_l = \varnothing$ by the disjointness of the sets comprising J(or K), so that $J \cap K$ is a disjoint union of measurable rectangles, and $\mathcal{N}_0$ is closed under intersection. (A rectangle is empty if and only if one of its components is empty.) This completes the proof.

We next study "sections" of measurable sets and measurable functions.

Definition 5.11.2. Let $A \in \mathcal{N}$, and let f be a real-valued, measurable function on N, i.e., $f^{-1}(B) \in \mathcal{N}$ for all Borel sets B.

For $x \in L$ define $A_x = \{y \in M: (x, y) \in A\}$.
For $y \in M$ define $A^y = \{x \in L: (x, y) \in A\}$.

The sets $A_x \subseteq M$ and $A^y \subseteq L$ are the *sections* of A cut by x and y, respectively.

For $x \in L$ define $f_x(y) = f(x, y)$.
For $y \in M$ define $f^y(x) = f(x, y)$.

The functions f_x and f^y are the sections of f cut by x and y, respectively; f_x has domain M and f^y has domain L.

The following lemma answers some immediate questions.

Lemma 5.11.3. For $A \in \mathcal{N}$ and f measurable on N,

(a) the sets A_x and A^y are in $\mathcal{M}$ and $\mathcal{L}$, respectively.
(b) the functions f_x and f^y are measurable on M and L, respectively.

Proof. (a) Consider the class $\mathcal{N}_1$ of subsets of N such that for $A \in \mathcal{N}_1$ the sets A_x and A^y are in $\mathcal{M}$ and $\mathcal{L}$, respectively, for all $x \in L$ and $y \in M$. Clearly $\mathcal{N}_1$ contains all measurable rectangles. It is also evident that $\mathcal{N}_1$ is a σ-algebra of subsets of N. It follows that $\mathcal{N}_1 \supset \mathcal{N}$ and the proof of (a) is finished.

(b) Let B be a Borel set in R and let f be a measurable function on N, The measurability of f_x and f^y follows from (a) and the relations

$$(f^y)^{-1}(B) = [f^{-1}(B)]^y \quad \text{and} \quad f_x^{-1}(B) = [f^{-1}(B)]_x.$$

This completes the proof of the lemma.

We have not yet defined a measure on the σ-algebra $\mathcal{N}$. For $A \in \mathcal{N}$ we consider the functions $f(x) = \mu(A_x)$ and $g(y) = \mu(A^y)$. Assuming they are measurable, we then attempt to define $v(A) = \int f \, d\lambda = \int g \, d\mu$; since f and g are nonnegative, the two integrals are defined, but their equality must be proved. This then is our program; however, some technical difficulties arise which necessitate the following digression.

Definition 5.11.4. A nonempty collection of sets is called a *monotone class* if

(a) $\bigcup_{n=1}^{\infty} A_n$ is in the class whenever $\{A_n\}$ is a sequence in the class with
$A_1 \subseteq A_2 \subseteq A_3 \subseteq \cdots$.

(b) $\bigcap_{n=1}^{\infty} A_n$ is in the class whenever $\{A_n\}$ is a sequence in the class with
$A_1 \supseteq A_2 \supseteq A_3 \cdots$.

The notation $\underset{n\to\infty}{\text{limit}}\, A_n$ will be used to designate the sets in (a) and (b) of the definition. (See exercise, 3, section 2 for related matters.)

Since the collection of all subsets of a set X is a monotone class, and since the intersection of a family of monotone classes is a monotone class, the notion of the smallest monotone class containing a given collection $\mathscr{S}$ of sets is meaningful; it is the intersection of all monotone classes containing $\mathscr{S}$. Clearly, a *σ-ring is a monotone class*. More interesting is the fact that *a ring of sets which is a monotone class is a σ-ring*. For, if $\{A_n\}$ is a sequence of sets in the ring, then

$$\bigcup_{n=1}^{\infty} A_n = A_1 \cup [A_1 \cup A_2] \cup \cdots \cup [A_1 \cup \cdots \cup A_n] \cdots$$

is the limit of a monotone sequence in the ring, and is thus also in the ring. We will need the following basic lemma on monotone classes.

Lemma 5.11.5. Let $\mathscr{R}$ be a ring of sets and let $\mathscr{S}$ be the smallest σ-ring containing $\mathscr{R}$. If $\mathscr{T}$ is a monotone class containing $\mathscr{R}$, then $\mathscr{T} \supseteq \mathscr{S}$.

Proof. Let $\mathscr{T}_0$ be the smallest monotone class containing the ring $\mathscr{R}$. Then $\mathscr{T} \supseteq \mathscr{T}_0 \supseteq \mathscr{R}$ and it suffices to show that $\mathscr{T}_0$ is a σ-ring, for in that case $\mathscr{T} \supseteq \mathscr{T}_0 \supseteq \mathscr{S} \supseteq \mathscr{R}$. But by the above remarks we need only show that $\mathscr{T}_0$ is a ring.

Thus the proof is reduced to the problem of showing that the smallest monotone class $\mathscr{T}_0$ containing a ring $\mathscr{R}$ is itself a ring.

To show that $\mathscr{T}_0$ is closed under the operations of set union and differences, we define $\mathscr{G}(A) = \{B: \; A \cup B, A - B, B - A \in \mathscr{T}_0\}$; $\mathscr{T}_0$ will then be a ring if $\mathscr{G}(B) \supseteq \mathscr{T}_0$ for every $B \in \mathscr{T}_0$.

We now show that for $A \in \mathscr{T}_0$, $\mathscr{G}(A)$ is a monotone class; let $\{A_n\}$ be a monotone sequence in $\mathscr{G}(A)$. Then

$$(\underset{n\to\infty}{\text{limit}}\, A_n) - A = \underset{n\to\infty}{\text{limit}}\,(A_n - A) \in \mathscr{T}_0, \; A - \underset{n\to\infty}{\text{limit}}\, A_n = \underset{n\to\infty}{\text{limit}}\,(A - A_n) \in \mathscr{T}_0,$$

and

$$(\underset{n\to\infty}{\text{limit}}\, A_n) \cup A = \underset{n\to\infty}{\text{limit}}\,(A_n \cup A) \in \mathscr{T}_0$$

by virtue of the fact that $\mathscr{T}_0$ is a monotone class. Hence, for $A \in \mathscr{T}_0$, $\mathscr{G}(A)$ is a monotone class, if it is nonempty. In particular, if $A \in \mathscr{R}$, then $\mathscr{G}(A) \supseteq \mathscr{R}$ and hence $\mathscr{G}(A) \supseteq \mathscr{T}_0$, because $\mathscr{T}_0$ is the smallest monotone class containing $\mathscr{R}$.

Now, from the definition of $\mathscr{G}(A)$ we see that $B \in \mathscr{G}(A)$ if and only if $A \in \mathscr{G}(B)$. Hence we have

$$(B \in \mathcal{T}_0, A \in \mathcal{R}) \Rightarrow (B \in \mathcal{G}(A)) \Rightarrow (A \in \mathcal{G}(B))$$

and so $\mathcal{T}_0 \subseteq \mathcal{G}(B)$ for all $B \in \mathcal{T}_0$. This completes the proof.

Corollary. For any ring of sets $\mathcal{R}$, the smallest σ-ring containing $\mathcal{R}$ is identical with the smallest monotone class containing $\mathcal{R}$.

We are now ready to establish the results promised earlier.

Theorem 5.11.6. If $A \in \mathcal{N}$ then the nonnegative functions $f(x) = \mu(A_x)$ and $g(y) = \lambda(A^y)$ are measurable on L and M, respectively, and $\int f \, d\lambda = \int g \, d\mu$.

Proof. Let $\mathcal{N}_1$ be the collection of all sets in $\mathcal{N}$ for which the theorem holds. If $C = A \times B$ is a nonempty, measurable rectangle then $f(x) = \mu(B) f_A(x)$ where $f_A(x)$ is the characteristic function of A, and $g(y) = \lambda(A) g_B(y)$ where $g_B(y)$ is the characteristic function of B. Hence f and g are measurable and $\int f \, d\lambda = \int g \, d\mu = \lambda(A) \cdot \mu(B)$. It follows that $\mathcal{N}_1$ contains all measurable rectangles.

We now verify that $\mathcal{N}_1$ is closed under the formation of countable disjoint unions. Let $C = \bigcup_{n=1}^{\infty} C_n$ where $C_i \cap C_j = \varnothing$ and $C_n \in \mathcal{N}_1$ for all n. If $f(x) = \mu(C_x)$ and $f_n(x) = \mu(C_{nx})$ then the sets $\{C_{nx}\}$ are disjoint and measurable in M so that

$$f(x) = \mu(C_x) = \mu\left(\bigcup_{n=1}^{\infty} C_{nx}\right) = \sum_{n=1}^{\infty} \mu(C_{nx}) = \sum_{n=1}^{\infty} f_n$$

is measurable, since it is the pointwise limit of a sequence of measurable functions (the partial sums of the series); similarly, the function $g(y) = \lambda(C_y)$ is measurable. Moreover,

$$\int f \, d\lambda = \int \left[\sum_{n=1}^{\infty} f_n\right] d\lambda = \sum_{n=1}^{\infty} \left[\int f_n \, d\lambda\right] = \sum_{n=1}^{\infty} \left[\int g_n \, d\mu\right]$$
$$= \int \left[\sum_{n=1}^{\infty} g_n\right] d\mu = \int g \, d\mu;$$

this is justified by our hypothesis that the sets $\{C_n\}$ are in $\mathcal{N}_1$. It follows that $\mathcal{N}_1$ is closed under the formation of countable disjoint unions, as claimed.

From the above results, we see that $\mathcal{N}_1$ contains $\mathcal{N}_0$, the ring of all finite disjoint unions of measurable rectangles. The strategy now is to attempt to show that $\mathcal{N}_1$ is a monotone class and to apply the preceding lemma, which would yield $\mathcal{N} \subseteq \mathcal{N}_1$ as desired. Unfortunately, some difficulty arises and we must proceed as follows. The σ-finiteness of L and M yields easily the fact that $N = \bigcup_{n=1}^{\infty} D_n$ where the sets $\{D_n\}$ are disjoint measurable rectangles with

$D_n = A_n \times B_n$ and $\lambda(A_n) < \infty$, $\mu(B_n) < \infty$. Since any $C \in \mathcal{N}_1$ can be written $C = \bigcup_{n=1}^{\infty} (C \cap D_n)$, a countable disjoint union, and since $\mathcal{N}_1$ is closed under the formation of such unions, it suffices to prove that any $E \in \mathcal{N}$ which is contained in a finite measurable rectangle must also be in $\mathcal{N}_1$. In other words, we can assume henceforth that $\lambda(L) < \infty$ and $\mu(M) < \infty$.

It remains to be shown that $\mathcal{N}_1$ is a monotone class in the case that $\lambda(L) < \infty$, $\mu(M) < \infty$. Let $\{E_n\}$ be a monotone sequence in $\mathcal{N}_1$ with limit E_0; then by the definition of $\mathcal{N}_1$ the functions $f_n(x) = \mu(E_{nx})$ and $g_n(y) = \lambda(E_n^y)$ are measurable and $\int f_n \, d\lambda = \int g_n \, d\mu$. If $\lim_{n \to \infty} E_n = E_0$ then clearly $f_0(x) = \lim_{n \to \infty} f_n(x) = \mu(E_{0x})$ and $g_0(y) = \lim_{n \to \infty} g_n(y) = \lambda(E_0^y)$ due to the finiteness of $\lambda(L)$ and $\mu(M)$, so that f_0 and g_0 are measurable. The Lebesgue dominated convergence theorem now applies, again due to the finiteness of $\lambda(L)$ and $\mu(M)$, and we have

$$\int f_0 \, d\lambda = \lim_{n \to \infty} \int f_n \, d\lambda = \lim_{n \to \infty} \int g_n \, d\mu = \int g_0 \, d\mu,$$

so that $E_0 \in \mathcal{N}$ as desired. This completes the entire proof.

The set function $\nu(C) = \int \lambda(C^y) \, d\mu = \int \mu(C_x) \, d\lambda$ is now defined on $\mathcal{N}$, and for measurable rectangles we have $\nu(A \times B) = \lambda(A) \cdot \mu(B)$. This takes care of part of the following theorem.

Theorem 5.11.7. If $(L, \mathcal{L}, \lambda)$ and $(M, \mathcal{M}, \mu)$ are σ-finite measure spaces and $\mathcal{N}$ is the smallest σ-algebra containing $\mathcal{N}_0$, the algebra of all finite disjoint unions of measurable rectangles in $N = L \times M$, then the set function

$$\nu(C) = \int \lambda(C^y) \, d\mu = \int \mu(C_x) \, d\lambda$$

is a σ-finite measure on $\mathcal{N}$, and

$$\nu(A \times B) = \lambda(A) \cdot \mu(B)$$

for every measurable rectangle $A \times B$.

Corollary. The function

$$\nu_1 \left[\bigcup_{n=1}^{m} (A_n \times B_n) \right] = \sum_{n=1}^{m} \lambda(A_n) \mu(B_n),$$

where the measurable rectangles $\{A_n \times B_n\}$ are disjoint, defines a measure on $\mathcal{N}_0$, and the measure $\nu(C)$ defined in the theorem is the unique extension of ν_1 from $\mathcal{N}_0$ to $\mathcal{N}$.

The demonstration that v is a measure requires application of the Lebesgue monotone convergence theorem; the details of the proof of the theorem and corollary are left as an exercise, where the question of the completeness of v is also considered.

Having defined the product measure, we come to the problem of reducing double integration to the calculation of an iterated integral. We begin with terminology.

If $f(x, y)$ is measurable on $N = L \times M$ and if $\int f \, dv$ is defined (i.e., if either f^+ or f^- is in $L^{(1)}(N, \mathcal{N}, v)$, then it is called the *double integral* of f. If $f_x(y)$ and $f^y(x)$ are integrable, and if $F_1(x) = \int f_x(y) \, d\mu$ and $F_2(y) = \int f^y(x) \, d\lambda$ are also integrable, then the expressions

$$\int \left[\int f_x(y) \, d\mu \right] d\lambda \quad \text{and} \quad \int \left[\int f^y(x) \, d\lambda \right] d\mu,$$

or, more usually,

$$\iint f(x, y) \, d\mu \, d\lambda \quad \text{and} \quad \iint f(x, y) \, d\lambda \, d\mu,$$

are called *iterated integrals*.

Theorem 5.11.8. If $f(x, y)$ is a nonnegative measurable function on $N = L \times M$, then

$$\int f \, dv = \iint f \, d\lambda \, d\mu = \iint f \, d\mu \, d\lambda.$$

Proof. If f is the characteristic function of some $A \in \mathcal{N}$ then the conclusion is valid by our definition of product measure and theorem 5.11.7. Secondly, by the linearity of the integration process, the theorem is valid in the case that f is a simple function. In the general case, we can find an increasing sequence $\{f_n\}$ of nonnegative simple functions converging pointwise to f and, by the Lebesgue monotone convergence theorem, we have $\int f_n \, dv \rightarrow \int f \, dv$.

We also have $\int f_n \, dv = \iint (f_n)_x \, d\mu \, d\lambda$ since f_n is simple. Now, clearly $(f_n)_x \rightarrow f_x$ in a nondecreasing way, so that the measurable functions $\int (f_n)_x \, d\mu$ converge to $\int f_x \, d\mu$ in a nondecreasing way. Thus, we have

$$\int f_n \, dv = \iint (f_n)_x \, d\mu \, d\lambda \rightarrow \iint f_x \, d\mu \, d\lambda.$$

Hence $\int f \, dv = \iint f \, d\mu \, d\lambda$ and one of the desired equalities is proved. The other one is treated in an identical fashion. This completes the proof.

The next theorem is a result of Fubini.

Theorem 5.11.9. (Fubini.) If $f(x, y) \in L^{(1)}(N, \mathcal{N}, v)$ then the functions $f_1(x) = \int f(x, y)\, d\mu$ and $f_2(y) = \int f(x, y)\, d\lambda$ are finite valued almost everywhere, integrable, and satisfy

$$\int f\, dv = \int f_1\, d\lambda = \int f_2\, d\mu.$$

Proof. Since $f \in L^{(1)}(N, \mathcal{N}, v)$ if and only if its positive and negative parts are in $L^{(1)}(N, \mathcal{N}, v)$, it suffices to assume $f \geq 0$. But in this case the equality is given by the preceding theorem; moreover, the finiteness of $\int f\, dv$ implies that f_1 and f_2 are finite valued almost everywhere. This completes the proof.

We close this section with the remark that the results extend without difficulty to the case of a product of n σ-finite measure spaces.

EXERCISES

1. Prove the corollary to lemma 5.11.5.

2. (a) Show that if $C = \bigcup\limits_{n=1}^{m} (A_n \times B_n)$ and $\bigcup\limits_{k=1}^{p} (C_k \times D_k)$ are two representations of a set C in $\mathcal{N}_0$ as a finite union of disjoint rectangles, then $v(C) = \sum\limits_{n=1}^{m} \lambda(A_n) \cdot \mu(B_n) = \sum\limits_{k=1}^{p} \lambda(C_k) \cdot \mu(D_k)$ so that the set function $v(C)$ is unambiguously defined on $\mathcal{N}_0$.
    (b) Without using the theorems of this section, show that $v(C)$ is a measure on $\mathcal{N}_0$.

3. Prove theorem 5.11.7 and its corollary.

4. Let $H(x, y)$ be a simple function on $N = L \times M$. Show that every section of H is a simple function.

5. (a) Consider the product $R \times R$ and the measure induced on it by using Lebesgue measure on the components. If S is a nonmeasurable subset of R and x_0 is a fixed point in R, show that the set $\{x_0\} \times S$ is nonmeasurable in $R \times R$, and hence that a product measure may not be complete even if the two component measures are complete.
    (b) Show that theorems 5.11.8 and 5.11.9 remain valid if v is replaced by its completion.

6. Show that all Borel sets in $R \times R$ are measurable in the product measure on the plane. Does theorem 5.2.14 extend to this case?

7. Show that if $\lambda(L) < \infty$, $\mu(M) < \infty$, and if $f(x)$ and $g(y)$ are integrable on L and M, respectively, then the product $f(x)g(y)$ is integrable on N and $\int fg \, dv = \int f \, d\lambda \cdot \int g \, d\mu$. What happens if the finiteness hypotheses are dropped?

Section 12 SIGNED MEASURES; THE RADON-NIKODYM THEOREM

Let $(X, \mathcal{M}, \mu)$ be a measure space and let $x \in L^{(1)}(X, \mathcal{M}, \mu)$. We recall from section 5 that the set function $v_1(A) = \int_A |x| \, d\mu$ is a measure on $\mathcal{M}$ and that it is absolutely continuous with respect to μ. Obviously, $v_1(A)$ does not determine $x(t)$, but we can consider the set function $v_2(A) = \int_A x \, d\mu$ and attempt to recover x from it. (Our knowledge of the situation for Lebesgue measure on R is encouraging, for in that case $v_2(A)$ determines, and is determined by, $F(t) = \int_a^t x(s) \, ds$, and x can essentially be recovered from F by differentiation.) Since $v_2(A)$ is not always nonnegative, we need the following definition.

Definition 5.12.1. A *signed measure* is an extended real-valued set function $v(A)$ on a σ-algebra $\mathcal{M}$ of subsets of a set X such that

(a) $v(\varnothing) = 0$.
(b) v assumes at most one of the values $+\infty$, $-\infty$.
(c) v is countably additive, that is, $v\left(\bigcup_{n=1}^{\infty} A_n\right) = \sum_{n=1}^{\infty} v(A_n)$ for any disjoint sequence $\{A_n\}$.

The condition (b) is necessary to avoid the meaningless symbol $\infty - \infty$. In condition (c) it is understood that the series $\sum_{n=1}^{\infty} v(A_n)$ either converges absolutely or diverges properly (to $+\infty$ or $-\infty$, independent of the order of the terms). Note that a signed measure is subtractive, but may not be monotone.

Our first goal is to show that a signed measure v can always be written as the difference of two measures. Now, the particular signed measure $v_2(E) = \int_E x \, d\mu$ can always be written in the form $v_2(E) = \int_E x^+ \, d\mu - \int_E x^- \, d\mu$. If in this situation we let $A^+ = \{t: \ x(t) \geq 0\}$ and $A^- = \{t: \ x(t) \leq 0\}$ then we have $v_2(E) = v_2(E \cap A^+) - v_2(E \cap A^-)$.

Guided by this model, we define a set $A \in \mathcal{M}$ to be *positive* if $(B \in \mathcal{M}, B \subseteq A) \Rightarrow (\nu(B) \geq 0)$, and *negative* if $(B \in \mathcal{M}, B \subseteq A) \Rightarrow (\nu(A) \leq 0)$. A set which is both positive and negative is called a *null* set. The method then of decomposing a signed measure into positive and negative parts is to find sets A^+ and A^-.

Theorem 5.12.2 (The Hahn Decomposition Theorem). If ν is a signed measure on a σ-algebra $\mathcal{M}$ of subsets of X, then there is a positive set A^+ and a negative set A^- such that $X = A^+ \cup A^-$ and $A^+ \cap A^- = \varnothing$.

Proof. We assume $+\infty$ is the infinite value not taken on by ν, but there exists a set E_0 such that $0 < \nu(E_0) < \infty$. (If $\nu(E) \leq 0$ for all $E \in \mathcal{M}$ then take $\varnothing = A^+$, $K = A^-$.)

The first step is to show that E_0 has a positive subset A with $\nu(A) > 0$. If E_0 is not positive, then it contains a subset E_1 with $\nu(E_1) < 0$. Let $\alpha_1 = \inf \{\nu(B): B \subseteq E_0\}$; then $-\infty < \alpha_1 < 0$ and we can assume E_1 was chosen so that $\nu(E_1) < \frac{1}{2}\alpha_1$. If the set $E_0 - E_1$ is not positive, then let $\alpha_2 = \inf \{\nu(B): B \subseteq E_0 - E_1\}$ and choose E_2 so that $\nu(E_2) < \frac{1}{2}\alpha_2$ (note that $-\infty < \alpha_2 < 0$). We thus arrive at a disjoint sequence $\{E_n\}$ with

$$E_n \subseteq E_0 - \bigcup_{k=1}^{n-1} E_k$$

$$\nu(E_n) < \tfrac{1}{2}\alpha_n = \tfrac{1}{2}\inf \left\{\nu(B): \quad B \subseteq \left(E_0 - \bigcup_{k=1}^{n-1} E_k\right)\right\}.$$

If for any n the set $E_0 - \bigcup_{k=1}^{n} E_k$ is positive with $\nu\left(E_0 - \bigcup_{k=1}^{n} E_k\right) > 0$, then we are finished. Otherwise, we let $A = E_0 - \bigcup_{k=1}^{\infty} E_k$; then

$$E_0 = A \cup \left(\bigcup_{k=1}^{\infty} E_k\right), \quad \text{disjoint,}$$

and

$$\nu(E_0) = \nu(A) + \sum_{k=1}^{\infty} \nu(E_k)$$

where the series converges absolutely. Since $\alpha_k < \nu(E_k) < \frac{1}{2}\alpha_k$, we see that the series $\sum_{k=1}^{\infty} \alpha_k$ converges absolutely and thus $\alpha_k \to 0$. Now, if A contained any set B with $\nu(B) < 0$, then $0 > \alpha_k > \nu(B)$ for sufficiently large k; but this is impossible by the nature of the construction of $\{E_n\}$. Thus A is a positive set contained in E_0. Finally since $\nu(E_k) \leq 0$ we cannot have $\nu(A) = 0$, for the relation $\nu(E_0) = \nu(A) + \sum_{k=1}^{\infty} \nu(E_n)$ would then yield $\nu(E_0) \leq 0$ in contradiction to our hypothesis.

Once X is known to have a positive subset it is not difficult to find a maximal positive set to serve as A^+. Let $\beta = \sup \{v(P): \ P \text{ positive}\}$; then choose $\{P_n\}$, a sequence of positive sets, such that $v(P_n) \to \beta$. Let $Q_1 = P_1$, $Q_2 = P_2 - Q_1, \ldots, Q_n = P_n - \bigcup_{k=1}^{n-1} Q_k, \ldots$ so that $\{Q_n\}$ is a sequence of disjoint positive sets ($Q_n \subseteq P_n$, and any subset of a positive set is positive). If $Q_0 = \bigcup_{n=1}^{\infty} Q_n$, then Q_0 is positive, for any $C \subseteq Q_0$ can be written $C = \bigcup_{n=1}^{\infty} (C \cap Q_n)$ and $v(C) = \sum_{n=1}^{\infty} v(C \cap Q_n) \geq 0$ by the positivity of the Q_n's. (This shows that the union of any sequence of positive sets is positive.) Now, $\beta \geq v\left(\bigcup_{k=1}^{n} P_k\right) = v\left(\bigcup_{k=1}^{n} Q_k\right) \geq v(P_n)$ and hence $v(Q_0) = \beta > 0$.

Let $A^+ = Q_0$ and $A^- = \tilde{Q}_0$; we must now show that A^- is negative. If there is a set $B \subset A^-$ with $v(B) > 0$ then, by our argument above, B contains a positive set B_0 with $v(B_0) > 0$. Now, $A^+ \cup B_0$ is positive, and $v(A^+ \cup B_0) > v(A^+) = \beta$, since $B_0 \cap A^+ = \varnothing$, in contradiction to the definition of β, Hence $v(B) \leq 0$ for all $B \subseteq A^-$ and A^- is negative. This completes the proof.

The Hahn decomposition is not unique, for clearly any null set contained in A^+ can be transferred to A^- and vice versa. However, this is the extent of the nonuniqueness (see the exercises).

The *Jordan decomposition* of a signed measure comes as a corollary of the preceding theorem. We define two measures v_1 and v_2 to be *mutually singular*, in symbols, $v_1 \perp v_2$, if X can be written $X = A \cup B$, disjoint, where $v_1(A) = v_2(B) = 0$. If we are given a signed measure v, and if $X = A^+ \cup A^-$ is the corresponding Hahn decomposition, then we define

$$v^+(E) = \quad v(E \cap A^+),$$
$$v^-(E) = -v(E \cap A^-).$$

Clearly v^+ and v^- are mutually singular measures on X and $v = v^+ - v^-$. This decomposition of v is unique (see the exercises). Thus we have the following theorem:

Theorem 5.12.3. If v is a signed measure on a σ-algebra $\mathcal{M}$ of sets, then there are two uniquely determined mutually singular measures v^+ and v^- such that $v = v^+ - v^-$.

The measures v^+ and v^- are called the positive and negative parts of v, while the measure $|v|$, defined by

$$|v|(E) = v^+(E) + v^-(E),$$

is called the *absolute value of v*. At most, one of the parts of v can assume an infinite value; if both are finite, then v is called a *finite signed measure*.

We recall that we want to specify exactly which signed measures on a measure space $(X, \mathcal{M}, \mu)$ are indefinite integrals. Absolute continuity is necessary, but it has only been defined for finite-valued set functions (definition 5.5.6); we also need to extend the notion of mutual singularity to signed measures.

Definition 5.12.4. Let μ and v be two signed measures on a σ-algebra $\mathcal{M}$ of subsets of X.

(a) μ and v are said to be *mutually singular* if $|\mu| \perp |v|$.

(b) v is said to be *absolutely continuous* with respect to μ if $|v|(E) = 0$ whenever $|\mu|(E) = 0$. Symbolically, $v \ll \mu$.

In the case where μ is nonnegative and v is finite valued, (b) is actually equivalent to definition 5.5.6, so that our present definition is a valid generalization. We prove this now.

Lemma 5.12.5. If μ is a measure and v is a finite-valued signed measure, then the following conditions are equivalent:

(a) $\mu(E) = 0 \Rightarrow |v|(E) = 0$,

(b) For any $\epsilon > 0$ there is a $\delta(\epsilon)$ such that $|v(E)| \leq \epsilon$ whenever $\mu(E) \leq \delta(\epsilon)$.

Proof. We first show that (b) is equivalent to

(b') For any $\epsilon > 0$ there is a $\delta(\epsilon)$ such that $|v|(E) \leq \epsilon$ whenever $\mu(E) \leq \delta(\epsilon)$.

Clearly, (b') $\Rightarrow$ (b), for $|v(E)| = |v^+(E) - v^-(E)| \leq v^+(E) + v^-(E) = |v|(E)$. Conversely, choose $\delta(\epsilon/2)$ from (b). Then

$$\mu(E) \leq \delta(\epsilon/2) \Rightarrow \mu(E \cap A^+) \leq \delta(\epsilon/2) \quad \text{and} \quad \mu(E \cap A^-) \leq \delta(\epsilon/2),$$

where A^+ and A^- are sets defining a Hahn decomposition for v. Hence

$$|v(E \cap A^+)| = v^+(E) \leq \frac{\epsilon}{2}$$

and

$$|v(E \cap A^-)| = v^-(E) \leq \frac{\epsilon}{2}$$

from which we have $|v|(E) \leq \epsilon$. Thus (b') $\Leftrightarrow$ (b).

We now show (b') $\Leftrightarrow$ (a):

(b') $\Rightarrow$ (a) is clear. We prove (a) $\Rightarrow$ (b') by a contrapositive argument. If (b') is false, then for some η there exists a sequence of sets $\{E_n\}$ such that $\mu(E_n) \leq 2^{-n}$ and $|\nu|(E_n) \geq \eta$. Letting $D_n = E_n \cup E_{n+1} \cup \cdots$ we have

$$\mu(D_n) \leq \sum_{k=n}^{\infty} 2^{-k} = 2^{-n+1} \quad \text{while} \quad |\nu|(D_n) \geq |\nu|(E_n) \geq \eta.$$

Since $\mu(D_1)$ and $|\nu|(D_1)$ are finite, we have

$$\mu\left(\bigcap_{n=1}^{\infty} D_n\right) = \operatorname*{limit}_{n \to \infty} \mu(D_n) = 0 \quad \text{while} \quad |\nu|\left(\bigcap_{n=1}^{\infty} D_n\right) \geq \eta.$$

Hence (a) is false, and the proof is complete.

We proceed now to our main theorem.

Theorem 5.12.6 (Radon-Nikodym). Let $(X, \mathcal{M}, \mu)$ be a σ-finite measure space and let ν be a signed measure on $\mathcal{M}$, with $\nu \ll \mu$. Then there exists a measurable function x_0 such that

$$E \in \mathcal{M} \Rightarrow \nu(E) = \int_E x_0 \, d\mu.$$

This function is unique almost everywhere.

Proof. The uniqueness part is a consequence of the integration theory already developed, for if $\int_E x \, d\mu = \int_E y \, d\mu$ for all $E \in \mathcal{M}$, then $x(t) = y(t)$ almost everywhere. Because of the Jordan decomposition theorem, we can assume that ν is a measure; also we exclude the trivial case where ν is identically zero.

CASE 1. $\mu(X) < \infty, \nu(X) < \infty, \nu \ll \mu$. Consider the nonempty family

$$\mathscr{F} = \left\{ 0 \leq y \in L^{(1)}(X, \mathcal{M}, \mu) \colon \int_E y \, d\mu \leq \nu(E) \text{ for all } E \in \mathcal{M} \right\};$$

then $\lambda = \sup \{\int y(t) \, d\mu \colon y(t) \in \mathscr{F}\} < \infty$. Now choose a sequence $\{y_n\}$ from $\mathscr{F}$ such that $\{\int y_n \, d\mu\}$ has limit λ. Since $\{y_n\}$ may not be nondecreasing, we construct $\{x_n\}$ by letting

$$x_1 = y_1, x_2 = y_1 \cup y_2, \ldots, x_n = y_1 \cup \cdots \cup y_n, \ldots.$$

Clearly $x_n \in L^{(1)}(X, \mathcal{M}, \mu)$; to see that $x_n \in \mathscr{F}$ we note that any $E \in \mathcal{M}$ can be written $E = \bigcup_{i=1}^{n} E_i$, disjoint, where

$$E_1 = \left\{ t \in E: \quad y_1(t) = \max_{i=1}^{n} y_i(t) \right\}$$

$$E_2 = \left\{ t \in E: \quad y_2(t) = \max_{i=1}^{n} y_i(t) \right\} - E_1, \ldots,$$

and

$$E_n = \left\{ t \in E: \quad y_n(t) = \max_{i=1}^{n} y_i(t) \right\} - \bigcup_{i=1}^{n-1} E_i;$$

then

$$\int_E x_n \, d\mu = \sum_{i=1}^{m} \int_{E_i} y_i \, d\mu \le \sum_{i=1}^{m} \nu(E_i) = \nu(E).$$

Hence $\{x_n\}$ is an increasing sequence of nonnegative functions in $\mathscr{F}$ and $\int x_n \, d\mu \to \lambda < \infty$. By the Lebesgue monotone convergence theorem, the function $x_0(t) = \lim_{n \to \infty} x_n(t)$ satisfies $\int x_0 \, d\mu = \lambda$, and x_0 is almost everywhere finite valued. Since $\int_E x_n \, d\mu \le \nu(E)$ for all n, we see that $x_0 \in \mathscr{F}$.

The function x_0 is the one whose existence is asserted in the theorem, and we must show that $\int_E x_0 \, d\mu = \nu(E)$ for all E, or, that the measure $\pi(E) = \nu(E) - \int_E x_0 \, d\mu$ is identically zero. Note that $\pi(X) < \infty$ and that π is absolutely continuous with respect to μ.

Let (A_n^+, A_n^-) be the Hahn decomposition of the signed measure $\pi - (1/n)\mu$. We assert that, for some N, $\mu(A_N^+) > 0$. To see this, let $A_0^+ = \bigcup_{n=1}^{\infty} A_n^+$ and $A_0^- = \bigcap_{n=1}^{\infty} A_n^-$; then $A_0^+ \cup A_0^- = X$ and $A_0^+ \cap A_0^- = \varnothing$. Since $A_0^- \subseteq A_n^-$, we have $\pi(A_0^-) - (1/n)\mu(A_0^-) \le 0$, or $\pi(A_0^-) \le (1/n)\mu(A_0^-)$ for all n, and so $\pi(A_0^-) = 0$. Hence, if π is not identically zero, we have $\pi(A_0^+) > 0$ and, by absolute continuity, $\mu(A_0^+) > 0$. The subadditivity of μ now shows that $\mu(A_N^+) > 0$ for some N.

Finally, let $z(t)$ be $(1/N)$ times the characteristic function of A_N^+. Then,

$$\int_E [x_0 + z] \, d\mu = \int_E x_0 \, d\mu + \frac{1}{N}\mu(A_N^+ \cap E)$$

$$= \int_{E - A_N^+} x_0 \, d\mu + \left[\int_{E \cap A_N^+} x_0 \, d\mu + \frac{1}{N}\mu(A_N^+ \cap E) \right]$$

$$\le \nu(E - A_N^+) + \nu(E \cap A_N^+) = \nu(E),$$

since $E \cap A_N^+$ is a positive set for $\pi - (1/N)\mu$; thus $x_0 + z \in \mathscr{F}$. But $\int [x_0 + z] \, d\mu = \lambda + (1/N)\mu(A_N^+) > \lambda$. This contradicts the definition of λ, and we are forced to the conclusion that π is identically zero. The proof in case 1 is now complete.

CASE 2. $\mu(X) < \infty$, ν σ-finite, $\nu \ll \mu$. Here we can write $X = \bigcup_{k=1}^{\infty} E_k$, disjoint, where $\nu(E_k) < \infty$ for all k. Using case 1, we have, for any $E \in \mathcal{M}$,

$$\nu(E \cap E_k) = \int_{E \cap E_k} x_k \, d\mu$$

and hence

$$\nu(E) = \sum_{k=1}^{\infty} \nu(E \cap E_k) = \sum_{k=1}^{\infty} \int_{E \cap E_k} x_k \, d\mu = \int_E \left[\sum_{k=1}^{\infty} x_k \right] d\mu = \int_E x_0 \, d\mu$$

as desired.

CASE 3. $\mu(X) < \infty$, $\nu \ll \mu$, $\nu(X) = \infty$. Let $\mathcal{F}$ be the nonempty collection of all sets E in $\mathcal{M}$ such that $\nu(F)$ is 0 or ∞ whenever $F \in \mathcal{M}$, $F \subseteq E$, and let $\alpha = \sup \{\mu(E); E \in \mathcal{F}\}$, If $\alpha > 0$, then let $\{E_n\}$ be a sequence in $\mathcal{F}$ such that $\mu(E_n) \to \alpha$ and let $E_0 = \bigcup_{n=1}^{\infty} E_n$. For $F \subseteq E_0$ we have $F = \bigcup_{n=1}^{\infty} (E_n \cap F)$ and $\nu(F)$ is 0 or ∞ so that $E_0 \in \mathcal{F}$ and $\mu(E_0) \leq \alpha$. But $\mu(E_0) \geq \mu(E_n)$ for all n; and hence $\mu(E_0) \geq \alpha$; it follows that $\mu(E_0) = \alpha$. In the case $\alpha = 0$, let E_0 be the empty set $\varnothing$.

We now show that $\tilde{E}_0$ is a countable union of sets of finite ν measure. If $\nu(\tilde{E}_0) < \infty$, there is nothing to prove. If $\nu(\tilde{E}_0) = \infty$ then $\mu(E_0) > 0$ (by absolute continuity) and some subset of E_0 must have finite positive ν measure. (Otherwise, $\tilde{E}_0 \in \mathcal{F}$, $\tilde{E}_0 \cup E_n \in \mathcal{F}$, and $\mu(\tilde{E}_0 \cup E_n) > \alpha$ for sufficiently large n; this is a contradiction.) Let $\beta = \sup \{\mu(G): G \subseteq \tilde{E}_0, \nu(G) < \infty\}$; then $\beta > 0$ by absolute continuity, and we can find a sequence $\{G_n\}$ such that $G_n \subseteq \tilde{E}_0$, $\mu(G_n) \to \beta$, $\nu(G_n) < \infty$.

We now define $G_0 = \bigcup_{n=1}^{\infty} G_n$, $H_0 = \tilde{E}_0 - G_0$, and show $\mu(H_0) = 0$. Firstly, $\mu(H_0) > 0$ is incompatible with $\nu(H_0) < \infty$, for in that case a contradiction to the definition of β could easily be obtained. Secondly, $\mu(H_0) > 0$ is incompatible with $\nu(H_0) = \infty$, again by the definition of α and β. Thus ν is σ-finite on $\tilde{E}_0$ as claimed.

We now return to E_0. If ν vanishes on any subset A of E_0 with $\mu(A) > 0$, then $0 < \gamma = \sup \{\mu(A): \nu(A) = 0, A \subseteq E_0\}$. Let $\{A_n\}$ be a sequence of such sets with $\mu(A_n) \to \gamma$, and let $A_0 = \bigcup_{n=1}^{\infty} A_n$; it is then easy to show that $\nu(A_0) = 0$, $\mu(A_0) = \gamma$, and that ν does not vanish on any subsets of $E_0 - A_0$ of positive μ measure. (If ν does not vanish on any subsets of E_0 of positive measure, we let $A_0 = \varnothing$.)

The partition $X = \tilde{E}_0 \cup A_0 \cup (E_0 - A_0)$ now leads to the conclusion of case 3. Define $x_0(t)$ on $\tilde{E}_0$ as in case 2, and let $x_0(t)$ be 0 and ∞ for $t \in A_0$ and $E_0 - A_0$, respectively. It is left as an exercise to show $\nu(E) = \int_E x_0 \, d\mu$.

CASE 4. μ σ-finite, $\nu < <\mu$. The demonstration in case 4 proceeds from case 3 exactly as 2 followed from 1. This completes the entire proof.

An interesting application of the Radon-Nikodym theorem is the Lebesgue decomposition theorem, as follows.

Theorem 5.12.7. Let $(X, \mathcal{M}, \mu)$ be a σ-finite measure space and ν a σ-finite measure on $\mathcal{M}$. Then there exists a unique decomposition $\nu = \nu_1 + \nu_2$ where ν_1 and ν_2 are σ-finite measures and $\nu_1 \perp \mu$, $\nu_2 \ll \mu$.

Proof. The measure $\lambda = \mu + \nu$ is σ-finite and obviously $\mu \ll \lambda$, $\nu \ll \lambda$. It follows that there exist functions x and y such that

$$\mu(E) = \int_E x \, d\lambda \quad \text{and} \quad \nu(E) = \int_E y \, d\lambda.$$

Let $A_0 = \{t: \ x(t) = 0\}$ and $A_+ = \{t: \ x(t) > 0\}$, and define

$$\nu_1(E) = \nu(E \cap A_0), \qquad \nu_2(E) = \nu(E \cap A_+).$$

Clearly, $\nu = \nu_1 + \nu_2$ and since $\nu_1(A_+) = \mu(A_0) = 0$ we have $\nu_1 \perp \mu$. To see that $\nu_2 \ll \mu$ we assume $\mu(E_0) = 0$ and show $\nu_2(E_0) = 0$.

Since $0 = \mu(E_0) = \int_{E_0} x \, d\lambda$ we have $\lambda(E_0 \cap A_+) = 0$ and hence $\nu(E_0 \cap A_+) = 0$. But $\nu(E_0 \cap A_+) = \nu_2(E_0)$, and the existence part of the proof is finished. The uniqueness part of the proof is left as an exercise.

Corollary. The theorem is valid if ν is a σ-finite signed measure.

EXERCISES

1. Investigate the uniqueness of the Hahn decomposition.

2. Show that the Jordan decomposition is unique.

3. Show that the following are equivalent for any signed measures μ and ν

    (a) $\nu \ll \mu$
    (b) $\nu^+ \ll \mu$, $\nu^- \ll \mu$
    (c) $|\mu|(E) = 0 \Rightarrow \nu(E) = 0$.

4. Complete the proof of theorem 5.12.6, part 3.

5. (a) Show that if μ and ν are signed measures with $\mu \perp \nu$, $\nu \ll \mu$, then ν is identically zero.
    (b) Prove the uniqueness part of theorem 5.12.7.

6.   Prove the corollary to theorem 5.12.7.

7.   Prove lemma 5.3.3 for signed measures.

8.   (a) Show that if $|\mu(A)| < \infty$ for all $A \in \mathcal{M}$, then the signed measure μ is bounded, i.e., $|\mu(A)| \leq M$ for all $A \in \mathcal{M}$.
   (b) Let V be the set of all bounded signed measures on $\mathcal{M}$. Show that V is a real vector space and that $|\alpha\mu| = |\alpha||\mu|$, $|\mu + \nu| \leq |\mu| + |\nu|$.
   (c) For $\mu \in V$ define $\|\mu\| = |\mu|(X)$; show that V is a normed vector space. Is it a Banach space?

9.   Give an example of a Lebesgue-Stieltjes measure which is singular with respect to Lebesgue measure.

10.   Show that the Radon-Nikodym theorem remains valid if μ is a σ-finite signed measure.

Section 13   THE RIEMANN-STIELTJES INTEGRAL

In this section we discuss the Riemann-Stieltjes integral, which is a generalization of the Riemann integral, but is not the same as the Lebesgue-Stieltjes integral.

As in section 5.1 we consider partitions

$$\pi = \{0 = t_0, t_1, \ldots, t_{n-1}, t_n = 1\}$$

of $[0, 1]$, and let $\sigma = \{t'_1, \ldots, t'_n\}$ where $t_{k-1} \leq t'_k \leq t_k$. For any two real-valued functions x and y on $[0, 1]$, we can then write sums of the type

$$S[\pi, \sigma] = \sum_{k=1}^{n} x(t'_k) [y(t_k) - y(t_{k-1})]$$

and ask if they approach some limit as the partitions become "finer" in some sense. To make this precise, we let

$$|\pi| = \max \{t_k - t_{k-1}; k = 1, 2, \ldots, n\}$$

Definition 5.13.1. $\lim_{|\pi| \to 0} S[\pi, \sigma] = I$ if and only if, for any $\epsilon > 0$, there exists a $\delta(\epsilon)$ such that $|S[\pi, \sigma] - I| \leq \epsilon$ whenever $|\pi| \leq \delta(\epsilon)$, for all σ. In this case the number I is denoted by $\int_0^1 x \, dy$ and is called the Riemann-Stieltjes integral of x with respect to y.

We collect some basic properties of the Riemann-Stieltjes integral in the following lemma, whose proof is left as an exercise.

Lemma 5.13.2. If x_1 and x_2 are integrable with respect to y, and if x is integrable with respect to y_1 and y_2 then

$$\int_0^1 [\alpha_1 x_1 + \alpha_2 x_2] dy = \alpha_1 \int_0^1 x_1 dy + \alpha_2 \int_0^1 x_2 \, dy$$

$$\int_0^1 x \, d[\alpha_1 y_1 + \alpha_2 y_2] = \alpha_1 \int_0^1 x \, dy_1 + \alpha_2 \int_0^1 x \, dy_2,$$

where the existence of the integrals on the left is implied by the hypotheses. Further, if $\int_0^1 x \, dy$ exists and $0 < c < 1$, then

$$\int_0^1 x \, dy = \int_0^c x \, dy + \int_c^1 x \, dy.$$

We now prove the integration-by-parts theorem for Riemann-Stieltjes integrals.

Theorem 5.13.3. If either $\int_0^1 x \, dy$ or $\int_0^1 y \, dx$ exists, then the other exists, and

$$\int_0^1 x \, dy + \int_0^1 y \, dx = x(1)y(1) - x(0)y(0).$$

Proof. We have

$$\sum_{k=1}^n x(t_k')[y(t_k) - y(t_{k-1})] = - \sum_{k=1}^{n-1} y(t_k)[x(t_k') - x(t_{k-1}')]$$
$$- x(t_1')y(t_0) + x(t_n')y(t_n),$$

where the right member can be rewritten

$$x(1)y(1) - x(0)y(0) - \Big\{ [x(t_1') - x(t_0)]y(t_0) + [x(t_n) - x(t_n')]y(t_n)$$
$$- \sum_{k=1}^{n-1} y(t_k)[x(t_k') - x(t_{k-1}')] \Big\}.$$

Now, if $\int_0^1 x \, dy$ exists, then the limit as $|\pi| \to 0$ of the right member must exist; but from its form we see that this limit is $\int_0^1 y \, dx$. Thus the integrals exist simultaneously, and

$$\int_0^1 x \, dy + \int_0^1 y \, dx = x(1)y(1) - x(0)y(0).$$

This completes the proof.

We now consider the problem of the existence of the integral; the most interesting and useful case is that in which x is continuous and y is of bounded variation. Making these assumptions, we define

$$m(\delta) = \sup \{|x(t_1) - x(t_2)|: \ |t_1 - t_2| \le \delta\}$$

and note that $m(\delta)$ is a nonnegative, nondecreasing function on $[0, 1]$ such that limit $m(\delta) = 0$, due to the uniform continuity of x on $[0, 1]$.
$$\delta \to 0$$

Lemma 5.13.4. If $|\pi_1| \le \delta$ and $|\pi_2| \le \delta$, then

$$|S[\pi_1, \sigma_1] - S[\pi_2, \sigma_2]| \le 2m(\delta)V_0^1(y).$$

Proof. Consider first the case $\pi_1 = \pi_2 = \pi$, where $|\pi| \le \delta$ but $\sigma_1 \ne \sigma_2$. Let $\pi = \{t_0, t_1, \ldots, t_{n-1}, t_n\}$, $\sigma_1 = \{u_1, \ldots, u_n\}$, and $\sigma_2 = \{v_1, \ldots, v_n\}$, where $t_{k-1} \le u_k, v_k \le t_k$. Then

$$|S[\pi, \sigma_1] - S[\pi, \sigma_2]| \le \sum_{k=1}^{n} |x(u_k) - x(v_k)|\,|y(t_k) - y(t_{k-1})|$$

$$\le m(\delta) \sum_{k=1}^{n} |y(t_k) - y(t_{k-1})| \le m(\delta)V_0^1(y).$$

Next we assume that π_2 is a refinement of π_1, that is, every point of π_1 is also a point of π_2 and $|\pi_1| \le \delta$. Let $[t_{k-1}, t_k]$ be an interval in π_1, and let $t_{k-1} = s_{k,0}, s_{k,1}, \ldots, s_{k,m} = t_k$ be the points of π_2 in this interval. If $\sum_k$ is the part of $S[\pi_1, \sigma_1]$ contributed by $[t_{k-1}, t_k]$, then $\sum_k = x(u_k)[y(t_k) - y(t_{k-1})]$. If $\sum_k'$ denotes the terms in $S[\pi_2, \sigma_2]$ from this interval, then

$$\sum_k' = \sum_{j=1}^{m} x(s_{k,j})[y(s_{k,j}) - y(s_{k,j-1})]$$

and

$$\left|\sum_k - \sum_k'\right| = \left|\sum_{j=1}^{m} [x(u_k) - x(s_{kj}')][y(s_{k,j}) - y(s_{k,j-1})]\right|$$

$$\le m(\delta)V_{t_{k-1}}^{t_k}(y).$$

Thus

$$|S[\pi_1, \sigma_1] - S[\pi_2, \sigma_2]| \le m(\delta) \sum_{k=1}^{n} V_{t_{k-1}}^{t_k}(y) = m(\delta)V_0^1(y).$$

Finally, if π_1 and π_2 are arbitrary partitions with $|\pi_1|, |\pi_2| \le \delta$, then, letting π be their common refinement, we have

$$|S[\pi_1, \sigma_1] - S[\pi_2, \sigma_2]| \le |S[\pi_1, \sigma_1] - S[\pi, \sigma]| + |S[\pi, \sigma] - S[\pi_2, \sigma_2]|$$

$$\le 2m(\delta)V_0^1(y).$$

This completes the proof.

Theorem 5.13.5. If $x \in C[0, 1]$ and $y \in BV[0, 1]$, then the Riemann-Stieltjes integral $\int_0^1 x\, dy$ exists.

Proof. Let A_n be the closure of the set of real numbers which occur as Riemann-Stieltjes sums $S[\pi, \sigma]$ with $|\pi| \leq 1/n$. By the preceding lemma, this set is bounded, and if its diameter is denoted by λ_n, then $\lambda_n \leq 2m(1/n)\,V_0^1(y)$, a quantity which tends to zero as $n \longrightarrow \infty$. Since $A_n \supseteq A_{n+1}$, the closed sets in the sequence $\{A_n\}$ are nested, and by the completeness of the real-numbers system, there is a unique number I in $\bigcap_{n=1}^{\infty} A_n$. Now, for $\epsilon > 0$, choose n so that $\lambda_n < \epsilon$; then

$$|\pi| < \frac{1}{n} \Rightarrow |I - S[\pi, \sigma]| \leq \lambda_n \leq \epsilon.$$

This completes the proof.

For $x \in C[0, 1]$ and $y \in BV[0, 1]$ we have the following two inequalities:

$$\left| \int_0^1 x\, dy \right| \leq \int_0^1 |x|\, dV_0^t(y)$$

$$\left| \int_0^1 x\, dy \right| \leq \|x\|\, V_0^1(y).$$

The proofs are left for the exercises.

There is an analogue to theorem 5.1.4 for the Riemann-Stieltjes integral. For completeness we state the result, without proof.

Definition 5.13.6. A set E is said to have y-measure zero, where $y \in BV[0, 1]$, if for any $\epsilon > 0$ there is a sequence $\{I_n\}$ of intervals such that $\bigcup_{n=1}^{\infty} I_n \supseteq E$ and $\sum_{n=1}^{\infty} V_{I_n}(y) \leq \epsilon$

Theorem 5.13.7. For $x \in B[0, 1]$, $y \in BV[0, 1]$ the integral $\int_0^1 x\, dy$ exists if and only if the set of discontinuities of x has y-measure zero.

EXERCISES

1. Prove lemma 5.13.2.

2. Show that $\int_0^c x\, dy$ and $\int_c^1 x\, dy$ may both exist while $\int_0^1 x\, dy$ does not exist.

3. For $x \in C[0, 1]$, $y \in BV[0, 1]$, prove the inequalities

   a. $\left| \int_0^1 x \, dy \right| \leq \int_0^1 |x| \, dV_0^t(y)$

   b. $\left| \int_0^1 x \, dy \right| \leq ||y|| V_0^1(y)$.

4. Let $x \in C[0, 1]$ and $y' \in C[0, 1]$. Prove that $\int_0^1 x \, dy$ exists and has value $\int_0^1 xy' \, dt$.

5. Let y be a step function with jumps at $t_1, t_2, \ldots, t_n$ where $0 \leq t_1 < t_2 < \cdots < t_n \leq 1$. Show that for $x \in C[0, 1]$,

   $$\int_0^1 x \, dy = \sum_{k=1}^n x(t_k) \, [y(t_k^+) - y(t_k^-)]$$

6. Regard the Riemann-Stieltjes integral as a function from the product space $C[0, 1] \times BV[0, 1]$ onto R, the real numbers. Show that it is continuous.

7. Using exercise 6, generalize exercise 5 to the case where y is the limit, in the norm of $BV[0, 1]$, of step functions. (See exercises 5 and 6 of section 4.8.)

8. Using exercise 7, show how to write $\int_0^1 x \, dy$ as the sum of an integral $\int_0^1 x \, dz$, where z is a continuous function of bounded variation, and a series.

6 Linear Functionals: an introduction

DEFINITIONS AND GENERALITIES

We begin with some considerations which are purely algebraic. Let V be any vector space.

Definition 6.1.1. A linear functional on V is a linear transformation T of V into the field of scalars (see definition 4.1.13).

If T_1 and T_2 are linear transformations with domain V, we define

$$(T_1 + T_2)x = T_1(x) + T_2(x),$$
$$(\alpha T)x = \alpha[T(x)].$$

It is easy to show that $T_1 + T_2$ and αT are again linear. In fact, we have:

Lemma 6.1.2. The set of all linear transformations with domain V and range in a vector space W is itself a vector space under the above definitions of addition and multiplication by scalars.

Let T be any linear transformation with domain V and let

$$N = \{x \mid T(x) = 0\};$$

then $x_1,\ x_2 \in N \Rightarrow \alpha_1 x_1 + \alpha_2 x_2 \in N$, for $T(\alpha_1 x_1 + \alpha_2 x_2) = \alpha_1 T(x_1) + \alpha_2 T(x_2) = 0$. Thus the *null space* N of T is a subspace of the domain of T.

252

Among the subspaces of V, the null spaces of linear functionals form a special class which we now proceed to characterize. We will henceforth use the notation

$$\langle x \rangle = \{\alpha x : \alpha \text{ is a scalar}\}$$

for the one-dimensional subspace determined by the vector x.

Definition 6.1.3. A subspace S of a vector space V is said to have finite co-dimension k if V can be written $V = S \oplus W$ (a *direct* sum) where dim $W = k < \infty$.

It is left as an exercise to show that the co-dimension of a subspace S of finite co-dimension is uniquely determined. A subspace of co-dimension one is called a *hyperspace*, while a set of the form $x_0 + H = \{x_0 + h : h \in H\}$ where H is a hyperspace, is called a *hyperplane*.

Lemma 6.1.4. The null space N of any nonzero linear functional on V is a hyperspace in V. Conversely, every hyperspace in V is the null space of a linear functional F on V. Moreover, F is uniquely determined by its null space to within a scalar multiple.

Proof. Let F be a nonzero linear functional on V with null space N and let x_0 be any vector in V for which $F(x_0) \neq 0$. The sum

$$N + \langle x_0 \rangle$$

is clearly direct, and since for any z in V we have

$$z = \left[z - \frac{F(z)}{F(x_0)} x_0 \right] + \frac{F(z)}{F(x_0)} x_0,$$

N is a hyperspace.

Conversely, if H is a hyperspace of V then $V = H \oplus \langle x_0 \rangle$ for some x_0, so that any $z \in V$ can be written uniquely

$$z = x + \alpha_z x_0, \qquad x \in H.$$

The mapping

$$F(z) = \alpha_z$$

defines a linear functional on V with null space H, and $F(x_0) = 1$.

Finally, let F_1 and F_2 be two linear functionals with null space N; then $F_1(x_0) \neq 0 \Leftrightarrow F_2(x_0) \neq 0$ so that for such an x_0 we have $V = N \oplus \langle x_0 \rangle$ and any z in V can be written $z = n + \alpha x_0$. Hence $F_1(z) = \alpha F_1(x_0)$, $F_2(z) = \alpha F_2(x_0)$, and F_2 is a scalar multiple of F_1. This completes the proof.

Corollary. If $F_1, \ldots, F_n$ are n linearly independent functionals on V with null spaces $N_1, \ldots, N_n$ respectively, then $N_1 \cap \cdots \cap N_n$ is a subspace of V co-dimension n. Conversely every subspace of V of co-dimension n can be represented in this way.

We come now to some remarks which enable us to reduce the study of complex-valued linear functionals on a complex vector space to the case of real-valued linear functionals on a real vector space. Note first that a vector space V of over C, the complex numbers, can be regarded as a vector space over R, the real numbers. If this is done, however, the vectors x and ix become independent (for $x \neq 0$) so that in general an n-dimensional vector space over C is a $2n$-dimensional space over R.

Now let F be a complex-valued linear functional on the complex vector space V. Then

$$F(x) = F_1(x) + iF_2(x)$$

and it is a routine matter to verify that F_1 and F_2 are real-valued linear functionals on V, regarded as a real vector space. Now

$$F(ix) = F_1(ix) + iF_2(ix),$$

but also

$$F(ix) = iF(x) = iF_1(x) - F_2(x).$$

Hence

$$F_2(x) = -F_1(ix)$$

for all x, so that

$$F(x) = F_1(x) - iF_1(ix).$$

Thus, a complex linear functional on a complex vector space is determined by its real part.

We now consider a *normed* vector space V and restrict our attention to the *continuous* linear functionals.

Lemma 6.1.5. A linear functional on a normed vector space V is continuous if and only if its null space is closed.

Proof. If F is continuous then $N = \{x: F(x) = 0\}$ is the inverse image of the closed set $\{0\}$ and is consequently closed.

Conversely, assume that the null space N of a nonzero linear functional F is closed. By our general results there is a vector $x_0 \notin N$ such that any vector x in V can be written $x = x_1 + F(x)x_0$ for some $x_1 \in N$. Let H be the closed hyperplane $x_0 + N$; then $0 < d = \inf \{\|x_0 + x_1\|: x_1 \in N\}$. We claim that $|F(x)| < 1$ whenever $\|x\| < d$. Otherwise there exists a vector z such that $F(z) = \alpha$, where $|\alpha| \geq 1$, and $\|z\| < d$. But then $F\left(\dfrac{1}{\alpha}z\right)$ $= 1$ and $\left\|\dfrac{1}{\alpha}z\right\| < d$. These relations are incompatible, for no vector y can be in the hyperplane $x_0 + N$ (i.e., can satisfy $F(y) = 1$) and simultaneously satisfy $\|y\| < d$. It follows that F sends the sphere $\{x: \|x\| < d\}$ onto a bounded set of scalars, and thus F is continuous.

The following corollary, whose proof is left as an exercise, presents an interesting view of the lemma.

Corollary. A nonzero linear functional F on a normed vector space V is discontinuous if and only if its null space is dense in V.

In the case that V is a normed vector space we define V^* to be the collection of all *continuous* linear functionals on V. Thus $x^* \in V^*$ if and only if there exists an M such that

$$|x^*(x)| \leq M \|x\|$$

for all $x \in V$. (See section 3 of chapter 4.) We define

$$\|x^*\| = \inf \{M: |x^*(x)| \leq M \|x\| \text{ for all } x\}$$

or equivalently (see the exercises),

$$\|x^*\| = \sup \{|x^*(x)|: \|x\| = 1\}.$$

V^* is called the *conjugate*, or *dual*, space to V, and $|x^*(x)| \leq \|x^*\| \|x\|$ always.

Lemma 6.1.6. V^*, as defined above, is a Banach space.

Proof. Since the sum of two members of V^* and the product of a member of V^* by a scalar are continuous and linear, it is clear that V^* is a vector space. It is equally clear that

$$\|x^*\| \geq 0, \|\alpha x^*\| = |\alpha| \|x^*\| \quad \text{and} \quad \|x^*\| = 0 \Rightarrow x^* = 0.$$

For $\|x\| = 1$ we have

$$|x^*(x) + y^*(x)| \leq |x^*(x)| + |y^*(x)| \leq \|x^*\| + \|y^*\|$$

so that we obtain

$$\|x^* + y^*\| \le \|x^*\| + \|y^*\|,$$

and thus V^* is a normed vector space.

Now let $\{x_n^*\}$ be a Cauchy sequence in V^*; from the inequality

$$|x_n^*(x) - x_m^*(x)| \le \|x_n^* - x_m^*\| \, \|x\|$$

we see that $\{x_n^*(x)\}$ is a Cauchy sequence of scalars for every x, and hence $\{x_n^*(x)\}$ has a limit which we call $x^*(x)$. Now, since Cauchy sequences are bounded, we have $\|x_n^*\| \le M$ for all n, and thus $|x_n^*(x)| \le M\|x\|$ for all n; it follows that $|x^*(x)| \le M\|x\|$. The linearity of x^* is a consequence of the fact that a pointwise limit of linear functions is again linear.

EXERCISES

1. (a) Show that the set V^* of all linear functionals on a vector space V is itself a vector space under the operations given in this section.
   (b) Complete the proof of lemma 6.1.6.

2. (a) Show that the co-dimension of a subspace W of finite co-dimension in V is uniquely determined.
   (b) Prove the corollary to lemma 6.1.4.

3. Prove the corollary to lemma 6.1.5.

4. (a) Let V be a normed algebra and let M^* be the subset of V^* consisting of those functionals satisfying

   $$F(xy) = F(x)F(y).$$

   Show that M^* is a closed subalgebra of V^*. Such functionals are said to be multiplicative.
   (b) Let s be a point in the compact metric space K. Define the functional F_s in $C^*[K]$ by $F_s(x) = x(s)$. Show that F_s is multiplicative and that the set of all multiplicative functionals of this type is closed in $C^*[K]$.

5. Prove that a hyperspace in a normed vector space V is either closed or dense.

Section 2 EXAMPLES

In this section we exhibit the dual spaces of several Banach spaces.

A. $l^{(p)}*$; $1 < p < \infty$

The Hölder inequality is central to the discussion of $l^{(p)}*$. For $x = \{\alpha_n\} \in l^{(p)}$ and $y = \{\beta_n\} \in l^{(q)}$, where $1/p + 1/q = 1$, we have

$$\sum_{n=1}^{\infty} |\alpha_n \beta_n| \leq \|\{\alpha_n\}\|_p \|\{\beta_n\}\|_q.$$

This implies at once that every $y = \{\beta_n\} \in l^{(q)}$ defines an $F_y \in l^{(p)}*$ by

$$F_y(x) = \sum_{n=1}^{\infty} \alpha_n \beta_n$$

where $x = \{\alpha_n\} \in l^{(p)}$; moreover $\|F_y\| \leq \|y\|_q$.
The mapping of $l^{(q)}$ into $l^{(p)}*$ given by

$$L: y \longrightarrow F_y$$

is clearly linear. It is our goal to show that it is also an isometry of $l^{(q)}$ *onto* $l^{(p)}*$.

To show that L is an isometry we note first that $\|0\| = \|F_0\|$ so that we need consider only nonzero vectors $y = \{\beta_k\} \in l^{(q)}$. Corresponding to such a vector we define

$$x_N = \{\alpha_1, \ldots, \alpha_n; 0 \ldots 0 \ldots\}$$

where, for $i = 1, 2, \ldots, N$,

$$\alpha_i = 0 \quad \text{whenever} \quad \beta_i = 0, \text{ and}$$

$$\alpha_i = |\beta_i|^{q-1} \frac{\bar{\beta_i}}{|\beta_i|} \quad \text{otherwise.}$$

Then

$$\|x_N\|_p = \left(\sum_{i=1}^{N} |\alpha_i|^p\right)^{1/p} = \left(\sum_{i=1}^{N} |\beta_i|^{pq-p}\right)^{1/p} = \left(\sum_{i=1}^{N} |\beta_i|^q\right)^{1/p}$$

and

$$|F_y(x_N)| = \sum_{i=1}^{N} |\beta_i|^q = \left(\sum_{i=1}^{N} |\beta_i|^q\right)^{1/q} \left(\sum_{i=1}^{N} |\beta_i|^q\right)^{1/p} = \left(\sum_{i=1}^{N} |\beta_i|^q\right)^{1/q} \|x_N\|_p.$$

Thus, for N sufficiently large to ensure $\| x_N \|_p \neq 0$, we have

$$\left| F_y\left(\frac{x_N}{\| x_N \|_p} \right) \right| = \left(\sum_{i=1}^{N} |\beta_i|^q \right)^{1/q}.$$

From the definition $\| F_y \| = \sup \{ |F_y(x)| : \| x \|_p = 1 \}$ we now obtain at once the inequality

$$\| F_y \| \geq \| y \|_q$$

which, coupled with the reverse inequality proved above, yields the desired relation $\| F_y \| = \| y \|_q$. Thus, L is a linear isometry, and hence is one to one.

It remains to be shown that L is an *onto* mapping. To this end we let

$$u_k = (0, \ldots, 0, 1, 0, \ldots)$$

where the 1 is in the kth place, and for a given $F \in l^{(p)*}$ we denote $F(u_k)$ by β_k. Defining the vectors x_N in terms of the β_k's exactly as above, we have, as before,

$$|F(x_N)| = \left| F\left(\sum_{k=1}^{N} \alpha_k u_k \right) \right| = \left| \sum_{k=1}^{N} \alpha_k \beta_k \right| = \sum_{k=1}^{N} |\beta_k|^q,$$

$$\| x_N \|_p = \left(\sum_{k=1}^{N} |\beta_k|^q \right)^{1/p},$$

$$\left(\sum_{k=1}^{N} |\beta_k|^q \right) \leq \| F \| \left(\sum_{k=1}^{N} |\beta_k|^q \right)^{1/p}.$$

If $\beta_k = 0$ for all k then $\{\beta_k\} \in l^{(q)}$; otherwise for N sufficiently large we can rewrite this inequality as

$$\left(\sum_{k=1}^{N} |\beta_k|^q \right)^{1/q} \leq \| F \|$$

and it again follows that $y = \{\beta_k\} \in l^{(q)}$.

We now claim that $F = F_y$. This conclusion is reached by noting that F and F_y agree on the subspace V of $l^{(p)}$ consisting of all (finite) linear combinations of the vectors u_k. Since V is dense in $l^{(p)}$ and F and F_y are (uniformly) continuous, we then must have $F = F_y$ (see theorem 3.3.7).

This completes the proof of the following.

Theorem 6.2.1 For $1 < p < \infty$, $1/p + 1/q = 1$, the space $l^{(p)*}$ is isomorphic and isometric to $l^{(q)}$.

The proof of the theorem can be slightly modified to show that $l^{(1)*} = l^{(\infty)}$; this task is relegated to the exercises.

B. $L^{(p)}* (X, \mathcal{M}, \mu)$

We assume the measure space is σ-finite, and that $1 < p < \infty$. Again, the Hölder inequality is the key to the problem. For $x \in L^{(p)}$, $y \in L^{(q)}$ where $1/p + 1/q = 1$ we have $xy \in L^{(1)}$ and

$$\int |xy|\, d\mu \leq \|x\|_p \|y\|_q.$$

Thus y defines a member F_y of $L^{(p)}*$ according to

$$F_y(x) = \int xy\, d\mu$$

and the mapping

$$T : y \longrightarrow F_y$$

is a linear function from $L^{(q)}$ into $L^{(p)}*$ such that $\|F_y\| \leq \|y\|_q$. To show that T is an isometry we let

$$x(t) = \begin{cases} |y(t)|^{q-1} \dfrac{y(t)}{|y(t)|} & \text{for } y(t) \neq 0 \\ 0 & \text{for } y(t) = 0 \end{cases}$$

Then $|x(t)|^p = |y(t)|^{pq-p} = |y(t)|^q$, $x \in L^{(p)}$ and $\|x\|_p = (\int |y|^q\, d\mu)^{1/p}$. Finally,

$$|F_y(x)| = \left| \int xy\, d\mu \right| = \int |y|^q\, d\mu$$
$$= \left(\int |y|^q\, d\mu \right)^{1/q} \left(\int |y|^q\, d\mu \right)^{1/p} = \|y\|_q \|x\|_p,$$

and it follows that $\|F_y\| \geq \|y\|_q$. Combined with the reverse inequality proved earlier, this yields $\|F_y\| = \|y\|_q$ as desired. Hence T is a one-to-one linear isometry of $L^{(q)}$ into $L^{(p)}*$.

To show that T is an *onto* mapping we first consider the case $\mu(X) < \infty$. Let $F \in L^{(p)}*$ and define for any $A \in \mathcal{M}$

$$\nu(A) = F(\chi_A)$$

where χ_A is the characteristic function of A. The set function $\nu(A)$ is certainly additive; to see that it is countably additive, let $A_0 = \bigcup_{k=1}^{\infty} A_k$ where $A_i \cap A_j$

$= \varnothing$ for $i \neq j$, and let $B_n = \bigcup_{k=1}^{n} A_k$. Then $\|\chi_{A_0} - \chi_{B_n}\|_p \rightarrow 0$ and thus, by the continuity of F, we have $F\left(\sum_{k=1}^{n} \chi_{A_k}\right) \rightarrow F(A_0)$; in other words, ν is countably additive. Moreover, $\mu(A) = 0$ implies $F(\chi_A) = \nu(A) = 0$, and ν is absolutely continuous with respect to μ. (The *linear* mapping F sends the zero function into zero.) We can now apply the Radon-Nikodym theorem to obtain a function $f \in L^{(1)}$ $(X, \mathcal{M}, \mu)$ such that

$$F(\chi_A) = \nu(A) = \int_A f d\mu = \int \chi_A f \, d\mu$$

for all $A \in \mathcal{M}$.

It is evident now that for any simple function s we also have

$(*)$
$$F(s) = \int sf \, d\mu.$$

Now let g be any nonnegative function in $L^{(p)}$ and let $\{s_n\}$ be a nondecreasing sequence of nonnegative simple functions converging pointwise to g. Thus $\{s_n^p\}$ converges pointwise to g^p and the Lebesgue monotone convergence theorem implies that $\int |s_n^p - g^p| d\mu \rightarrow 0$. But $(s_n - g) \in L^{(p)}$ and $\int |s_n - g|^p \, d\mu \leq \int |s_n^p - g^p| \, d\mu$ so that $\|s_n - g\|_p \rightarrow 0.\dagger$ Let M be a bound for $\{\|s_n\|_p\}$.

We wish to extend the relation $(*)$ to hold for all nonnegative g in $L^{(p)}$. Let h^+ be the characteristic function of $\{t : f(t) > 0\}$; then

$$\left| \int s_n f^+ \, d\mu \right| = \left| \int (s_n h^+) f \, d\mu \right| = |F(s_n h')|$$

$$\leq \|F\| \|s_n h^+\|_p \leq \|F\| \|s_n\|_p \leq \|F\| M.$$

Since $s_n f^+$ is a nondecreasing sequence in $L^{(1)}$ converging pointwise to gf^+, we can apply the monotone convergence theorem to establish that $gf^+ \in L^{(1)}$ and $\int s_n f^+ \, d\mu \rightarrow \int gf^+ \, d\mu$. Similarly, we can show that $gf^- \in L^{(1)}$ and $\int s_n f^- \, d\mu \rightarrow \int gf^- \, d\mu$. Combining these results we see that $gf \in L^{(1)}$ and $\int s_n f \, d\mu \rightarrow \int gf \, d\mu$. The relation $F(s_n) = \int s_n f \, d\mu$ and the continuity of F now yields, for all nonnegative $g \in L^{(p)}$,

$$F(g) = \int gf \, d\mu.$$

Finally, this obviously extends to all $g \in L^{(p)}$ by virtue of the fact that $g \in L^{(p)}$ if and only if both g^+ and g^- are in $L^{(p)}$.

$\dagger$If $p \geq 1$, $0 \leq a \leq b$, then $(b - a)^p \leq b^p - a^p$.

It remains to be shown that the function f defined by $F \in L^{(p)}*$ is actually in $L^{(q)}$ and not merely in $L^{(1)}$. (Recall that for $\mu(X) < \infty$ we have $L^{(1)} \supseteq L^{(q)}$.) To do this we define

$$f_n(t) = \begin{cases} |f(t)|^{q-1} \dfrac{f(t)}{|f(t)|} & \text{for } |f(t)|^{q-1} \leq n \\[2mm] n\dfrac{f(t)}{|f(t)|} & \text{for } |f(t)|^{q-1} > n. \end{cases}$$

Then f_n, as a bounded measurable function, is in $L^{(p)}$ and

$$\left| \int f_n f \, d\mu \right| = |F(f_n)| \leq \|F\| \left(\int |f_n|^p \, d\mu \right)^{1/p}.$$

But

$$f_n f = |f_n| \, |f| \geq |f_n| \, |f_n|^{1/(q-1)} = |f_n|^p$$

so that

$$\left(\int |f_n|^p \, d\mu \right) \leq \|F\| \left(\int |f_n|^p \, d\mu \right)^{1/p}$$

and hence

$$\left(\int |f_n|^p \, d\mu \right)^{1/q} \leq \|F\|.$$

But $\{|f_n|^p\}$ converges pointwise to $|f|^q$, the monotone convergence theorem applies, and we have $(\int |f|^q \, d\mu)^{1/q} \leq \|F\|$, so that $f \in L^{(q)}$ and $\|F\| = (\int |f|^q \, d\mu)^{1/q}$ as desired.

The σ-finite case can be handled easily by writing X as the countable disjoint union of sets of finite measure; the necessary argument is left to the exercises.

We sum up this discussion in the following theorem:

Theorem 6.2.2. For $1 < p < \infty$ the space $L^{(p)}*(X, \mathcal{M}, \mu)$, where X is σ-finite, is isomorphic and isometric to $L^{(q)}$.

It is also true that $L^{(1)}* = L^{(\infty)}$; discussion of this result is left to the exercises.

C. $(C_0)*$

We consider the problem of characterizing the space $(C_0)*$, where (C_0) is the Banach space consisting of all sequences $\{\alpha_n\}$ of scalars which converge to zero, and $\|\{\alpha_n\}\| = \sup_n |\alpha_n|$. Let

$$x_n = (0, \ldots 0, 1, 0, \ldots)$$

where the 1 is in the nth position. It is then a simple matter to show that for any $x = \{\alpha_n\} \in C_0$ we have

$$x = \{\alpha_n\} = \sum_{n=1}^{\infty} \alpha_n x_n$$

where the partial sums converge in norm to x. Indeed, $\left\| x - \sum_{k=1}^{n} \alpha_k x_k \right\|$ $= \sup_{j>n} |\alpha_j| \to 0$, since $\{\alpha_n\}$ is a null sequence.

Now let $F \in (C_0)^*$ and let $F(x_n)$ be denoted by β_n; by the continuity and linearity of F we see at once that

$$F(x) = F\left(\sum_{n=1}^{\infty} \alpha_n x_n \right) = \sum_{n=1}^{\infty} \alpha_n \beta_n,$$

so that any $F \in (C_0)^*$ is defined by the sequence of scalars $\{F(x_n)\} = \{\beta_n\}$. Let $y_k = \{\delta_n\}$ where

$$\delta_n = \begin{cases} 0 \text{ for } n > k, \\ 0 \text{ for } \beta_n = 0, n \le k, \\ \dfrac{\beta_n}{|\beta_n|} \text{ for } \beta_n \ne 0, n \le k. \end{cases}$$

Then $y_k \in (C_0)$ and for all k we have

$$\sum_{n=1}^{k} |\beta_n| = \left| \sum_{n=1}^{k} \delta_n \beta_n \right| = |F(y_k)| \le \|F\| \|y_k\| \le \|F\|,$$

so the $\{\beta_n\} \in l^{(1)}$, and $\|\{\beta_n\}\|_1 \le \|F\|$. On the other hand,

$$|F(\{\alpha_n\})| = \left| \sum_{n=1}^{\infty} \alpha_n \beta_n \right| \le (\sup_n |\alpha_n|) \sum_{n=1}^{\infty} |\beta_n| = \|\{\alpha_n\}\| \|\{\beta_n\}\|_1, \quad \text{and}$$

$$\|F\| \le \|\{\beta_n\}\|_1.$$

Thus, to every $F \in (C_0)^*$ there corresponds a sequence $\{\beta_n\}$ in $l^{(1)}$ such that $\|F\| = \|\{\beta_n\}\|_1$ and $F(\{\alpha_n\}) = \sum_{n=1}^{\infty} \alpha_n \beta_n$. Conversely, by the calculation just given, every $y = \{\beta_n\}$ in $l^{(1)}$ defines a member F_y of $(C_0)^*$ by $F_n(\{\alpha_n\})$ $= \sum_{n=1}^{\infty} \alpha_n \beta_n$. This completes the proof of the following theorem.

Theorem 6.2.3 The space $(C_0)^*$ is isometric and isomorphic to $l^{(1)}$.

A discussion of $(C)^*$, the dual of the space (C) of all convergent sequences, will be found in the exercises.

EXERCISES

1. Show that $l^{(1)*} = l^{(\infty)}$.

2. Prove the σ-finite case of theorem 6.2.2.

3. Prove that $L^{(1)*} = L^{(\infty)}$.

4. Let (C) be the Banach space of all convergent sequences of scalars with $\|\{\alpha_n\}\| = \sup_n |\alpha_n|$. Characterize the dual space $(C)^*$. [*Hint:* Besides the vectors x_n, $n = 1, 2, \ldots$, introduce the vector $x_0 = (1, 1, \ldots, 1, \ldots)$ and show that the linear combinations of the vectors $x_0, x_1, \ldots, x_n, \ldots$ are dense in (C).]

5. Let V be a finite-dimensional normed vector space. Show that every linear functional on V is continuous, and characterize V^*.

6. Let $\mathcal{M}$ be a σ-algebra of subsets of X, let μ and ν be finite measures on $\mathcal{M}$, and define $\lambda = \mu + \nu$.

    (a) Show that $x \in L^{(2)} (X, \mathcal{M}, \mu) \Rightarrow x \in L^{(1)} (X, \mathcal{M}, \mu)$ and that $F(x) = \int x \, d\mu$ is a bounded linear functional on $L^{(2)} (X, \mathcal{M}, \lambda)$.

    (b) Obtain a function $y \in L^{(2)} (X, \mathcal{M}, \lambda)$ such that $F(x) = \int xy \, d\lambda$, and $0 \leq y(t) \leq 1$ for all t.

    (c) Show that $\mu(A) = \int_A y \, d\lambda$ and $\nu(A) = \int_A [1 - y] \, d\lambda$.

    (d) Show that if $E \subseteq \{t : y(t) = 0\}$, then $\mu(E) = 0$.

    (e) Show that if ν is absolutely continuous with respect to μ, then λ is absolutely continuous with respect to μ, and hence

$$\lambda(A) = \int_A y^{-1} \, d\mu$$

$$\left[\text{Consider } A_k = \left\{ t \in A : \frac{k}{n+1} < y(t) \leq \frac{k}{n} \right\} \text{ and show}\right.$$

$$\left. \frac{n}{k}\mu(A_k) \leq \lambda(A_k) \leq \frac{n+1}{k}\mu(A_k), \text{ etc.} \right]$$

    (f) Assuming ν is absolutely continuous with respect to μ, show that $\nu(A) = \int_A [y^{-1} - 1] \, d\mu$.

Section 3 LINEAR FUNCTIONALS ON HILBERT SPACE

In section 2 we characterized $l^{(p)*}$ and $L^{(p)*}$. The case $p = 2$ is special, for there the corresponding q is also 2 and we have

$$l^{(2)*} = l^{(2)} \quad \text{and} \quad L^{(2)*} = L^{(2)},$$

where the equality sign indicates both isometry and isomorphism. In characterizing $l^{(2)*}$ and $L^{(2)*}$ the mappings

$$\varnothing_1 : (\{\alpha_n\}, \{\beta_n\}) \longrightarrow \sum_{n=1}^{\infty} \alpha_n \beta_n$$

and

$$\varnothing_2 : (f, g) \longrightarrow \int fg \, d\mu$$

played a central role. In this section we introduce an abstract generalization of $l^{(2)}$ and $L^{(2)}$, and characterize its conjugate space.

Definition 6.3.1. A vector space V is called an inner-product space if there is a scalar-valued function on $V \times V$ having properties

(1) $(x,y_1) + (x,y_2) = (x,y_1 + y_2)$
(2) $(\alpha x, y) = \alpha(x,y)$
(3) $(x,y) = \overline{(y,x)}$

(If V is a real vector space, the bar is omitted.)

(4) $(x,x) \geq 0;$ $\quad (x,x) = 0 \quad$ only if $x = 0$.

It is left as an exercise to check that $l^{(2)}$ and $L^{(2)}$ with $\varnothing_1$ and $\varnothing_2$ satisfy (1)–(4). The scalar (x,y) is called the *inner product* of the vectors x and y. It is easily shown to have the additional properties

$$(0,x) = (x,0) = 0,$$
$$(x_1 + x_2, y) = (x_1, y) + (x_2, y),$$
$$(x, \beta y) = \bar{\beta}(x, y),$$

and indeed

$$\left(\sum_{i=1}^{m} \alpha_i x_i, \sum_{j=1}^{n} \beta_j y_j \right) = \sum_{i,j=1}^{m,n} \alpha_i \bar{\beta}_j (x_i, y_j).$$

By analogy with $l^{(2)}$ and $L^{(2)}$ we define

$$\|x\| = (x,x)^{1/2};$$

this is justified by the following lemma.

Lemma 6.3.2. If (x,y) is an inner product on a vector space V, then the function $||x|| = (x,x)^{1/2}$ is a norm on V, and for all x and y in V,

$$|(x,y)| \le ||x|| \, ||y|| \qquad \text{(Schwarz inequality)}.$$

Proof. The properties $||\alpha x|| = |\alpha| \, ||x||$ and $[||x|| = 0] \Rightarrow [x = 0]$ are immediate. The inequality $|(x,y)| \le ||x|| \, ||y||$ is true if either $x = 0$ or $y = 0$; otherwise it can be written $|(x \, ||x||^{-1}, \, y \, ||y||^{-1})| \le 1$, and thus it suffices to show that $|(x,y)| \le 1$ whenever $||x|| = ||y|| = 1$.

Now, for any α, we have

$$0 \le (x + \alpha y, \, x + \alpha y) = (x,x) + \alpha(y,x) + \bar{\alpha}(x,y) + \alpha\bar{\alpha}(y,y)$$

or

$$0 \le 1 + |\alpha|^2 + \alpha(\overline{x,y}) + \bar{\alpha}(x,y).$$

Choosing $\alpha = -(x,y)$, we obtain

$$0 \le 1 - |(x,y)|^2$$

or $|(x,y)| \le 1$, as desired.

The triangle inequality is now easily established; we have

$$||x + y||^2 = (x + y, \, x + y) = (x,x) + (x,y) + (y,x) + (y,y)$$
$$= ||x||^2 + 2\,\text{Re}\,(x,y) + ||y||^2$$
$$\le ||x||^2 + 2\,|(x,y)| + ||y||^2$$
$$\le ||x||^2 + 2\,||x|| \, ||y|| + ||y||^2 = (||x|| + ||y||)^2$$

so that $||x + y|| \le ||x|| + ||y||$.

Corollary: The scalar-valued function (x,y) is continuous on $V \times V$.

The continuity of the inner product is a consequence of the Schwarz inequality, for if $x_n \to x$ and $y_n \to y$ we can write

$$|(x_n, y_n) - (x,y)| = |(x_n, y_n) - (x_n, y) + (x_n, y) - (x,y)|$$
$$= |(x_n, y_n - y) + (x_n - x, y)|$$
$$\le |(x_n, y_n - y)| + |(x_n - x, y)|$$
$$\le ||x_n|| \, ||y_n - y|| + ||x_n - x|| \, ||y||,$$

a quantity which approaches zero as $n \to \infty$, due to the boundedness of $\{||x_n||\}$.

Definition 6.3.3. An inner-product space which is complete is called a *Hilbert space*.

Thus $l^{(2)}$ and $L^{(2)}$ are Hilbert spaces. It is our aim to characterize the conjugate space of an arbitrary Hilbert space. The notion of orthogonality is essential.

Definition 6.3.4. A set $\{x_\alpha\}$ in an inner-product space V is called *orthogonal* if $(x_\alpha, x_\beta) = 0$ for every pair of (distinct) vectors in the set. The set is orthonormal if in addition $\|x_\alpha\| = 1$ for all α.

An orthogonal set of nonzero vectors is linearly independent, for, by an elementary computation,

$$\left\| \sum_{k=1}^n \alpha_k x_k \right\|^2 = \sum_{k=1}^n |\alpha_k|^2 \, \|x_k\|^2$$

whenever $(x_i, x_j) = 0$, $i \neq j$. Thus, $\sum_{k=1}^n \alpha_k x_k = 0$ if and only if $\alpha_1 = \alpha_2 = \cdots = \alpha_n = 0$.

For notational convenience we shall use orthonormal sets henceforth. The following theorem shows how such sets may be useful.

Theorem 6.3.5. Let $A = \{x_\alpha\}$ be an orthonormal set in an inner-product space V. Then assertions (a), (b), and (c) below are equivalent:

(a)   For every $x \in V$, $x = \sum_\alpha (x, x_\alpha) x_\alpha$

(b)   For every $x \in V$, $\|x\|^2 = \sum_\alpha |(x, x_\alpha)|^2$

(c) $\overline{S(A)} = V$

Proof: We first note that for any given $x \in V$ at most countably many members of the set $\{(x, x_\alpha)\}$ are nonzero. For, if $y_1, y_2, \ldots, y_n$ are orthonormal, then

$$0 \leq \left\| x - \sum_{k=1}^n (x, y_k) y_k \right\|^2 = \left(x - \sum_{k=1}^n (x, y_k) y_k, \, x - \sum_{j=1}^n (x, y_j) y_j \right).$$

A little computation yields

$$0 \leq \|x\|^2 - \sum_{k=1}^n |(x, x_k)|^2.$$

Applying this to the set $\{x_\alpha\}$ we find

$$\sum_\alpha |(x, x_\alpha)|^2 \leq \|x\|^2$$

so that the series converges. Thus, for any N, at most $N - 1$ terms $|(x,x_\alpha)|^2$ are greater than $N^{-1}\|x\|^2$, so that only countably many can be nonzero. The equivalence of (a) and (b) now follows from the equality

$$\left\| x - \sum_{k=1}^{n} (x,x_k)x_k \right\| = \|x\|^2 - \sum_{k=1}^{n} |(x,x_k)|^2,$$

where the set $x_1, x_2, \ldots, x_n, \ldots$ is the subset of $\{x_\alpha\}$ consisting of those vectors for which $(x,x_\alpha) \neq 0$.

The implication (a) $\Rightarrow$ (c) is clear.

To prove the reverse implication we first note that, for *any* scalars $r_1, \ldots, r_n,$

$$\left(x - \sum_{k=1}^{n} (x,x_k)x_k, \ \sum_{k=1}^{n} r_k x_k \right) = 0$$

(the vectors $x_1, \ldots, x_n$ are as above). In particular,

$$\left(x - \sum_{k=1}^{n} (x,x_k)x_k, \ \sum_{k=1}^{n} [(x,x_k) - \beta_k]x_k \right) = 0$$

for $\beta_1, \ldots, \beta_n$ arbitrary. If we now apply the Pythagorean theorem (see exercise 5)

$$(y,z) = 0 \Leftrightarrow \|y + z\|^2 = \|y\|^2 + \|z\|^2,$$

to the vectors

$$y = x - \sum_{k=1}^{n} (x,x_k)x_k \quad \text{and} \quad z = \sum_{k=1}^{n} [(x,x_k) - \beta_k]x_k$$

we obtain

$$\left\| x - \sum_{k=1}^{n} \beta_k x_k \right\|^2 = \left\| x - \sum_{k=1}^{n} (x,x_k)x_k \right\|^2 + \left\| \sum_{k=1}^{n} [(x,x_k) - \beta_k]x_k \right\|^2$$

$$= \left\| x - \sum_{k=1}^{n} (x,x_k)x_k \right\|^2 + \sum_{k=1}^{n} |(x,x_k) - \beta_k|^2.$$

Thus, for any scalars $\beta_1, \ldots, \beta_n,$

$$\left\| x - \sum_{k=1}^{n} \beta_k x_k \right\|^2 \geq \left\| x - \sum_{k=1}^{n} (x,x_k)x_k \right\|^2$$

and so

$$\left\| x - \sum_{k=1}^{n} (x,x_k)x_k \right\| \leq \left\| x - \sum_{k=1}^{n} \beta_k x_k \right\|,$$

where the equality holds if and only if $\beta_k = (x, x_k)$ for $k = 1, 2, \ldots, n$.

Now, if $\overline{S(A)} = V$ then linear combinations $\sum_{k=1}^{n} \beta_k x_k$ can be found arbitrarily close to x; thus, by the above inequality, a subsequence of the sequence of partial sums of the series $\sum_{k=1}^{\infty} (x, x_k) x_k$ converges to x. But the equality

$$\left\| \sum_{k=m+1}^{m} (x, x_k) x_k \right\|^2 = \sum_{k=m+1}^{m} |(x, x_k)|^2,$$

together with the convergence of the series $\sum_{k=1}^{\infty} |(x, x_k)|^2$, implies that these partial sums form a Cauchy sequence in V. Thus the series converges to x, and the proof is complete.

The inequality

$$\sum_{\alpha} |(x, x_\alpha)|^2 \leq \|x\|^2$$

is called the *Bessel inequality*, while the equality

$$\sum_{\alpha} |(x, x_\alpha)|^2 = \|x\|^2$$

is called the *Parseval equality*. The vector

$$z_n = \sum_{k=1}^{n} (x, x_k) x_k$$

is called the *projection* of x onto the n-dimensional subspace V_n of V having $x_1, \ldots, x_n$ as an orthonormal basis.

We will need the notion of a projection onto an arbitrary subspace of V.

Definition 6.3.6. Let W be a subspace of an inner-product space V, and x a vector in V but not in W. A vector x_1 in W is called the *projection of x onto W* if $(x - x_1, y) = 0$ for all $y \in W$.

The reader may verify that the vector z_n above is indeed the projection of x onto the subspace V_n, according to the definition.

The *uniqueness* of the projection can be proved as follows. Suppose x_1 and y_1 are both projections of x onto W. Then $x_1 - y_1$ is in W and $(x - x_1, x_1 - y_1) = (x - y_1, y_1 - x_1) = 0$. The Pythagorean theorem, applied twice, then says

$$\|x - y_1\|^2 = \|x_1 - y_1\|^2 + \|x - x_1\|^2$$
$$= \|x_1 - y_1\|^2 + [\|x - y_1\|^2 + \|y_1 - x_1\|^2]$$

from which it follows that $\|x_1 - y_1\|^2 = 0$, or $x_1 = y_1$.

Projections do not always exist, but we now prove a theorem giving a general result in this direction.

Theorem 6.3.7 (Projection Theorem). If W is a complete subspace of an inner-product space V, then any vector x in V has a projection on W.

Proof: Let $d = \inf \{\|x - w\| : w \in W\}$, and let $\{w_n\}$ be a sequence in W such that $\|x - w_n\| \to d$. Using the definition of the norm in terms of the inner product, it is easily verified that

$$\|(w_m - x) - (w_n - x)\|^2 + \|(w_m - x) + (w_n - x)\|^2$$
$$= 2[\|w_m - x\|^2 + \|w_n - x\|^2].$$

(In fact, for any two vectors y and z in an inner-product space V, $\|y + z\|^2 + \|y - z\|^2 = 2[\|y\|^2 + \|z\|^2]$; this is known as the *parallelogram identity*.) In other words,

$$\|w_m - w_n\|^2 = 2[\|w_m - x\|^2 + \|w_n - x\|^2 - 2\|\tfrac{1}{2}(w_m + w_n) - x\|^2].$$

Now, since $\tfrac{1}{2}(w_m + w_n) \in W$ we have

$$d \leq \|\tfrac{1}{2}(w_m + w_n) - x\| = \tfrac{1}{2}\|(w_m - x) + (w_n - x)\|$$
$$\leq \tfrac{1}{2}[\|w_n - x\| + \|w_m - x\|],$$

a quantity which approaches d as $m, n \to \infty$. With the aid of this fact we can see that $\{w_n\}$ is a Cauchy sequence in W; let its limit be x_1.

To show that x_1 is the projection of x onto W, we must prove that $(x - x_1, w) = 0$ for all $w \in W$. Since

$$\|x - x_1\| = \inf \{\|x - x_0\| : x_0 \in W\},$$

we have, for all $w \in W$, and all α,

$$\|x - x_1\| \leq \|x - (x_1 - \alpha w)\|.$$

Squaring and expanding, we find

$$(x - x_1, x - x_1) \leq ([x - x_1] + \alpha w, [x - x_1] + \alpha w).$$
$$\leq (x - x_1, x - x_1) + |\alpha|^2 \|w\|^2 + \alpha \overline{(x - x_1, w)}$$
$$+ \bar{\alpha}(x - x_1, w)$$

and thus

$$0 \leq |\alpha|^2 \|w\|^2 + \alpha \overline{(x - x_1, w)} + \bar{\alpha}(x - x_1, w).$$

Letting

$$\alpha = -\|w\|^{-2}(x - x_1, w),$$

we obtain

$$0 \leq -\|w\|^{-2}|(x - x_1, w)|^2.$$

Hence $(x - x_1, w) = 0$ for all $w \in W$ and the proof is complete.

A corollary to the theorem is the fact that if W is a closed (and therefore complete) subspace of a Hilbert space H, then there is a vector $x \in H$ which is orthogonal to every vector in W. Thus the set

$$W^\perp = \{x : (x, w) = 0 \text{ for all } w \in W\}$$

is nonempty. It is called the *orthogonal complement* of W and is easily seen to be a closed subspace of H; moreover, $H = W \oplus W^\perp$.

We can now return to our project of studying continuous linear functionals on a Hilbert space.

If F is a nonzero continuous linear functional on a Hilbert space H, then N, the null space of F, is a closed proper subspace of H, and by the projection theorem there is a nonzero vector x_0 orthogonal to N. We can assume without loss that $F(x_0) = 1$. Now,

$$H = N \oplus \langle x_0 \rangle,$$

and

$$x = \frac{(x, x_0)}{(x_0, x_0)} x_0 + n$$

for any x in H, where $n \in N$. Thus,

$$F(x) = \left(x, \frac{x_0}{\|x_0\|^2}\right),$$

and we see that every F in H^* determines a vector y_0 in H such that for all x in H,

$$F(x) = (x, y_0).$$

(If F is the zero functional, let $y_0 = 0$.) The uniqueness of y_0 is easy to prove, for if $(x, y_0) = (x, z_0)$ for all x in H, then $(x, y_0 - z_0) = 0$ for all x in H, and in particular

$$(y_0 - z_0, y_0 - z_0) = \|y_0 - z_0\|^2 = 0 \qquad \text{so that } y_0 = z_0.$$

From the Schwarz inequality we have

$$|F(x)| = |(x, y_0)| \leq ||y_0|| \, ||x||$$

so that $||F|| \leq ||y_0||$. But $F\left(\dfrac{y_0}{||y_0||}\right) = \left(\dfrac{y_0}{||y_0||}, y_0\right) = ||y_0||$ so that actually $||F|| = ||y_0||$. This completes the proof of the following theorem.

Theorem 6.3.8. If $F \in H^*$, where H is a Hilbert space, then there exists a unique vector $y_0 \in H$ such that $F(x) = (x, y_0)$ for all $x \in H$, and $||F|| = ||y_0||$.

Moreover, if we let F_y be the functional $x \longrightarrow (x,y)$, then the correspondence

$$T : y \longrightarrow F_y$$

is an isomorphism and an isometry of the Hilbert space H onto H^*. Only the linearity of T remains unproved, and this is left as an exercise.

EXERCISES

1. (a) Show that in any inner-product space the norm satisfies the parallelogram law

$$||x + y||^2 + ||x - y||^2 = 2[||x||^2 + ||y||^2].$$

   (b) Show that the inner product can be recovered from the norm by the formula

$$(x,y) = \tfrac{1}{4}[||x + y||^2 - ||x - y||^2]$$
$$+ \tfrac{1}{4}i[||y - ix||^2 - ||y + ix||^2]$$

   the imaginary part being omitted if real scalars are used.

2. Show that in the Banach space $C[a,b]$ the parallelogram law does not hold.

3. Show that

$$||x|| = ||y|| \Rightarrow (x + y, x - y) = 0$$

   in a real inner-product space, but not in a complex inner product space.

4. (a) Show that for nonzero vectors x and y the Schwarz inequality becomes an equality if and only if

$$x ||x||^{-1} = e^{i\theta} y ||y||^{-1}.$$

(b)   Show that for nonzero x and y

$$||x + y|| = ||x|| + ||y||$$

if and only if $x = \alpha y$, where α is real and positive.

5.   Prove the Pythagorean theorem:

$$(x,y) = 0 \Leftrightarrow ||x + y||^2 = ||x||^2 + ||y||^2.$$

6.   Let $A = \{x_1, x_2, \ldots, x_n \ldots\}$ be a denumerable, linearly independent set in an inner-product space V. Define

$$y_1 = x_1$$

$$y_2 = x_2 - \frac{(x_2, y_1)}{(y_1, y_1)} y_1$$

.
.
.

$$y_n = x_n - \sum_{k=1}^{n-1} \frac{(x_n, y_k)}{(y_k, y_k)} y_k$$

.
.
.

Show that the set $B = \{y_1, y_2, \ldots, y_n \ldots\}$ is orthogonal and that
$S(A) = S(B)$.

7.   Show that in every finite-dimensional inner-product space there
is an orthonormal basis (see exercise 6).

8.   Show that in any separable inner-product space V there is a
countable orthonormal set A such that $\overline{S(A)} = V$. (*Hint:* From
a countable dense set B construct a countable, linearly independent
set C such that $\overline{S(C)} = V$; then apply exercise 6.)

9.   Let $A = \{x_n\}$ be an orthonormal set in an inner-product space V.

(a)   Show that if $x = \sum_{k=1}^{\infty} \alpha_k x_k$, then $\alpha_k = (x, x_k)$ for all k.

(b)   Show that if $x = \sum_{k=1}^{\infty} \alpha_k x_k$ and $y = \sum_{k=1}^{\infty} \beta_k x_k$ then $(x,y) =$
$\sum_{k=1}^{\infty} \alpha_k \overline{\beta_k}$.

(c)   In the case that V is a Hilbert space show that a series $\sum_{k=1}^{\infty} \alpha_k x_k$

converges in V if and only if $\sum_{k=1}^{\infty} |\alpha_k|^2 < \infty$.

(d)   Show that any separable Hilbert space is isomorphic and
isometric to $l^{(2)}$ (see exercise 8).

10. Let K be a closed convex subset of a Hilbert space H.

$$(x, y \in K \Rightarrow \alpha x + (1 - \alpha)y \in K \quad \text{for } 0 \leq \alpha \leq 1.)$$

Show that the projection theorem holds if the closed subspace W is replaced by K. Show that the uniqueness assertion also continues to be valid.

11. Consider the space $L^{(2)} [-\pi, \pi]$ of complex-valued square integrable functions on $[-\pi, \pi]$. Show that the set $A = \{e^{int} : n = 0, \pm 1, \pm 2, \ldots\}$ is orthogonal and that $\overline{S(A)} = L^{(2)} [-\pi, \pi]$.

Section 4 THE DUAL SPACE OF $C[0,1]$

From section 11 of chapter 5 we know that for $x \in C[0,1]$, $y \in BV[0,1]$ the Riemann-Stieltjes integral, $\int_0^1 x \, dy$, exists and

$$\left| \int_0^1 x \, dy \right| \leq \sup \{ |x(t)| : \ 0 \leq t \leq 1 \} V_0^1[y].$$

The mapping

$$F_y : x \longrightarrow \int_0^1 x \, dy$$

is linear, and from the preceding inequality we infer that $F_y \in C[0,1]^*$ with

$$\| F_y \| \leq V_0^1[y]$$

Since $V_0^1[y]$ can be increased by redefining y at a single point (thus leaving $\int_0^1 x \, dy$ unchanged) the equality sign does not hold in general; moreover, different y's can define the same functional. We examine this question more closely in the following lemma.

Lemma 6.4.1. If $z \in BV[0,1]$ and $\int_0^1 x \, dz = 0$ for all $x \in C[0,1]$, then $z(0) = z(t^-) = z(t^+) = z(1)$ for all $t \in [0,1]$. [$z(0^-)$ and $z(1^+)$ are not defined.] Conversely, if a function z in $BV[0,1]$ satisfies $z(0) = z(t^+) = z(t^-) = z(1)$ for all $t \in [0,1]$, then $\int_0^1 x \, dz = 0$ for all $x \in C[0,1]$.

Proof. For $x(t) \equiv 1$, we have $0 = \int_0^1 dz = z(1) - z(0)$, so that $z(0) = z(1)$.

Now let t_0 be a point at which $z(t)$ is continuous, and consider the function

$$x(t) = \begin{cases} 1 & 0 \le t \le t_0 \\ 1 - \frac{1}{h}(t - t_0) & t_0 \le t \le t_0 + h \\ 0 & t_0 + h \le t. \end{cases}$$

Then $0 = \int_0^1 x \, dz = \int_0^{t_0} 1 \, dz + \int_{t_0}^{t_0+h} \left[1 - \frac{1}{h}(t - t_0)\right] dz + \int_{t_0+h}^1 0 \, dz$

$$= z(t_0) - z(0) + \int_{t_0}^{t_0+h} \left[1 - \frac{1}{h}(t - t_0)\right] dz$$

and hence

$$|z(0) - z(t_0)| = \left| \int_{t_0}^{t_0+h} \left[1 - \frac{1}{h}(t - t_0)\right] dz \right|$$

$$\le \int_{t_0}^{t_0+h} \left| 1 - \frac{1}{n}(t - t_0) \right| dV \le V_{t_0}^{t_0+h}$$

where V is the variation of z. Since the variation function is continuous wherever z is continuous, we see that $z(t_0) = z(0) = z(1)$ at all points t_0 where z is continuous.

Finally, z has at most countably many discontinuities, so that its points of continuity are dense. It follows that for all $t \in [0,1]$, $z(0) = z(t^+) = z(t^-) = z(1)$.

The converse part of the lemma also follows from the fact that the continuity points are dense; in fact, the sums approximating the integral $\int_0^1 x \, dz$ can be formed using only these points, and these sums have value zero. Hence $\int_0^1 x \, dz = 0$ whenever $z \in BV[0,1]$, and $z(0) = z(t^+) = z(t^-) = z(1)$; the proof is now complete.

We now let N be the collection of all functions z of bounded variation on $[0,1]$ satisfying the condition $z(0) = z(t^-) = z(t^+) = z(1)$ for all $t \in [0,1]$, and define, for $y_1, y_2 \in BV[0,1]$,

$$y_1 * y_2 \Leftrightarrow y_1 - y_2 \in N.$$

This is clearly an equivalence relation on $BV[0,1]$. Moreover, if $y_1 * y_2$, then $y_1 - y_2$ is constant on the complement of a countable set. It follows that there is only one function in each equivalence class satisfying the conditions

$$y(t^+) = y(t) \qquad 0 \le t < 1$$
$$y(0) = 0.$$

Such functions are said to be of normalized bounded variation, and the set of all of these functions will be labeled $NBV[0,1]$.

Consider an equivalence class and let y_0 be the unique function of normalized bounded variation in it. We assert that if $y * y_0$, then

$$V_0^1[y_0] \leq V_0^1[y].$$

To see this, note that, apart from the addition of a constant, y can be obtained from y_0 by changing the values of y_0 on a countable set. Such a change cannot decrease the variation. (A change at a point of continuity of y_0 increases the variation, while a change at a discontinuity point t_1 would increase or leave unaffected the variation, depending on whether or not

$$|y(t_1) - y_0(t_1)| + |y(t_1) - y_0(t_1^-)| > |y_0(t_1) - y_0(t_1^-)|.)$$

It follows from this discussion that the mapping

$$T : y \longrightarrow F_y$$

of $NBV[0,1]$ into $C[0,1]^*$ is linear and one to one. It is our object to show that T is also *onto* and that $V_0^1[y] = \| F_y \|$. The major part of this information is contained in the following theorem.

Theorem 6.4.2. For any $F \in C[0,1]^*$ there is a $y \in BV[0,1]$ such that for all $x \in C[0,1]$,

$$F(x) = \int_0^1 x \, dy$$

and

$$\| F \| = V_0^1[y].$$

Proof. Let F be a fixed member of $C[0,1]^*$. For any $x \in C[0,1]$ we recall that the Bernstein polynomials

$$P_n(t) = \sum_{k=0}^{n} x\left(\frac{k}{n}\right)\binom{n}{k}t^k(1 - t)^{n-k}$$

converge uniformly to x. It follows that

$$\lim_{n \to \infty} F(P_n) = F(x).$$

Using the linearity of F, we calculate

$$F(P_n) = \sum_{k=0}^{n} x\left(\frac{k}{n}\right)F\left[\binom{n}{k}t^k(1 - t)^{n-k}\right].$$

If we introduce the abbreviation

$$\alpha_k^n = F\left[\binom{n}{k}t^k(1 - t)^{n-k}\right]$$

then we have

$$F(P_n) = \sum_{k=0}^{n} x\left(\frac{k}{n}\right)\alpha_k^n$$

or

$$F(P_n) = \sum_{k=0}^{n} x\left(\frac{k}{n}\right)[(\alpha_k^n + \alpha_{k-1}^n + \cdots + \alpha_0^n) - (\alpha_{k-1}^n + \cdots + \alpha_0^n)].$$

This suggests the definition

$$y_n(t) = \begin{cases} 0 & 0 = t \\ \sum_{0 \le k/n < t} \alpha_k^n & 0 < t < 1 \\ \sum_{k=0}^{n} \alpha_k^n & t = 1. \end{cases}$$

The functions y_n are step functions, and, by the properties of the Riemann-Stieltjes integral, we have

$$F(P_n) = \int_0^1 x \, dy_n$$

for any x in $C[0,1]$, where $\{P_n\}$ is the sequence of Bernstein polynomials uniformly approximating x. Thus we also have

$$F(x) = \lim_{n \to \infty} \int_0^1 x \, dy_n.$$

for *all* x in $C[0,1]$. The vital fact in this relation is that the functions y_n depend only on F and not on x.

Before proceeding further, we need to investigate some properties of the sequence $\{y_n\}$. Define

$$s_k^n = \begin{cases} +1 & \text{if } \alpha_k^n \ge 0 \\ -1 & \text{if } \alpha_k^n < 0. \end{cases}$$

Then we have

$$V_0^1[y_n] = \sum_{k=0}^{n} s_k^n \alpha_k^n = \sum_{k=0}^{n} s_k^n F\left[\binom{n}{k} t^k (1-t)^{n-k}\right]$$

$$= F\left[\sum_{k=0}^{n} s_k^n \binom{n}{k} t^k (1-t)^{n-k}\right]$$

$$\le \|F\| \sup\left\{\left|\sum_{k=0}^{n} s_k^n \binom{n}{k} t^k (1-t)^{n-k}\right| : 0 \le t \le 1\right\}$$

$$\le \|F\|,$$

since

$$\left| \sum_{k=0}^{n} s_k^n \binom{n}{k} t^k (1 - t)^{n-k} \right| \le \sum_{k=0}^{n} \binom{n}{k} t^k (1 - t)^{n-k} = 1.$$

Thus, for all n,

$$V_0^1[y_n] \le \|F\|$$

and the functions y_n are *uniformly* of bounded variation; this implies, of course, that they are also uniformly bounded. (Recall that $y_n(0) = 0$ for all n.)

We now interrupt the proof of theorem 6.4.2 with a lemma on such sequences.

Lemma 6.4.3. If $\{y_n\}$ is any sequence of functions on [0,1] such that, for all n,

$$V_0^1[y_n] \le M$$

and

$$\sup \{|y_n(t)| : 0 \le t \le 1\} \le M$$

then there exists a function $\hat{y}(t)$ satisfying these same conditions and a sub-sequence $\{\hat{y}_{n(k)}\}$ of $\{y_n\}$ such that

$$\lim_{k \to \infty} \hat{y}_{n(k)}(t) = \hat{y}(t)$$

for all t in [0,1]

Proof of Lemma 6.4.3. Consider first the case where the functions y_n are all nondecreasing. Let $D = \{t_m\}$ be a denumerable dense set in [0,1]. Since the sequence $\{y_n(t_1)\}$ is bounded, it has a convergent subsequence $\{y_{n1}(t_1)\}$. The sequence $\{y_{n1}(t_2)\}$ is also bounded, and also has a convergent subsequence $\{y_{n2}(t_2)\}$. If we continue in this way, we obtain the array of functions

$$
\begin{array}{ll}
y_{11}, y_{21}, \cdots & , y_{n1}, \cdots \cdots \\
y_{12}, y_{22}, \cdots & , y_{n2}, \cdots \cdots \\
\quad \cdot & \\
\quad \cdot & \\
\quad \cdot & \\
y_{1n}, y_{2n}, \cdots & , y_{nn}, \cdots \cdots \\
\quad \cdot & \\
\quad \cdot & \\
\quad \cdot &
\end{array}
$$

where

$$\lim_{n \to \infty} y_{nk}(t_k)$$

exists for every k, and each row in the array contains only members of the preceding row. The diagonal sequence $\{y_{nn}\}$ has the property that

$$\lim_{n \to \infty} y_{nn}(t_k)$$

exists for all k.

We denote this diagonal sequence by $\{y_{n(k)}\}$; it is a subsequence of $\{y_n\}$ and for every $t \in D$

$$\lim_{k \to \infty} y_{n(k)}(t)$$

exists. Thus we can say that the sequence $\{y_{n(k)}\}$ converges pointwise on D; let the limit function be denoted by y. As a function on D, y is also nondecreasing (see exercise 8, section 8, chapter 4).

For an everywhere defined monotone function z, the limits $z(t^+)$ and $z(t^-)$ exist at every t, and $z(t^+) \neq z(t^-)$ on at most a countable set C. These facts are also true for any monotone function defined on a dense set. Thus the function y on D can be extended to the set $\sim (C \cup D)$ by defining

$$y(t) = y(t^+) = y(t^-)$$

for t in $\sim (C \cup D)$.

We now assert that for $t_0 \in \tilde{C}$

$$\lim_{k \to \infty} y_{n(k)}(t_0) = y(t_0),$$

i.e., the sequence $\{y_{n(k)}\}$ is pointwise convergent on $\tilde{C}$.

To prove this we first find t_1 and t_2 in D such that $t_1 < t_0 < t_2$, and

$$0 \leq y(t_2) - y(t_0) \leq \epsilon$$
$$0 \leq y(t_0) - y(t_1) \leq \epsilon.$$

Next, find $k(\epsilon)$ such that for $k \geq k(\epsilon)$

$$|y_{n(k)}(t_1) - y(t_1)| \leq \epsilon$$

and

$$|y_{n(k)}(t_2) - y(t_2)| \leq \epsilon.$$

We then have

$$y(t_0) \geq y(t_2) - \epsilon \geq [y_{n(k)}(t_2) - \epsilon] - \epsilon \geq y_{n(k)}(t_0) - 2\epsilon$$
$$y(t_0) \leq y(t_1) + \epsilon \leq [y_{n(k)}(t_1) + \epsilon] + \epsilon \leq y_{n(k)}(t_0) + 2\epsilon$$

Therefore

$$|y(t_0) - y_{n(k)}(t_0)| \leq 2\epsilon$$

whenever $k \geq k(\epsilon)$, and our assertion is established.

Since C is countable and the functions in the sequence $\{y_{n(k)}\}$ are bounded, we can repeat the *diagonal process* argument used previously to obtain a subsequence $\{\hat{y}_{n(k)}\}$ of $\{y_{n(k)}\}$ converging everywhere. The limit, $\hat{y}$, of $\{\hat{y}_{n(k)}\}$ is the function we seek. The facts that

$$V_0^1[\hat{y}] \leq M \text{ and sup } \{|y(t)| : 0 \leq t \leq 1\} \leq M$$

follow at once from the corresponding properties of the functions $\hat{y}_{n(k)}$ and the monotonicity of $\hat{y}$.

This completes the proof of lemma 6.4.3 for the case where the functions in the sequence $\{y_n\}$ are all nondecreasing. The general case is handled by writing each y_n as the difference of two nondecreasing functions. The details are relegated to the exercises at the end of the section.

Let us return to the proof of theorem 6.4.2. For simplicity the sequence $\{\hat{y}_{n(k)}\}$ found in lemma 6.4.3 is renamed $\{\hat{y}_n\}$. We know that

$$F(x) = \lim_{n \to \infty} \int_0^1 x \, dy_n$$

and therefore

$$F(x) = \lim_{n \to \infty} \int_0^1 x \, d\hat{y}_n.$$

We will now show that

$$\lim_{n \to \infty} \int_0^1 x \, d\hat{y}_n = \int_0^1 x \, d\hat{y}$$

for all x in $C[0,1]$.

To begin with, we apply lemma 4.8.3 and obtain

$$V_0^1[\hat{y}] \leq \lim_{n \to \infty} V_0^1[\hat{y}_n] \leq \|F\|.$$

So that $\hat{y}$ is of bounded variation. Letting

$$z_n = \hat{y}_n - \hat{y}$$

we see that it suffices to show

$$\operatorname*{limit}_{n \to \infty} I_n = \operatorname*{limit}_{n \to \infty} \int_0^1 x \, dz_n = 0.$$

At this point we need a basic inequality from section 5.13. For $x \in C[0,1]$, $z \in BV[0,1]$, and $|\pi| \le \delta$ we have

$$\left| \int_0^1 x \, dz - \sum_{k=1}^n x(\sigma_k)[z(t_k) - z(t_{k-1})] \right| \le 2m(\delta)V_0^1(z),$$

where

$$m(\delta) = \sup \{|x(t') - x(t'')| : |t' - t''| \le \delta\},$$

the modulus of uniform continuity of x on $[0,1]$. Since

$$V_0^1[z_n] = V_0^1[\hat{y}_n - \hat{y}] \le 2 \, \|F\|,$$

we have, for $|\pi| \le \delta$,

$$\left| I_n - \sum_{k=1}^m x(\sigma_k)[z_n(t_k) - z_n(t_{k-1})] \right| \le 4 \, \|F\| \, m(\delta)$$

for all n. Now, $m(\delta) \to 0$ as $\delta \to 0$, and thus we can choose a δ_0 such that

$$(*) \qquad \left| I_n - \sum_{k=1}^m x(\sigma_k)[z_n(t_k) - z_n(t_{k-1})] \right| \le \frac{\epsilon}{2}$$

whenever $|\pi| \le \delta_0$. For such a π, say $\pi = \{t_0, t_1, \ldots, t_m\}$, we can, by the pointwise convergence of $\{z_n\}$ to the zero function, find an $N(\epsilon)$ such that $n \ge N(\epsilon)$ implies

$$|z_n(t_k)| \le \frac{\epsilon}{4Cm}$$

for $k = 0, 1, 2, \ldots, m$, where $C = \sup \{|x(t)| : 0 \le t \le 1\}$. It then follows that

$$(**) \qquad \left| \sum_{k=1}^m x(\sigma_k)[z_n(t_k) - z_n(t_{k-1})] \right| \le C \cdot 2m \frac{\epsilon}{4Cm} = \frac{\epsilon}{2}.$$

Upon combining (*) and (**) we obtain

$$|I_n| \le \epsilon$$

whenever $n \geq N(\epsilon)$; thus

$$I_n \longrightarrow 0 \quad \text{as} \quad n \longrightarrow \infty,$$

as desired.

With this we have fully established that

$$F(x) = \int_0^1 x \, d\hat{y}$$

for all x in $C[0,1]$. The equality

$$\|F\| = V_0^1[\hat{y}]$$

is a consequence of the previously demonstrated inequalities

$$\|F\| \leq V_0^1[\hat{y}] \quad \text{and} \quad V_0^1[\hat{y}] \leq \|F\|.$$

The entire proof of theorem 6.4.2 is now complete. The main theorem of this section follows easily.

Theorem 6.4.4. The space $C[0,1]^*$ is isometric and isomorphic to the space $NBV[0,1]$.

Proof: We have already established that the mapping

$$T : y \longrightarrow F_y$$

of $NBV[0,1]$ into $C[0,1]^*$ is linear and one to one. Theorem 6.4.2. ensures that for every F in $C[0,1]^*$ there is a y in $BV[0,1]$ such that

$$F(x) = \int_0^1 x \, dy = F_y(x).$$

If this y is replaced by the equivalent y_0 in $NBV[0,1]$, it is still true that

$$F(x) = \int_0^1 x \, dy_0$$

and thus T is an *onto* mapping. Moreover, since

$$\|F\| \leq V_0^1[y_0] \leq V_0^1[y] = \|F\|$$

we have

$$\|F\| = V_0^1[y_0].$$

But $y_0(0) = 0$ and so $V_0^1[y_0]$ is the norm of y_0 in $NBV[0,1]$. This completes the proof.

EXERCISES

1. Complete the proof of lemma 6.4.3.

2. Prove that $NBV[0,1]$ is a closed subspace of $BV[0,1]$.

3. Let y be a nondecreasing function in $NBV[0,1]$, and let F_y be the corresponding member of $C[0,1]^*$. Show that

$$x \geq 0 \Rightarrow F(x) \geq 0.$$

Investigate the converse.

4. Consider $C_R[M]$, the space of all continuous real-valued functions on a compact metric space M. Let G be a linear mapping of $C[M]$ into R such that for all x and y in $C[M]$

$$G(xy) = G(x)G(y).$$

Show that $G \in C[M]^*$ and that there exists a point t_0 in M such that

$$G(x) = x(t_0)$$

for all x in $C[M]$. (See exercise 4 of section 1)

5. Suppose

$$F(x) = x(t_0)$$

for all x in $C[0,1]$. Describe the function y in $NBV[0,1]$ for which

$$F(x) = \int_0^1 x \, dy.$$

6. Suppose

$$F(x) = \sum_{k=1}^{m} \alpha_k x(t_k)$$

where $t_1, \ldots, t_m$ and $\alpha_1, \ldots, \alpha_m$ are fixed. Describe the function y in $NBV\ [0,1]$ such that

$$F(x) = \int_0^1 x \, dy.$$

Appendix

Zorn's Lemma, Axiom of Choice, and Applications

Zorn's lemma and the axiom of choice are equivalent set theoretic principles. The latter is readily acceptable on intuitive grounds, at least to most mathematicians. We will not here prove their equivalence, but will content ourselves with statements and applications.

AXIOM OF CHOICE. If $\{S_\alpha\}$ is any collection of sets, then there exists a set S such that $S \subseteq \bigcup_\alpha S_\alpha$ and, for each α, $S \cap S_\alpha$ has exactly one member.

The statement of Zorn's lemma involves the notions of partially ordered set and simply ordered set. (See section 4 of chapter 1.) An element x of a partially ordered set is said to be *maximal* if $y \geq x$ implies $y = x$. An element z is an *upper bound* of a subset A if $z \geq a$ for all $a \in A$.

ZORN'S LEMMA. If every simply ordered subset of a partially ordered set E has an upper bound in E, then E has a maximal element.

We begin by giving several applications of Zorn's lemma to the theory of cardinal numbers (see chapter 1, section 5).

Theorem A.1. Let A and B be any two nonempty sets. Then either there exists a one-to-one mapping of A into B, or there exists a one-to-one mapping of B into A.

Proof. A one-to-one mapping between a subset of A and a subset of B

283

is specified by a collection of pairs (a, b) with $a \in A$, $b \in B$, such that no two distinct pairs have either the same first or second member; thus, the mapping is determined by its graph. Let $\{G_\alpha\}$ be the set of all such graphs, and let $\{G_\alpha\}$ be partially ordered by set inclusion. If $\{G_\beta\}$ is a simply ordered subset of $\{G_\alpha\}$, then $\bigcup_\beta G_\beta$ is in $\{G_\alpha\}$, so that Zorn's lemma applies and $\{G_\alpha\}$ has a maximal member, G. By its maximality, G defines the mapping whose existence is asserted in the theorem.

The theorem fills the gap left in section 5 of chapter 1, where it was asserted that cardinal numbers form a simply ordered set under the relation $\leq$, given in definition 1.5.2.

We are now in a position to examine further properties of cardinal numbers. Letting A and B be disjoint sets, we define

$$c(A) + c(B) = c(A \cup B)$$
$$c(A)c(B) = c(A \times B).$$

(The definition can be applied to nondisjoint sets simply by replacing one of them by a duplicate of itself which is disjoint from the other.)

Theorem A.2. If either A or B is infinite, then

$$c(A) + c(B) = c(A)c(B) = \max\,[c(A), c(B)].$$

Proof. We can assume that B is infinite and that $c(A) \leq c(B)$. From the inequalities

$$c(B) \leq c(A) + c(B) \leq c(B) + c(B)$$
$$c(B) \leq c(A)c(B) \leq c(B)c(B),$$

we see that it will suffice to prove that

$$c(B) = c(B) + c(B) = c(B)c(B)$$

for any infinite set B.

If the digit 2 is used to represent a set with two elements, we have

$$c(2 \times B) = c(B) + c(B),$$

so that $c(B) = c(B) + c(B)$ is equivalent to $c(B) = c(2 \times B)$. To prove this equation, we recall that the infinite set B contains a denumerable subset I. [If B is itself denumerable, then we know already from chapter 1 that $c(B) = c(2 \times B)$.] Let f be a one-to-one mapping of I onto $2 \times I$, and consider the set of all extensions of f to one-to-one mappings of S onto $2 \times S$ where

$B \supseteq S \supseteq I$. Let this collection of mappings be partially ordered by set inclusion of their graphs; then Zorn's lemma applies (as in the proof of the previous theorem), and a maximal extension F of f is obtained. Let B_0 be the domain of F; if $B_0 = B$ we are finished. Otherwise, $B - B_0$ must be finite, for if $B - B_0$ were infinite, it would contain a denumerable subset J, and an extension of F could be constructed because $c(J) = c(2 \times J)$. Now, the set $(B - B_0) \cup I$ is denumerable, and hence $c\{2 \times [(B - B_0) \cup I]\} = c[(B - B_0) \cup I]$. Further, $c(B_0 - I) = c[2 \times (B_0 - I)]$, for the restriction of F to $(B_0 - I)$ is a one-to-one correspondence between them. Thus

$$B = \{(B - B_0) \cup I\} \cup (B_0 - I)$$

is in one-to-one correspondence with

$$2 \times B = \{2 \times [(B - B_0) \cup I]\} \cup \{2 \times (B_0 - I)\}$$

and $c(2 \times B) = c(B)$ as claimed.

It is an easy corollary that, for any infinite set B,

$$c(n \times B) = c(B),$$

where n represents a set with n elements (n is a positive integer). (See the exercises for proof.) We will need this later.

Finally, we come to the proof of the relation $c(B \times B) = c(B)$, where B is infinite. As before, B contains a denumerable set I, and there is a one-to-one mapping F of I onto $I \times I$. We can then consider the collection of all mappings which extend F to a one-to-one correspondence between sets B' and $B' \times B'$ where $B \supseteq B' \supseteq I$. Zorn's lemma again applies, so that there exists a maximal extension F_0 whose domain we shall call B_0. Thus $c(B_0) = c(B_0 \times B_0)$ and, while we cannot conclude that B_0 coincides with B, it will suffice to show that $c(B_0) = c(B)$.

Clearly, $c(B_0) \leq c(B)$; proceeding indirectly, we assume in fact that $c(B_0) < c(B)$. Then $c(B - B_0) > c(B_0)$, for otherwise,

$$c(B) = c(B_0) + c(B - B_0) = c(B_0) < c(B),$$

a contradiction. Thus $B - B_0$ contains a subset B_1 such that $c(B_0) = c(B_1)$. Using this fact, plus the breakdown

$$(B_0 \cup B_1) \times (B_0 \cup B_1) = (B_0 \times B_0) \cup (B_1 \times B_0) \cup (B_0 \times B_1)$$
$$\cup (B_1 \times B_1),$$

we can construct an extension F_1 of F_0 from $B_0 \cup B_1$ to $(B_0 \cup B_1) \times (B_0 \cup B_1)$, for

$$c(B_1) = c(B_1 \times B_1) = c[3 \times (B_1 \times B_1)]$$
$$= c[(B_1 \times B_0) \cup (B_0 \times B_1) \cup (B_1 \times B_1)].$$

But F_0 is maximal, and thus the assumption $c(B_0) < c(B)$ is not tenable. It follows that $c(B) = c(B_0)$ as desired, and the proof is complete.

We will need the following corollary, whose proof is left as an exercise.

Corollary. If B is the collection of all finite subsets of an infinite set A, then $c(B) = c(A)$.

As an application of these results on cardinality, we prove a theorem on bases in a vector space. Recall that a *basis* in a vector space V is simply an independent set of vectors in V which spans V [i.e., every vector $x \in V$ can be uniquely written as a (finite) linear combination of basis vectors.]

Theorem A.3.   Every nonzero vector space V has a basis, and any two bases for V have the same cardinality.

Proof. Let P be the (nonempty) collection of all independent sets of vectors in V, partially ordered by set inclusion. Zorn's lemma applies, and any maximal member of P is a basis for V.

To prove the second part of the theorem, it suffices to assume that V is infinite dimensional (otherwise, see chapter 4, section 1). Hence we are given two infinite bases $A = \{x_\alpha\}$ and $B = \{y_\beta\}$, and we are required to show $c(A) = c(B)$. Now, each $x_\alpha \in A$ determines a unique finite subset B_α of B such that x_α is a linear combination of the members of B_α (no zero coefficients). The mapping $\varnothing : x_\alpha \rightarrow B_\alpha$ has domain A and range a subset of the set of all finite subsets of B. The sets $A_\alpha = \varnothing^{-1}(B_\alpha)$ are finite subsets of A and form a *partition* of A. Hence the mapping $\Phi : A_\alpha \longleftrightarrow B_\alpha$ is one-to-one and $c(\{A_\alpha\}) = c(\{B_\alpha\})$. Using the corollary of theorem A.2, we find $c(A) = c(\{A_\alpha\}) = c(\{B_\alpha\}) \leq c(S) = c(B)$ where S is the collection of all finite subsets of B. A similar argument then yields $c(B) \leq c(A)$, so that $c(A) = c(B)$ by theorem 1.5.3. This completes the proof.

A basis in a vector space V is called a *Hamel* basis for V. The field of scalars for V need not be R or C but can be any field. In particular, the real-numbers system R can be considered an infinite dimensional vector space over the field Q of rational numbers. (See the exercises for proof of the infinite dimensionality.) R then contains a Hamel basis H, and it can be assumed that the unique rational number in H is 1.

An interesting application of a Hamel basis H in R is as follows. Recall that a function x from one vector space to another is *additive* if

$$x(t_1 + t_2) = x(t_1) + x(t_2)$$

for all t_1 and t_2, and that an additive function automatically satisfies

$$x\left(\frac{p}{q}t\right) = \frac{p}{q}x(t)$$

for all integers $q \neq 0$ and p. If the domain and range of x are R, we then see that an additive continuous function must have the form $x(t) = ct$ where $c = x(1)$.

Theorem A.4. There exist discontinuous additive functions from R into R.

Proof. Let H be a Hamel basis for R over Q with $1 \in H$. Define the function x on H arbitrarily, except for the requirements that $x(1) \neq 0$ and $x(h_1) = 0$, where h_1 is a fixed element chosen from H, and extend x to R linearly. Then x is certainly additive, but since $0 = x(h_1) \neq x(1)h_1$, it cannot be continuous.

We come finally to an important application of the choice axiom to the theory of Lebesgue measure on R.

Theorem A.5. Every subset A of R of positive outer measure contains a nonmeasurable set.

Proof. It suffices to assume that A is bounded, for if A is unbounded, then

$$A = \overset{\infty}{\underset{n=1}{\cup}} (A \cap [-n, n]),$$

where $\mu^*(A \cap [-N, N]) > 0$ for some N, by the subadditivity of μ^*. Hence, if the theorem is true for bounded sets, then it also holds for unbounded sets.

Accordingly, we assume $A \subseteq [-N, N]$ and $2N \geq \mu^*(A) > 0$. Define, for x and y in A,

$$x \sim y \Leftrightarrow [x - y \text{ is rational}].$$

It is trivial to verify that this is an equivalence relation; if $\{A_\alpha\}$ is the set of equivalence classes, then $A = \underset{\alpha}{\cup} A_\alpha$. By the axiom of choice, there exists a set E such that, for each α, $E \cap A_\alpha$ consists of a single number.

Let $0 = r_0, r_1, r_2, \ldots, r_n, \ldots$ be an enumeration of the rationals in $[-N, N]$, and define the sets $\{E_k\}$ by

$$E_0 = E, \qquad E_k = \{x + r_k; x \in E\}.$$

Since inner and outer Lebesgue measures are invariant under translation, we have, for all k,

$$\mu^*(E_k) = \mu^*(E)$$
$$\mu_*(E_k) = \mu_*(E).$$

Moreover, the E_k's are disjoint. (If $x + r_k = y + r_l \in E_k \cap E_l$, then $x \sim y$, which is impossible since $x, y \in E$.)

Now, $A \subseteq \bigcup_{k=0}^{\infty} E_k$ and hence

$$0 < \mu^*(A) \leq \sum_{k=0}^{\infty} \mu^*(E_k),$$

from which we deduce that $\mu^*(E_k) = \mu^*(E) > 0$. On the other hand, $E_k \subseteq [-3N, 3N]$, and therefore

$$\bigcup_{k=0}^{\infty} E_k \subseteq [-3N, 3N],$$

which implies that

$$6N = \mu([-3N, 3N]) = \mu_*([-3N, 3N]) \geq \mu_*(\bigcup_{k=0}^{\infty} E_k) \geq \sum_{k=0}^{\infty} \mu_*(E_k).$$

It follows that $\mu_*(E_k) = \mu_*(E) = 0$. In particular, $\mu_*(E) < \mu^*(E)$, and the set E is not Lebesgue measurable.

Theorem A.5 enables us to exhibit a Lebesgue measurable set which is not a Borel set. Let $c(t)$ be the Cantor function and let $x(t) = \frac{1}{2}[t + c(t)]$. Then $x(t)$ and $x^{-1}(t)$ are continuous strictly increasing functions. The image of the Cantor set C_1 under $x(t)$ is easily calculated to be a set A of measure $\frac{1}{2}$, so that A contains a nonmeasurable set B. The set $L = x^{-1}(B)$ is contained in C_1 and, hence, is measurable with measure zero. Moreover, L cannot be a Borel set, because its inverse image under the measurable function $x^{-1}(t)$ is nonmeasurable.

EXERCISES

1. Prove or disprove each of the six possible arithmetic laws for cardinal numbers (commutativity, associativity, and distributivity).

2. Let $B_1, B_2, \ldots, B_n, \ldots$ be an infinite sequence of distinct copies of the infinite set B. Show that $C(\bigcup_{n=1}^{\infty} B_n) = C(B)$.

3. Prove the corollary of theorem A.2.

4. Construct a Lebesgue measurable set of positive measure which is not a Borel set.

5. A simply ordered set S is said to be *well-ordered* if every subset of S has a first member (i.e., if $S_1 \subseteq S$, then there exists an element x in S_1 such that $y < x$ implies $y \in \tilde{S}_1$). A simply ordered set S is said to satisfy the *transfinite induction principle* if S has a first element x_1, and if any subset S_1 of S, which contains x_1, and which contains x whenever it contains $\{y : y < x\}$, must coincide with S.

PROVE: A simply ordered set S is well-ordered if and only if it satisfies the transfinite induction principle.

6. Let x be a real-valued function on R such that $x(s + t) = x(s) + x(t)$.
  (a) Show that x is continuous everywhere if it is continuous at a single point.
  Now assume that x is measurable as well as additive. Let t_0 be any point.
  (b) Choose $\epsilon > 0$ and apply Lusin's theorem to obtain a continuous function $X(t)$ on R and a set S such that $\mu(\tilde{S}) < \epsilon$ and $x(t) = X(t)$ on S. Choose an interval (a, b) containing t_0 such that $b - a > 2\epsilon$, and an interval $[a_1, b_1] = [a - \eta, b + \eta]$ where $\eta > \epsilon$. Apply the uniform continuity of $X(t)$ on $[a_1, b_1]$ to find a $\delta(\epsilon) < \epsilon$ for $X(t)$. Show that for any h such that $|h| < \delta(\epsilon)$,
$$|x(t_0 + h) - x(t_0)| < \epsilon$$
  and hence that $X(t)$ is continuous at t_0.
  This exercise shows that a measurable additive function is continuous.

7. Prove that R is an infinite-dimensional vector space over Q.

8. Prove the following:
  (a) In a Hilbert space H there exists a maximal orthonormal subset.
  (b) Any two maximal orthonormal subsets in a Hilbert space H have the same cardinality.

Selected References

1. Buck, R. C. (ed.), *Studies in Modern Analysis*, Englewood Cliffs, New Jersey: Prentice-Hall, Inc., 1962.

2. Dieudonne, J., *Foundations of Modern Analysis*, New York: Academic Press, Inc., 1960.

3. Goffman, C. and Pedrick, G., *First Course in Functional Analysis*, Englewood Cliffs, N. J.: Prentice-Hall, Inc., 1965.

4. Goldberg, R. R., *Methods of Real Analysis*, New York: Blaisdell Publishing Co., 1964.

5. Halmos, P., *Finite Dimensional Vector Spaces*, 2nd Ed., Princeton, N. J.: D. Van Nostrand Co., Inc., 1958.

6. Kestleman, H. *Modern Theories of Integration*, New York: Oxford University Press, 1937.

7. Kolmogorov and Fomin, *Elements of the Theory of Functions and Functional Analysis*, Vols. I and II, Rochester, New York: Graylock Press, 1957, 1961.

8. Royden, H. L., *Real Analysis*, New York: The Macmillan Co., 1963.

9. Rudin, W., *Principles of Mathematical Analysis*, New York: McGraw-Hill Book Co., 1964.

Index of Symbols

In the list below, the frequently occurring symbols are given with a brief explanation and the number of the page on which they first appear.

Index